International Review of
Cytology

A Survey of
Cell Biology

VOLUME 166

International Review of Cytology
A Survey of Cell Biology

Edited by

Kwang W. Jeon

Department of Zoology
University of Tennessee
Knoxville, Tennessee

VOLUME 166

ACADEMIC PRESS
San Diego   New York   Boston   London   Sydney   Tokyo   Toronto

Copyright © 1996 by ACADEMIC PRESS, INC.

Academic Press, Inc.
A Division of Harcourt Brace & Company
525 B Street, Suite 1900, San Diego, California 92101-4495

United Kingdom Edition published by
Academic Press Limited
24-28 Oval Road, London NW1 7DX

International Standard Serial Number: 0074-7696

International Standard Book Number: 0-12-364570-0

PRINTED IN THE UNITED STATES OF AMERICA
96   97   98   99   00   01   EB   9   8   7   6   5   4   3   2   1

CONTENTS

Vitellogenin Receptors: Oocyte-Specific Members of the Low-Density Lipoprotein Receptor Supergene Family

Wolfgang Johann Schneider

Follicular Dendritic Cells and Germinal Centers

Yong-Jun Liu, Géraldine Grouard, Odette de Bouteiller, and Jacques Banchereau

Cell Biology of Kidney Glomerulus

Shinichi Ohno, Takeshi Baba, Nobuo Terada, Yasuhisa Fujii, and Hideho Ueda

Distribution and Functions of Parathyroid Hormone-Related Protein in Vertebrate Cells

Patricia M. Ingleton and Janine A. Danks

CONTRIBUTORS

Numbers in parentheses indicate the pages on which the authors' contributions begin.

Takeshi Baba (181), *Department of Anatomy, Yamanashi Medical University, Tamaho, Yamanashi 409-38 Japan*

Jacques Banchereau (139), *Schering-Plough, Laboratory for Immunological Research, 69571 Dardilly, France*

Janine A. Danks (231), *St. Vincent's Institute of Medical Research, Fitzroy, Victoria 3065, Australia*

Odette de Bouteiller (139), *Schering-Plough, Laboratory for Immunological Research, 69571 Dardilly, France*

Arnold De Loof (1), *Zoological Institute, 3000 Leuven, Belgium*

Yasuhisa Fujii (181), *Department of Anatomy, Yamanashi Medical University, Tamaho, Yamanashi 409-38, Japan*

Marie-Madeleine Giraud-Guille (59), *Histophysique et Cytophysique, Ecole Pratique des Hautes Etudes, Observatoire Océanologique, Université Pierre et Marie Curie and Centre National de la Recherche Scientifique, 66650 Banyuls-sur-Mer, France*

Géraldine Grouard (139), *Schering-Plough, Laboratory for Immunological Research, 69571 Dardilly, France*

Patricia M. Ingleton (231), *Institutes of Endocrinology and Cancer Studies, Sheffield University Medical School, Sheffield S10 2RX, United Kingdom*

Ine Janssen (1), *Zoological Institute, 3000 Leuven, Belgium*

Yong-Jun Liu (139), *Schering-Plough, Laboratory for Immunological Research, 69571 Dardilly, France*

Shinichi Ohno (181), *Department of Anatomy, Yamanashi Medical University, Tamaho, Yamanashi 409-38, Japan*

Wolfgang Johann Schneider (103), *Department of Molecular Genetics, University and Biocenter of Vienna, A-1030 Vienna, Austria*

Nobuo Terada (181), *Department of Anatomy, Yamanashi Medical University, Tamaho, Yamanashi 409-38, Japan*

Hideho Ueda (181), *Department of Anatomy, Yamanashi Medical University, Tamaho, Yamanashi 409-38, Japan*

Jozef Vanden Broeck (1), *Zoological Institute, 3000 Leuven, Belgium*

Hormones and the Cytoskeleton of Animals and Plants

Arnold De Loof, Jozef Vanden Broeck, and Ine Janssen
Zoological Institute of the K.U.Leuven, Naamsestraat 59, 3000 Leuven, Belgium

It is often overlooked that a cell can exert its specific functions only after it has acquired a specific morphology: function follows form. The cytoskeleton plays an important role in establishing this form, and a variety of hormones can influence it. The cytoskeletal framework has also been shown to function in a variety of cellular processes, such as cell motility (important for behavior), migration (important for the interrelationship between the endocrine and immune systems, e.g., chemotaxis), intracellular transport of particles, mitosis and meiosis, maintenance of cellular morphology, spatial distribution of cell organelles (e.g., nucleus and Golgi system), cellular responses to membrane events (e.g., endocytosis and exocytosis), intracellular communication including conductance of electrical signals, localization of mRNA, protein synthesis, and—more specifically in plants—ordered cell wall deposition, cytoplasmic streaming, and spindle function followed by phragmoplast function. All classes of hormones seem to make use of the cytoskeleton, either during their synthesis, transport, secretion, degradation, or when influencing their target cells. In this review special attention is paid to cytoskeleton-mediated effects of selected hormones related to growth, transepithelial transport, steroidogenesis, thyroid and parathyroid functioning, motility, oocyte maturation, and cell elongation in plants.

KEY WORDS: Abscisic acid, ACTH, Actin, Adrenes, Auxin, Chemotaxis, Cytoskeleton, Ecdysone, Gibberellin, Hormones, Intermediate filaments, Microfilaments, Microtubules, Myosin, Oocyte maturation, Plant hormones, Thyroid.

I. Introduction

A living entity can exert only those functions that are in line with its morphology. This fact is self-evident and well documented at the organismal level: wings—flying; gills—breathing in an aqueous environment, etc. Although usually less emphasized, the same situation prevails at the cellular level, as illustrated by the following tandems: huge concentrations of contractile filaments in muscle fibers—movement; axons with excitable membranes—signal transduction; flat leaf structure—photosynthesis; and epithelial cells—transport of water and solutes.

It is especially the cytoskeleton that bestows on cells a specific morphology, which is essential to their specific functioning. In multicellular systems, cell–cell contacts and the extracellular matrix are also important for the cell's architecture (Ben-Ze'ev, 1992). The dynamic three-dimensional cytoskeletal framework consists of a variety of proteins that have to be synthesized, assembled/disassembled, and interconnected in a specific way and at a specific time in development. The components of the cytoskeletal framework are believed to function in a variety of cellular processes, such as cell motility (important for behavior), migration (important for the interrelationship between the endocrine and immune systems, e.g., chemotaxis), intracellular transport of particles, mitosis and meiosis (Sawin and Endow, 1993), maintenance of cellular morphology (Avila, 1992; Remacle, 1992), spatial distribution of cell organelles (e.g., nucleus and Golgi system, De Loof, 1986; Karecla and Kreis, 1992), cellular responses to membrane events, intracellular communication including conductance of electrical signals (Lin and Cantiello, 1993), localization of mRNA (Singer, 1992; Surprenant, 1993; Bassell, 1993; Davies *et al.,* 1993), protein synthesis (Kirkeeide *et al.,* 1993), and—more specifically in plants—ordered cell wall deposition, cytoplasmic streaming, and spindle function followed by phragmoplast function (Seagull, 1989).

Effects of hormones on the cytoskeleton have already been reported in the literature (Hall, 1984), but they are scattered and quite often hidden in the bulk of other experimental data. A concise and unavoidably selective overview is given in this chapter.

II. Cytoskeleton

The cytoskeleton consists of a number of proteins that polymerize and connect with other proteins to form a complex three-dimensional network that runs from the plasma membrane to the chromosomes in the nucleus. The cytoskeleton can be subdivided into three interlinked parts (Fig. 1). The *membrane skeleton* is found just underneath the plasma membrane. The *cytoplasmic skeleton* is present all over the cytoplasm and is the best studied one. Its major constituents are microtubules, microfilaments, and intermediate filaments. In many cell types, the cytoplasmic skeleton keeps the nucleus anchored in a well-defined position in the cell, which may be

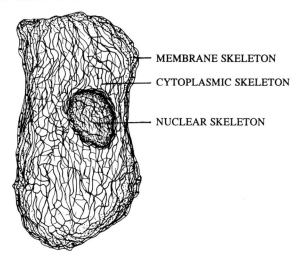

MEMBRANE SKELETON

CYTOPLASMIC SKELETON

NUCLEAR SKELETON

FIG. 1 Schematic representation of the cytoskeleton of a hypothetical cell.

important for control of gene expression (De Loof, 1986; Vanden Broeck *et al.,* 1992). It also makes special contacts with the membrane skeleton, the nuclear skeleton, and where this is relevant (e.g., in epithelial cells), with neighboring cells. The *nuclear skeleton* is probably by far the most complex one but little is as yet known about its functions. One should keep in mind that the three parts of the cytoskeleton form a continuum, at least in those cells where the nucleus cannot freely move in the cytoplasm.

A. Cytoskeleton and Origin of the Living State

In our opinion the presence of a primitive cytoskeleton was as important for the living state to arise as were nucleic acids for its continuation over successive generations. The reason is that the cytoskeleton plays a crucial role in cellular communication. Communication in turn is the key to understanding the very nature of "life" and "death" (De Loof and Vanden Broeck, 1995). In the era preceding the coming into existence of the first true cellular entity, a large number of different molecules able to carry out, in concert, a number of reactions had to be more or less orderly aggregated so that compartmentalization by a lipid membrane would result in a functional system. Early functions of the cytoskeleton in the primordial cell may have been: (i) generation of polarity; (ii) anchoring of DNA, some enzymes, ribosomes, and ion-transporting molecules in the plasma membrane, and (iii) movement (De Loof, 1992, 1993a). In eukaryotes such functions have been conserved up to the present day. Some are typical for the membrane cytoskeleton (Luna and Hitt, 1992), others for the cytoplasmic cytoskeleton, and still others for the nuclear skeletal framework. If this is indeed the case, it means that the cytoskeletal complex and at least some of its functions were already present long before some molecules acquired hormonal function.

B. Major Cytoskeletal Proteins of Eukaryotic Cells

We present only a very short overview to illustrate the complexity of the system and to facilitate the interpretation of the data presented in Tables I–XI.

1. The Membrane Skeleton

The major functions of the membrane cytoskeleton are to keep a variety of membrane compounds in place and to stabilize the lipid bilayer (Bretscher, 1991; Luna and Hitt, 1992; Small *et al.,* 1992; Hitt and Luna, 1994). These are essential for the functioning of epithelia and to establish proper contacts with the cytoplasmic cytoskeleton and with the extracellular matrix, if present.

Proteins closely related to spectrin, ankyrin, band 3, adducin, protein 4.1, protein 4.2 (pallidin), and dematin—to name a few—are associated with the plasma membrane of many cells and organisms ranging from humans to amoebae. Ankyrin and nonerythroid spectrin (fodrin) especially are often concentrated in membrane domains that may be specialized for mechanical stability and restricted diffusion of associated membrane proteins. Such flexible holdfasts include postsynaptic densities in the brain and neuromuscular junctions, infoldings of the basolateral membrane in epithelial cells, and the nodes of Ranvier and axon initial segments in myelined nerves. Actin and spectrin (α and β chain) provide contacts to a variety of proteins in the plasma membrane:

> *Ankyrin* attaches the β chain of spectrin to the anion exchanger of red blood cells, to other members of the anion exchanger gene family, to the voltage-gated Na^+ channel in the brain, to the α subunit of Na^+, K^+-ATPase, and probably to some other membrane proteins as well.
>
> *Protein 4.1* enhances the affinity of spectrin for actin, binds to transmembrane proteins and glycophorin C, and binds with low affinity to negatively charged phospholipids.
>
> *Adducin,* a heterodimeric calmodulin-associated protein, enhances spectrin binding to actin.

Tropomyosin and *dematin* (protein 4.9) bind along the sides of actin filaments.

Tropomodulin may control the length of the short actin filaments bound to the spectrin tetramers.

2. The Cytoplasmic Cytoskeleton

Microtubules

Tubulin Glycoprotein, monomeric unit (110 kDa) consisting of two nonidentical subunits, α and β. There are two binding sites for GTP per dimer. Tubulin forms complexes with microtubule-associated proteins (Map's).

Microfilaments

Actin Monomeric G-actin is a 42-kDa protein that can bind ATP and Ca^{2+}. Arrays of microfilaments are very dynamic, capable of changing organization and orientation in response to receptor-dependent or independent stimuli. Many proteins form complexes with actin, the best known ones being tropomyosin and dematin, and they exert a variety of functions (Hartwig and Kwiatkowski, 1991; Pollard *et al.*, 1994).

Intermediate filaments [five major classes (Ravindra and Grosvenor, 1990)]:

1. Tonofilaments consisting of cyto- or prekeratins (40–68 kDa)
2. Neurofilaments composed of triplet proteins (68, 160, 200 kDa)
3. Glial filaments with glial fibrillary proteins (51 kDa)
4. Desmin filaments made up of desmin (53 kDa)
5. Vimentin filaments with vimentin (53 kDa)

The intermediate filaments interact with microtubules and microfilaments; they are thought to reinforce the cytoskeleton. Subunits of the intermediate filaments may act as transmitters of signals from the cell periphery to the nucleus. An extensive system of highly cross-linked microtrabeculae has been visualized (Larson *et al.*, 1991).

3. The Nuclear Skeleton

The nucleus has its own specific skeleton, the nuclear matrix, which seems to be a site of organization for virtually all known nuclear processes, such as DNA loop attachment, DNA replication, transcription, RNA splicing and transport, hormone receptor function, carcinogen binding, oncogene function and protein phosphorylation. Furthermore, the nuclear matrix has linkages to intermediate filaments of the cytoskeleton (Vemuri *et al.*, 1993).

The following proteins were consistently found in isolated rat liver nuclear matrix (Nakayasu and Berezney, 1991):

Nuclear lamins A, B, and C
The nucleolar protein B-23
Residual components of core heterogeneous nuclear ribonucleoproteins
Eight matrins, namely, matrin 3 (125 kDa, slightly acidic), matrin 4 (105 kDa, basic), matrins D–G (60–75 kDa, basic), and matrins 12 and 13 (42–48 kDa, acidic). The matrins are not related to the lamins. Indirect immunofluorescence microscopy showed a strictly interior nuclear localization of the matrins in intact cells in contrast with the peripheral location of nuclear lamins.

The question of whether actin is present in substantial amounts in the nucleus has been debated for years. Recently, Sauman and Berry (1994) digested away DNA in ovarian cells of the moth *Manduca sexta* and then applied visualization methods for actin. They found that actin is a major component of the chromosomal skeleton, and they suggested that the binding of polymerase II to chromosomes may involve actin in addition to DNA.

There is evidence for physical connections between the nucleus and the cell periphery by means of cytoskeletal interactions (Pienta and Coffey, 1992). In plants there is growing evidence that the nuclear envelope may function as a microtubule-nucleating or microtubule-organizing center during spindle formation and that it may play an important role in controlling microfilament polymerization and organization (Seagull, 1989).

C. Cytoskeletal Diversity in Eukaryotic Multicellular Organisms

Theoretically, a given hormone may affect the cytoskeleton of a certain cell type in a given way and elicit a different effect or no effect on the cytoskeleton of another cell type. Several explanations are possible, of which only one is mentioned. Each cell type, and most probably even each individual cell, has its own unique cytoskeleton characterized by a specific morphology and a specific combination of a varying number of typical cytoskeletal proteins. The probability that two cells, even if they are neighbors in the same tissue, elaborate an identical cytoskeleton, is extremely low, if not nil. Thus, there is no such thing as "the" cytoskeleton.

Generating cytoskeletal diversity is an essential part of differentiation during embryogenesis. Indeed, all constituting cells of a differentiated organism are descendants of the very same zygote. With a few exceptions, which are not taken into consideration, they all have an identical genome but express it differentially. This enables the constituting cells to acquire a different morphology and engage in a variety of different functions. In our opinion the very basic rule underlying differentiation is: "Keep the genome constant but change again and again its macromolecular and/or ionic environment and realize this by continuously generating cells that differ in their plasma membrane–cytoskeletal complex." This difference is achieved by the double asymmetry principle: first, asymmetry is introduced in the plasma membrane–cytoskeletal complex of the zygote; next, this asymmetry is combined as soon as possible (preferably in the first cleavage, but not later than the third one) with asymmetrical cytokinesis (De Loof, 1993b).

D. The Cytoskeleton and Polarity

It is important to realize that in order to exert certain specific functions, cells have to be polarized and that some hormones influence this polarization. We briefly deal with polarization in three cell types, namely, single cells, muscle cells, and epithelial cells in which some hormones influence polarity.

1. Single Cells

Cells with a perfectly external and internal spherical symmetry do not seem to exist in nature. This follows from the architecture of their cytoskeleton. Indeed, the major cytoskeletal elements are threadlike proteinaceous polymers. Such linear structures are polarized by definition. If the primordial cytoskeletal elements had been spherical instead of threadlike, and if organisms with such structures had been viable, today's living world would be made up of spherical creatures instead of ones with the common bilateral symmetry. In our opinion perfect spherical symmetry and life are not compatible, for reasons that cannot be dealt with here. In the polarized state there is a front and a rear, or an apical and a basal side. However, at the moment that cells divide, the polarization of their cytoskeleton is largely lost. The daughter cells reconstruct a cytoskeleton that can be either similar to the parental one or drastically different. A typical example is the cytoskeleton of the spermatozoon as compared with that of primary spermatocytes. The typical form a cytoskeleton finally acquires depends on the interplay between genetics and factors present in the extra- and intracellular environments (epigenetics). The organization of functionally and structurally distinct plasma membrane domains depends on targeted delivery of transport vesicles between different membrane compartments and on protein sorting in the membranes of the Golgi complex and cell surface. Cytoskeletal components facilitate this process (Mays et al., 1994).

A variety of hormones can influence the polarity of individual cells. We deal with two examples, namely, the hormones that play a role in chemotaxis of leucocytes in the immune system (Section III,E) and the ones that cause maturation in oocytes (Section III,F).

2. Muscle Cells

All types of muscle cells have a very well developed—not to say "exaggerated"—and clearly polarized cytoskeleton that is organized in such a way (actin–myosin association) that contraction results in a substantial shortening of the cell. Not only some neurotransmitters but also a variety of myotropic factors (Section III,E) can influence the contractility of some muscles, often in a very specific way. Some hormones make the cytoskeleton of muscle cells develop (e.g., somatostatin; Section III,A); others cause degeneration (the molting hormone of insects causes programmed cell death of some larval muscles during metamorphosis).

3. Polarity in Epithelial Cells in Animals

Many hormones target epithelial cells. For example, vasopressin acts on the collecting ducts in the kidneys (Section III,B), progesterone on the chicken oviduct, ecdysone on the epidermis and salivary glands, and sex steroids on the skin and the epithelium of the uterus. Epithelial cells have to be polarized by definition to realize one of their major functions, namely, the transcellular transport of solutes and water. Their basal–apical polarization is due to the form of their cytoskeleton (Drenckhahn *et al.*, 1993). Specializations of this cytoskeleton (e.g., tight junctions) play a key role in making adjacent cells stick together so tightly that water and solutes cannot pass between the cells but have to be transported through the cells themselves. If the connections are not perfectly tight, the epithelium is said to be "leaky." The reason why epithelial cells are so important, especially in animals, resides in the very nature of animals themselves. In the Five Kingdoms classification system an animal is defined as an organism that develops from a blastula (De Loof, 1992). This simple definition harbors more information than one might at first sight think. Indeed, in histological terminology, a blastula is an epithelium. The definition thus says that in order to become an animal, a morula must organize itself into an epithelium-lined compartment, the blastula. Once this crucial event has occurred, the next developmental stage is the gastrula: the blastular epithelium invaginates, thereby forming the first inner epithelial compartment, namely, the archenteron or primitive gut. Later in development, more and more epithelial compartments are formed: the invaginations of the skin (e.g., the neural tube) and the appendages of the alimentary canal (lungs, liver, pancreas).

After gastrulation some cells escape from the ecto- or endodermal epithelial layers. They will give rise to mesodermal structures. Some cells retain their epithelial architecture (e.g., the cells lining the coelomic cavity), while others lose it and acquire different morphologies (muscle cells, blood cells, etc.). Thus, the basic cell type of any animal is the epithelial one; all other cellular architectures are secondarily acquired.

E. The Cytoskeleton and Signal Transduction

1. Intracellular Signaling Factors Affecting the Cytoskeleton

Intracellular signal transduction is initiated by specific binding of a certain hormone to its receptor protein. In a few cases the receptor protein can function as a transcription factor that is regulated by specific interaction with lipophilic ligands such as steroid hormones, ecdysteroids, thyroid hormone, or retinoic acid, but usually the functional receptor is found in the plasma membrane of the hormone's target cells, and cellular responses are elicited via rather complicated signaling pathways that are mainly controlled by activation/inactivation of a number of important molecular switches. Most, if not all, intracellular signaling mechanisms can directly or indirectly affect certain parts of the cytoskeleton. Indeed, many hormonal effects such as mitosis, phagocytosis, movement, contraction, or secretion are likely to be accompanied by short and/or long term changes of the cytoskeleton. In the case of cellular responses to growth and differentiation factors, certain functional and structural aspects of cytoskeletal (re-)organization are extremely important and will determine the nature of the final result.

Several intracellular signaling routes are able to evoke certain changes in the properties of the cytoskeleton. These changes include phosphorylations/ dephosphorylations of cytoskeletal components or of associated proteins, indicating that several protein Ser/Thr and/or Tyr kinases are involved in this kind of control. Src family protein tyrosine kinases play specific roles in regulating the cytoskeleton in developing mutant mouse embryos (Thomas *et al.*, 1995). In addition, evidence is accumulating for the existence of a direct (and indirect) control of the cytoskeleton by large (heterotrimeric) or small (Ras-related) GTP-binding proteins. The organization of actin filaments in activated mast cells appears to be regulated by both heterotrimeric and small (Rho and Rac) G proteins (Norman *et al.*, 1994). Rac is required for membrane ruffling and may be operating in invasive processes such as metastasis of tumor cells (Michiels *et al.*, 1995). Other GTP-ases are implicated in several types of intracellular trafficking: mRNA transport by the cell-cycle regulatory protein Ran/TC4 (e.g., Coutavas *et al.*, 1993); coated vesicle formation and fusion (e.g., Arf, Rothman, 1994;

XLαS, Kehlenbach *et al.,* 1994); exocytosis (e.g., Rab directs membrane traffic, Südhof, 1995); and endocytotic processes (e.g., dynamin, Hinshaw and Schmid, 1995). Cytoskeletal changes may also be induced by changes in intracellular ion concentrations; H^+ and Ca^{2+} especially are thought to play important regulatory roles. The special role of calcium is mainly due to the existence of quite a large number of regulatory calcium-modulated proteins, some of which are closely associated with cytoskeletal components. Protein kinase C, troponin C, calmodulin, and α-actinin are well known examples of proteins that can be regulated by changing Ca^{2+} concentrations.

2. Plasma Membrane Receptor Binding, Endocytosis, and Desensitization

The general rule is that a hormone must bind to a receptor in order to exert its function(s), but the binding should not last forever: a negative feedback pathway is usually required to switch off the activated receptor. Therefore, this mechanism is an important part of the agonist-induced cellular response. Hormone degradation/elimination is very important as well. When the molecular mass of the hormonal messenger is relatively high, as is the case for peptidic and proteinaceous hormones, the receptor is localized in the plasma membrane. The binding of a hormone to such a plasma membrane receptor sets off a cascade of events that starts in the region of the membrane skeleton and that, depending on the type of hormone, can end in the nucleus. A variety of signaling pathways can be involved, such as protein phosphorylation (see Section III,A and Table I: Effects of Insulin and Bombesin on Cytoskeletal Protein Phosphorylation) or secondary messengers (cAMP, Ca^{2+}, inositol polyphosphates, etc.). In cells where the plasma membrane receptors for hormones have to be specifically localized (e.g., in epithelial cells), anchoring mechanisms involving the cytoskeleton are likely to exist (Matus-Leibovitch *et al.,* 1993). It may be that some of these receptors use anchoring mechanisms similar to those of the epidermal growth factor receptor (EGFR), which is an actin-binding protein (den Hartig *et al.,* 1992).

Plasma membrane receptor–ligand complexes are internalized by endocytosis and then recycled. The ligand can be degraded or further transported to another destination. Endocytosis also occurs in cells that eliminate foreign substances from the body (hormones and the immunological system; Section III,E) and in oocytes that pinocytose vitellogenins, etc. Endocytosis requires the activity of the cytoskeleton just underneath the plasma membrane, probably in combination with elements of the membrane skeleton. Chemicals that disturb the action of the cytoskeletal elements involved in endocytosis can sometimes act as antihormones, as in the following example from insect endocrinology: Insects in which vitellogenesis takes place in the adult stage usually require juvenile hormone for pinocytotic uptake of vitellogenins. In some species, this process can be disturbed by benzodioxoles. Van Mellaert *et al.* (1983) demonstrated that benzodioxoles are true antijuvenile compounds, and Chang *et al.* (1994) showed that they act by inhibiting the polymerization of tubulin *in vitro* (see also Section IV).

The membrane cytoskeleton is also thought to play a role in desensitization to membrane-mediated cell inducers (Simantov *et al.*, 1980).

3. Effects of Steroid Hormones at the Level of the Plasma Membrane

For a long time there has been doubt whether steroid and other small lipophilic hormones (e.g., the terpenoid insect juvenile hormones) first interact with plasma membrane receptors as the peptide/protein hormones do. If they do, they must necessarily use the membrane skeleton for their internalization. Steroid hormone–mediated effects at the level of the plasma membrane, some of which occur within a few seconds after application of the hormone, have been reported in the literature since the early 1960s (De Loof, 1986), but until the 1990s, relatively little progress was made in the elucidation of the nature of the membrane molecules that are involved. Pappas *et al.* (1995) used confocal microscopy to show that antibodies directed against the nuclear estrogen receptor also recognize epitopes present in the plasma membrane. This suggests that the nuclear and the plasma membrane receptors are either very similar or identical.

The hydrophilic steroid molting hormone of insects, ecdysone, which served as another model in steroid hormone research, is still thought to have only a nuclear receptor. This nuclear receptor has been fully characterized in a few insect species, the best-documented one being *Drosophila* (Koelle *et al.*, 1991). Although there has been experimental evidence in favor of ecdysone-mediated plasma membrane effects in the dipteran *Chironomus* since 1963 (Kroeger, 1963), research to substantiate this further has never been very popular: a steroid hormone was supposed to be able to cross the membrane by diffusion without causing membrane effects. The findings that ecdysone activates a Na^+/H^+ exchanger, and that induced alkalinization of the cytoplasm in the absence of ecdysone causes substantial changes in nuclear volume (Wünsch *et al.*, 1993), urges for the identification of the membrane receptor of this model steroid and for an investigation of the role of the cytoskeleton in its signal transduction.

4. Intracellular Transport of Receptors

In the vagus nerve of rats corticotropin releasing factor (CRF) binding sites undergo axonal transport that may require some cytoskeletal elements

(Mercer et al., 1992). With respect to the role of the cytoskeleton in the translocation of steroid hormone receptors, not enough data are as yet available to present a coherent picture. In mouse hepatoma cells the 8S-glucocorticoid receptor binds to actin filaments through the 90-kDa heat shock protein (HSP90) (Yahara et al., 1991; Miyata and Yahara, 1991). In human gingival fibroblasts the receptor colocalizes with cytoplasmic microtubules (Akner et al., 1991a, 1991b). In L cells the association with the cytoskeleton may be required for translocation of the glucocorticoid receptor from its cytoplasmic site of synthesis to its nuclear site of action. The 9S complex is derived from this cytoskeleton-associated form. Association with the cytoskeleton can be a "docking" position (Pratt et al., 1989; Scherrer and Pratt, 1992). In ovariectomized adult guinea pigs colchicine causes an accumulation of the estrogen and progestin receptors in atypical brain areas, dendrites, and axon terminals (Blaustein and Olster, 1993). Perrot-Applanat et al. (1992) showed that in a permanent L cell line expressing a mutant progesterone receptor without karyophilic signal the cytoskeleton does not play a role in the translocation from cytoplasm to nucleus.

5. Cytoplasmic Anchoring

Cytoarchitectural components can regulate signal transduction by transiently anchoring regulatory molecules and by controlling their ordered transport to different locations in the cell. Some factors regulating nuclear activity (e.g., some protein kinase subunits or certain transcription factors) can be made temporarily unavailable to the nucleus by association with the cytoskeleton. It is well documented that an association between mRNA and the actin cytoskeleton may play a major role in mRNA targeting in the cell and in defining the site of translocation of proteins in a spatial context of the cell (Ben-Ze'ev, 1992). A variety of hormones have been reported to influence phosphorylation of certain cytoskeletal proteins (see tables later in this chapter).

6. Exocytosis

With respect to the mechanisms involved in peptide hormone secretion, the anterior pituitary has been used as a model for the past two decades. Pharmacological, morphological, and biochemical methods have been applied. The best-known hormones influencing peptide hormone secretion are the different releasing factors from the hypothalamus and the gonadal steroids that exert feedback control on the release of gonadotropins. The majority of hormonal peptides are packed into secretory vesicles (SV) in the Golgi apparatus, transported to the plasma membrane, and then secreted by exocytosis. Three oostatic peptide hormones have been found in insects that seem to originate from matrix proteins in the broad sense and that do not use the classical ER–Golgi pathway (De Loof *et al.*, 1995).

The secretory vesicles containing hormonal peptides to be secreted are formed in the Golgi apparatus. It has been shown that isolated Golgi complexes interact with microtubules *in vitro* and that this interaction depends on cytosolic factors and membrane-associated receptors but not on the microtubule-based motor protein, cytoplasmic dynein (Karecla and Kreis, 1992). After their release from the Golgi, the SV are transported to their release site. The association of SV with microtubules and the shift in equilibrium between soluble and polymerized tubulin pools during increased hormone secretion strongly suggests that microtubules are recruited for this transport.

With respect to the role of actin filaments in exocytosis of secretory granules, the rat parotid and submandibular gland acinar cells have been well studied with morphological methods by Segawa and Yamashina (1989). Here the microfilaments are localized mainly underneath the luminal plasma membrane and act as an obstacle to exocytosis when the cells are in the resting phase. During the secretory phase microfilaments allow exocytosis by disorganizing their barrier system and then, by encircling the discharged secretory membranes, provide forces for the extrusion of secretory products through the acto–myosin contractile system. Microfilaments are not detectable around the secretory granules that are stored in the cytoplasm but are clearly observed around those granules whose membranes are continuous with the luminal plasma membrane.

With respect to the biochemistry of SV exocytosis, a well-documented model system is chromaffin granule secretion by the medulla of the adrenes. The role of actin therein has been excellently reviewed by Trifaro *et al.* (1992), who propose the following sequence of events: under resting conditions cell viscosity is controlled through the formation of a mesh of actin microfilaments that are cross-linked and stabilized by fodrin and α-actinin. Both the secretory granule membranes and the plasma membrane contain a calmodulin-binding protein. At the Ca^{2+} concentrations found in resting cells (10^{-8} M), calmodulin and scinderin are not activated and there is a large percentage of nonfilamentous, nonphosphorylated myosin. Upon stimulation, Ca^{2+} enters the chromaffin cell and produces (i) a dissociation of actin from fodrin, (ii) patching of fodrin along the plane of the plasma membrane, and (iii) activation of scinderin with a consequent capping and shortening of the actin filaments. Calcium ion does not affect the binding of actin to granule membrane α-actinin. As a result of both (i) and (ii), the cytosol viscosity decreases, allowing the movement of granules toward the plasma membrane releasing sites. Whether an actin–myosin interaction (sliding mechanism) is also involved in granule movement remains to be determined. The intracellular Ca^{2+} concentration reached during stimulation is sufficient to activate calmodulin-dependent processes, including the binding of calmodulin to plasma granule membranes and plasma membranes (fusin) and the phosphorylation of myosin-like chains, a condition required for mysosin activation and bipolar filament formation. The secretion of peptide hormone–containing SV probably involves mechanisms similar to those operating in the adrenal medulla.

7. Actin Filaments: Biological "Electric Wires"?

A novel idea in signal transduction, in which actin filaments are of prime importance, has been suggested by Lin and Cantiello (1993). The essence of their hypothesis is that actin filaments have properties that enable them to function as biological electric wires that conduct electricity throughout the cell. The exciting idea is that an electrical signal, which is generated by ion pumps or channels in the plasma membrane that are linked to the membrane skeleton, does not immediately get dispersed in the cytoplasmic fluid around the filaments, as one would expect to happen in such a saline environment, but is conducted along filaments, perhaps to the interior of the nucleus. The underlying mechanism has been described as follows.

As charged polyelectrolytes, actin filaments may contain a proportion of their surrounding counterions in the form of a dense or "condensed" cloud about their surface that may be highly insensitive to large changes in the ionic strength conditions of the surrounding saline (Lin and Cantiello, 1993). This tightly "bound" ionic cloud, provided that it is electrically shielded from the bulk solution, will allow significant ionic movements along the polymer's length. The counterionic cloud about an actin filament could therefore be a highly conductive medium. As a consequence, electrically forced ions entering one end of the polymer will cause ions to exit the other end, with the ionic gradient lying along the polymer's length. Using a variation of the "patch clamp" technique applied to isolated actin filaments in solution, Lin and Cantiello (1993) studied ionic (= electrical) currents along the surface of electrically stimulated actin filaments. Such currents were observed about the polymer's surface under both high (100 mM KCl) and low (1 mM KCl) ionic strength conditions. Counterionic waves were highly nonlinear in nature and remained long after the electrical stimulation of the actin filaments had ceased. Upon pulse excitation, an input decomposed into a finite number of solitons and a low-amplitude oscillatory tail. The authors also observed that more net charge was transferred via actin filaments than via free solution. In solution, polyelectrolytes are known to electrically repel one another, thus minimizing the cross talk between neighboring polymers. It seems possible that actin filaments can exist in a complicated dynamic network and still preserve electrical screening from one another. According to Lin and Cantiello (1993) one can envision an electric field fluctuation, such as that elicited by a single channel opening to which an actin filament can be attached (Cantiello *et al.*, 1991) as a novel and efficient long-range intracellular signaling mechanism whose significance will be dictated by the overall status of its organization. A lot of hormones elicit electrogenic ion fluxes at the level of the plasma membrane. If the electricity is generated at a site that makes contact with an actin filament, one may expect that such electrically stimulated actin filaments generate soliton-like wave patterns based on the nonlinear wave behavior of ionic signals that are probably elicited by condensed ions around actin filaments. The ionic waves could perhaps play a relevant biological role as a novel intracellular signal transduction mechanism. One should bear in mind that at least in *Drosophila* (Agard and Sedat, 1983), the chromosomes are not distributed at random in the nucleus but occupy a very specific place. It may thus matter where in the nucleus a gene is located. Since there is also a complicated dynamic filamentous network present inside the nucleus, perhaps there are electrical connections between the actin filaments in the cytoplasm and the actin present in the chromosomal skeletons. The importance of electrical phenomena in signal transduction should not be underestimated: "life" as it can be defined today, is to a large extent an electrical phenomenon (De Loof and Vanden Broeck, 1995).

F. Chemical Tools in Cytoskeleton Research

To find out whether a given hormone requires the presence of actin, tubulin, or intermediate filaments in order to exert its typical effects, the hormone is usually administered in the presence of one of the following toxins that specifically act on the cytoskeleton:

Colchicine and *nocodazole* inhibit the assembly of tubulin monomers into microtubule monomers, finally causing the accumulation of tubulin in soluble pools.

Lumicolchicine is an inactive polymer of colchicine that does not inhibit microtubule function.

Vinblastine precipitates tubulin.

D_2O and *taxol* alter the equilibrium between the soluble and polymerized form of tubulin by overstabilizing them in their polymerized form.

Various *herbicides* and *carbamates* disrupt either microtubules or/and the ability of the microtubule-organizing center to produce organized arrays of microtubules (Seagull, 1989).

Cytochalasins in general inhibit the rate of actin polymerization.

Cytochalasin B preferentially blocks elongation of F-actin.

Phalloidin binds to actin and stabilizes it in the filamentous form. When
conjugated with a fluorescent dye such as rhodamine B, it is an excellent
marker for the microfilament cytoskeleton to be used in optical
methods.

Intermediate filaments are highly insoluble in physiological ionic solutions
and are extremely stable. As yet, only one compound is known that causes
their depolymerization, namely, β,β'-iminodipropionitrile (Feuilloley *et
al.*, 1994).

III. Cytoskeleton-Mediated Direct and Indirect Effects of Hormones

A. Growth, Growth Factors, Hormones

Following mitosis, cells will grow and realize a specific morphology. This
growth also requires a substantial reorganization of the cytoskeleton as *de
novo* synthesis of (some) cytoskeletal proteins at specific times. The major
cytoskeletal proteins, actin and tubulin, are polymers that can be assembled
or disassembled quickly, depending on the conditions. Cells usually contain
a pool of unpolymerized monomers that can be mobilized for quick actions.
In animal development the best-known factors controlling growth are
growth hormone (somatotropin) and growth factors: insulin-like growth
factor, epidermal growth factor, etc. Botanists invoke plant hormones for
elongation: auxin, gibberellins, etc. The borderline between growth factors
and hormones is sometimes very subtle or even nonexistent. Insulin itself
can act both ways. This hormone is the best documented one in the group
of peptides–proteins with respect to its possible effects on the cytoskeleton
(Table I). The list of functions exerted by insulin at the level of the cytoskeleton is already substantial (Al-Habori, 1993; Almås *et al.*, 1992a, 1992b).
Insulin has been reported to

Stimulate the synthesis of actin, certain cytokeratins, and glial fibrillary
acidic protein (GFAP)
Stimulate the polymerization of actin and the assembly of microtubules
Affect the general cellular morphology and to promote morphological
changes (e.g., membrane ruffling)
Cause phosphorylation of native tubulin and of fodrin and to stimulate
a serine–threonine kinase that phosphorylates microtubule-associated
protein 2 *in vitro* (MAP-2 kinase)
Shift ribosomes between cytoskeletal- and membrane-bound polysomes
Increase the actin content of the nuclear fraction

Cell growth and mitosis often go hand in hand. Growth factors may influence spindle function, as has been shown for bombesin in mitotic V 79 Chinese hamster cells (Önfelt, 1992). Development and growth are regulated by an interplay of growth stimulators and growth inhibitors (De Loof *et al.*, 1995). So far most attention has been paid to stimulators, but recently breakthroughs were realized in the discovery of growth inhibitors in insects. From two dipteran insects, the mosquito *Aedes aegypti* and the grey fleshfly *Neobellieria (Sarcophaga) bullata,* three peptides have been isolated that are potent inhibitors of oocyte growth, and one of them is a potent inhibitor of ecdysone biosynthesis (Borovsky *et al.*, 1990; Bylemans *et al.*, 1994, 1995; Hua *et al.*, 1994). These peptides do not show any resemblance to the activins and inhibins of vertebrates (Ackland *et al.*, 1992). These peptides seem to originate from matrix proteins in the broad sense, a novel source of compounds with hormonal action (De Loof *et al.*, 1995). Their exact role on the cytoskeleton remains to be identified.

The steroid 20-OH-ecdysone is the molting hormone of arthropods. Most of the research on its mode of action has dealt with effects at the level of the activation or inactivation of specific genes. Only few data are available on the cytoskeleton. However, ecdysone is a growth factor (Slama, 1982): it is exactly during molts that arthropods grow, and growth always involves changes in the cytoskeleton of the cells involved.

B. Transepithelial Transport: Vasopressin-Mediated Water Transport As an Example

As has already been mentioned, animals are to a large extent folded epithelia (De Loof, 1992). Many hormones therefore target epithelia and affect transepithelial transport. The best-documented hormone with respect to transepithelial transport is vasopressin, or antidiuretic hormone (ADH), with the amphibian urinary bladder serving as the favored model system (Table II). In general, investigators of epithelial transport refer to this ability of the cell to transport water as the *compliance* of the cell. It is possible that microfilaments may organize the cytoplasm in such a way as to facilitate the passage of water through the cell. Microtubules, on the other hand, may influence the permeability of surface membranes (Pearl and Taylor, 1983; Hall, 1984). It is not clear how the same hormone (vasopressin) applied by different research groups on the same epithelium (toad urinary bladder) can have exactly opposite effects on the actin skeleton. Holmgren *et al.* (1992) reported that ADH makes actin polymerize, while Ding *et al.* (1991) had described that it causes its depolymerization. Prat *et al.* (1993) reported that activation of epithelial Na^+ channels by protein kinase A requires actin filaments.

TABLE I

Interrelationship between the Cytoskeleton and Some Growth Influencing Vertebrate Hormones

Hormone and target	Functions	References
Growth hormone (somatostatin)		
Rat adipocytes	Stimulates the assembly of microtubules.	Soifer et al. (1971)
3T3 preadipose cells	Elicits alterations in vinculin expression and elevates levels of tubulin, whereas actin expression is unaffected.	Guller et al. (1991)
Insulin-like growth factor(s)		
Human neuroblastoma cells	Stimulates tubulin mRNA levels and neurite formation.	Mill et al. (1985)
Neonatal rat cardiomyocytes	Increases transcripts for skeletal α-actin but not for cardiac α-actin.	Ito et al. (1993)
Adult rat cardiomyocytes	Stimulates myofibril development and decreases smooth-muscle actin.	Donath et al. (1994)
Bovine chromaffin cells	Activates a microtubule-associated protein kinase.	Cahill and Perlman (1991)
Insulin		
Rat pancreas	Cytochalasin causes a decrease in the volume of beta cells and an increase in the secretion of insulin.	Orci et al. (1972)
Rat pancreas	Physiological and pharmacological stimulation of the secretion of insulin is accompanied by an increase in the proportion of tubulin in the polymerized form. Glucose induces the formation of microtubules in vitro.	Pipeleers et al. (1976, 1977)
Mouse brain cells	Stimulates neurite outgrowth.	Bhat (1983)
Human neuroblastoma cells	Stimulates tubulin mRNA levels and neurite formation.	Mill et al. (1985)
Rat adipocytes	Stimulates the assembly of microtubules.	Soifer et al. (1971)
Mouse 3T3 adipocytes	Causes breakdown or modification in microtubule organization.	Bockus and Stiles (1984)
Mouse 3T3 adipocytes	Rapidly stimulates a serine/threonine kinase that phosphorylates microtubule-associated protein 2 in vitro.	Ray and Sturgill (1987)
Rat pancreatic B-cell membrane	Binds preferentially on actin-rich microvilli.	Bendayan (1992)
Rat hepatocytes	Causes selective increase in synthesis of certain cytokeratins.	Leroux-Nicollet et al. (1983)
Embryonic human and mouse brain cells	Promotes neurite outgrowth.	Hooghe-Peeters et al. (1981)
		Recio-Pinto and Ishii (1988)
		Toran-Allerand et al. (1988)

Cell type	Effect	Reference
Porcine KB cells	Causes membrane ruffling, with more actin in association with membranes as result.	Goshima et al. (1984)
Porcine KB cells	No simultaneous changes in microtubule organization. Causes phosphorylation of fodrin (nonerythroid spectrin).	Kadowaki et al. (1985)
Human neuroblastoma cells	Stimulates tubulin mRNA levels and neurite formation.	Mill et al. (1985)
Rat hepatoma and human leukocytes	Increases the amount of G-actin.	Rao et al. (1985)
Pig erythrocytes	Reduces phosphorylation of spectrin.	Hesketh (1986)
3T3 fibroblasts	Reduces by 50% amount of actin released from detergent-insoluble cytoskeleton, suggesting that the hormone affects microfilament organization.	Hesketh and Pryme (1988)
	Increases the proportion of polysomes that are bound to the cytoskeleton.	Nanhua and Masters (1988)
Pig kidney cells	Increases the amount of G-actin.	Caron (1989)
Rat hepatocytes	Changes the stability of microtubule monomers, the level of nonpolymerized tubulin, and the rate of tubulin synthesis.	
Rat thyroid cells (FRTL5 cells)	Stimulates the phosphorylation of a cytoskeletal-associated phosphoprotein (pp 175).	Condorelli et al. (1989)
Krebs II ascites cells	24-hr stimulation causes extensive rearrangement of the microfilament network (cell adhesion and normal flattened morphology).	Pryme and Hesketh (1990)
Krebs II ascites cells	Changes the actin-binding properties of an already existing protein and the subcellular distribution of actin.	Vedeler et al. (1990)
	Promotes a shift of ribosomes into polysomes but also affects the distribution of ribosomes within the three polysome populations.	Vedeler et al. (1991)
Krebs II ascites cells	Increases actin content of nuclear fraction.	Almås et al. (1992a,b)
Krebs II ascites cells	Increases actin polymerization at the plasma membrane, independent of the synthesis of new cellular actin. Synthesis of new actin starts after 15 min.	Vedeler et al. (1991)
Krebs II ascites cells	Promotes entry of ribosomes into polysomes.	Kirkeeide et al. (1992)
Krebs II ascites cells	Changes pattern of proteins associated with cytoskeletal-bound polysome populations.	Moss et al. (1993)
Astroglia in cultures of E17 mouse cerebellum	Action on differentiation and expression of glial fibrillary acidic protein (GFAP) is dose dependent.	Toran-Allerand et al. (1991)

(continued)

TABLE I (*Continued*)

Hormone and target	Functions	References
3T3 preadipose cells	During differentiation important changes occur in actin and tubulin, and other proteins that make up the cytoskeleton, forming part of the lipid transportation within these cells.	O'Shea Alvarez (1991)
Rat muscle	Stimulates binding of phosphofructokinase and aldolase to muscle cytoskeleton.	Chen-Zion *et al.* (1992)
MPC-11 cells	Alters the pattern of proteins associated with ribosomes/mRNA in the three polysome preparations (free, cytoskeleton bound, and membrane bound).	Moss *et al.* (1991)
Diabetic rats	Strongly increases actin and desmin immunoreactivity, yet tubulin and vimentin immunoreactivity are unchanged.	Bestetti *et al.* (1992)
Rat H4IIE hepatoma cells (H4 cells)	Increases rapidly, but transiently, transcription of β-actin and α-tubulin genes.	Messina (1992)
Bombesin and bombesin-like peptides		
Normal human bronchial epithelial cells	Are growth factors.	Willey *et al.* (1984)
Human small cell lung cancer	Act as autocrine growth factors.	Cuttitta *et al.* (1985)
3T3 cells	Phosphorylates microtubule-associated protein kinase.	Miyasaka *et al.* (1992)
Quiescent 3T3 cells	Stimulation of p125 focal adhesion kinase/tyrosine phosphorylation requires integrity of the cytoskeleton.	Sinnett-Smith *et al.* (1993)

TABLE II

The Cytoskeleton, Hormones, and Transepithelial Transport: Vasopressin (ADH) As an Example

Model system and effects	References
Urinary bladder of the toad Bufo marinus	
Colchicine, vinblastine, podophyllotoxin, and cytochalasin B inhibit the action of vasopressin and cAMP on osmotic water movement across epithelium.	Taylor et al. (1973)
The initiation but not the maintenance of the intracellular transport of membrane-bound particles and their fusion is inhibited by colchicine, while cytocholasin B has only an effect on transport, not on fusion.	Kachadorian et al. (1979) Muller et al. (1980)
Cytochalasin B interferes with the pathway for water molecules and hence inhibits the response to ADH. Colchicine acts principally on the mucosal membrane, where microtubules enhance the permeability of the membrane to water.	Hardy and DiBona (1982)
ADH activates actin-containing filaments.	Davies et al. (1985)
ADH induces changes in the three-dimensional structure of apical surface.	Hartwig et al. (1987)
ADH depolymerizes F-actin.	Ding et al. (1991)
ADH induces polymerization of F-actin.	Holmgren et al. (1992)
Urinary bladder of frog Rana esculenta	
ADH causes intracellular transport of membrane-bound particles from internal but unidentified membrane to the cell surface membrane.	Bourquet et al. (1976)
Rat kidney collecting ducts	
ADH stimulates formation of coated pits.	Brown and Orci (1983)
Isolated rat hepatocytes	
ADH stimulates phosphorylation of myosin light chain 1.	Yamaguchi et al. (1991)

In recent years a number of diuretic peptides have been identified in insects. *Locusta*-vasopressin (Proux *et al.,* 1987) is one, but the best-documented ones belong to the corticotropin releasing factor family (Clottens *et al.,* 1994). Some myotropins also affect diuresis (Schoofs *et al.,* 1993). These peptides act on Malpighian tubules, which can easily be dissected out of the insect body and used in *in vitro* diuresis assays. The Malpighian tubule–diuretic hormone system of insects might be a good alternative to the urinary bladder of amphibians for studying the role of the cytoskeleton in water transport. Some of the changes the Malpighian tubules can undergo and some functional differences between primary and secondary cell types have been described by Meulemans and De Loof (1990a, 1990b; 1992).

There is still a lot of work to do in this domain. Despite many years of research, the transport of water through a cell and the role of the cytoskeleton therein are still poorly understood.

C. Role of the Cytoskeleton in Regulation of Steroidogenesis

The field of research into the role of the cytoskeleton in the regulation of steroidogenesis was opened by Crivello and Jefcoate (1978), who studied the effects of both a series of inhibitors of protein synthesis and a series of agents that inhibit the formation of microtubules and microfilaments on corticosterone synthesis by adrenals of female rats. Cytochalasin B, colchicine, and three other inhibitors of microtubule formation inhibited steroidogenesis without inhibiting protein synthesis and without any reversal effect with prolonged incubation. The effects were fully reversible. All inhibitors also inhibited the combination of cholesterol with cytochrome P-450$_{scc}$ in adrenal mitochondria, which correlated well with decreases in plasma corticosterone concentration. A labile protein is required for cholesterol transport to cytochrome P-450$_{scc}$. Additional data were collected, and the state of the art in this domain was recently summarized by Hall and Almahbobi (1992) (Table III).

Steroidogenesis begins in the inner mitochondrial membrane, where cholesterol is converted to pregnenolone. Cholesterol ester is stored in lipid droplets that are, just like the mitochondria, attached to the intermediate filament system. The first step, a slow one, involves the transport of cholesterol from the lipid droplets to the first enzyme in the pathway (P-450$_{scc}$). ACTH and cAMP stimulate this intracellular transport, but their effect can be inhibited by cytochalasins, by anti-actin antibodies, and by DNase I. This suggests that the response to ACTH requires a pool of monomeric (G-) actin that can be polymerized to F-actin. Thus, transport of cholesterol involves reorganization and contraction of actin filaments that may, in turn, cause movement of droplets and mitochondria together through their common attachment to intermediate filaments (Almahbobi *et al.,* 1992a, 1992b). Another well-studied model is the pituitary–gonad axis of vertebrates (Table IV). There seems to be a consensus about this model: We did not encounter substantial controversial data in the literature.

D. Thyroid and Parathyroid Functioning

There have been two waves in the research on the involvement of the cytoskeleton in thyroid hormone action. The first one started in 1970 with the discovery of Williams and Wolf that colchicine inhibits both the endocytotic accumulation of colloid in murine thyroid cells and the TSH-stimulated release of ^{131}I-labeled hormones. In the following years, until about 1985, the effects of a variety of cytoskeleton-disrupting agents were investigated. The second wave of research started around 1990 and was followed by the publication of much novel data on the effects of thyrotropin releasing hormone (TRH), thyrotropin stimulating hormone (TSH), thyroxine (T_4), and triiodothyronine (T_3) on different aspects of cytoskeletal organization and functioning (Table V). Cytoskeletal elements are needed for thyroxin formation (e.g., endocytotic accumulation of colloid in the thyroid cells) and release. A recent issue is the internalization of type II deiodinase. Little is known about the parathyroid hormone–cytoskeletal interrelationship (Table V).

E. Motility

1. Motility at the Cellular Level: Chemotaxis

All active movements carried out by animals ultimately rely on changes in the cytoskeleton of the cells involved. An amoeba moves by constantly remodeling its cytoskeleton: new assembly occurs at the tips of the forming pseudopods and depolymerization/breakdown occurs at the uroid site. Although to our knowledge this has not yet been well studied in any cell type involved in chemotaxis, it can be expected that a similar mechanism is used. Chemotaxis in animals is especially important in the context of immunology, as evidenced by the literature demonstrating a causal link between the endocrine and immunological systems (Table VI).

2. Myotropins: Control of Motility at the Muscular Level

As already mentioned, some hormones such as somatotropin (growth hormone), insulin, and some anabolic steroids have pronounced effects on muscle development. In the agricultural practice of some countries, some are used legally (e.g., bovine somatotropin) and others illegally (some steroids) for increasing meat (of which myosin and actin are major proteinaceous constituents) production in cattle and pork. Human somatostatin also stimulates muscle development (e.g., in elderly people).

Muscle contraction results from changing the relative positions of myosin and actin filaments. The role of Ca^{2+} in this process is well documented. The common situation is that muscles are innervated and that the release of an excitatory neurotransmitter causes depolarization of the sarcolemma, which is followed by a Ca^{2+} explosion, etc. In vertebrates the common neurotransmitter at neuromuscular junctions is acetylcholine, but especially in smooth muscles, a variety of peptides can also be involved, e.g., in vasoconstriction of blood vessels and in muscular activity of the gastrointestinal tract (gastrin, cholecystokinin, pancreatic polypeptide, etc.). Coelenterates have no true muscles: movements are realized by specializations of

TABLE III

Hormones, the Cytoskeleton, and Steroidogenesis: The Pituitary–Adrenals Axis and Vitamin D

Hormone and target	Functions	References
Adrenocorticotropic hormone (ACTH)		
Isolated adrenal cortex cells of female rats	Microfilaments play role in transport of cholesterol to mitochondria. Four inhibitors of microtubule formation inhibited steroidogenesis without inhibiting protein synthesis.	Crivello and Jefcoate (1978)
Human and rat adrenal cells	Causes extensive development of surface villi.	Armato et al. (1978)
Frog adrenals	Cytochalasin affects steroidogenesis.	Netchitailo et al. (1985)
Frog adrenals	Microfilaments play important role in steroidogenesis, but microtubules do not.	Feuilloley et al. (1986)
Frog adrenals	Required only for the coupling of the secretory response.	Feuilloley et al. (1988)
Rat adrenals	Increases expression of β-actin genes.	Imai et al. (1990)
Mouse adrenal tumor cells (Y-1 cells)	Cytochalasin B inhibits the action of ACTH and cAMP on steroidogenesis (cholesterol → pregnenolone inhibition).	Mrotek and Hall (1975, 1977)
Y-1 cells	Antimicrotubular agents stimulate steroid synthesis.	Ray and Scrott (1978)
Y-1 cells	Loading cells with anti-actin antibody prevents responses to both ACTH and cAMP by inhibiting transport of cholesterol to mitochondria.	Hall et al. (1979a)
Y-1 cells	Increases numbers of microvilli and pseudopodia; stress fibers are surrounded by disarranged microfilaments. High doses of cytochalasin B inhibit steroidogenesis.	Mattson and Kowal (1982)
Y-1 cells	Actin is involved in response to ACTH and cAMP.	Osawa et al. (1984)
Y-1 cells and cultured bovine fasciculata cells	Mitochondria and cholesterol-containing lipid droplets are attached to intermediate filaments.	Almahbobi et al. (1992a,b) Hall and Almahbobi (1992)
Pekin duck adrenal cells	Actin components are involved in the steroidogenic response to ACTH stimulation. Older cells have less well defined actin filaments and a lowered steroidogenic capacity.	Cronshaw et al. (1992)
Dexamethasone-treated rat adrenals	Increases expression of β-actin genes.	Ohno et al. (1992)
Human adrenal cells	Microtubules are required in a step preceding adenosine 3′,5′-cyclic monophosphate formation, whereas microfilaments are involved in a late and common step of adrenal steroidogenesis.	Feuilloley et al. (1994)
Adrenal cells	Causes reorganization of the actin cytoskeleton.	Raikhinstein and Hanukoglu (1994)
Rat zona fasciculata, not glomerulosa, cells	Increases peripheral cytoplasmic actin.	Loesser et al. (1994)

24

Gluco- and Corticosteroids

Perifused frog adrenals	Colchicine, nocodazole, and vinblastine do not affect spontaneous secretion of corticosterone and aldosterone. Vinblastine causes decrease in ACTH-induced stimulation of corticosterone and aldosterone production.	Feuilloley et al. (1986)
Rat brain	Acute and chronic treatment decreases mRNA of glial fibrillary acidic protein (GFAP).	Nichols et al. (1990)
Corticotrophs (AtT-20 cells) and human pneumocytes	Glucocorticoids stabilize actin filaments and inhibit corticotropin release by forming a stable open network near the plasma membrane that appears to hold secretory droplets in its interstices (modification of barrier theory).	Castellino et al. (1992)
Intact and adrenalized rats	Glucocorticoids regulate synthesis of GFAP.	O'Callaghan et al. (1991)
Mouse hepatoma cells	Glucocorticoids induced dissociation of the glucocorticoid binding protein from HSP90 and translocation of the receptor to the nucleus. HSP90 binds to actin filaments, and calmodulin or tropomyosin inhibits binding.	Miyata and Yahara (1991)
Rat C6 glioma cells	Hydrocortisone rearranges microtubules and modulates structure and function of the cell surface.	Armelin and Armelin (1983)
Adrenal tumor cells and primary cultures of bovine fasciculata cells	Intermediate filaments play a role in vectorial transport of lipid (cholesterol) droplets (encapsulated in a vimentin envelope) to mitochondria (site of steroidogenesis) which are also attached to such filaments.	Almahbobi et al. (1992a,b)
Mouse mammary epithelial cells	Glucocorticoids induce formation of tight junctions.	Zettl et al. (1992)

Angiotensin

Frog adrenal glands	Cytochalasin B inhibits basal corticosteroid output and its stimulation by corticotroph factors (e.g., angiotensin II).	Feuilloley et al. (1988)

Endothelin

Frog adrenocortical cells	This vasoconstrictor peptide is a potent stimulator of corticoid secretion. Colchicine antagonizes this effect. The intermediate filament inhibitor β,β'-iminodiproprionitrile has not effect on steroidogenesis.	Remy-Jouet et al. (1994)

1,25-dihydroxyvitamin D3

Mouse fibroblasts	Causes reorganization of the microfilament and microtubular systems.	Brackman et al. (1992)
Chick intestinal epithelium	Mediates alterations in microtubule proteins.	Nemere et al. (1991)

25

TABLE IV

Hormones, the Cytoskeleton, and Steroidogenesis: The Brain–Gonad Axis

Hormone and target	Functions	References
Gonadotropins (FSH, LH, HCG)		
Rat anterior pituitary cells in monolayer	Estradiol, TRH, 2-Br-α-ergocryptine, and colchicine affect morphology and prolactin secretion.	Antakly et al. (1979)
Rat Leydig cells	LH: microfilaments play role in transport of cholesterol to mitochondria.	Hall et al. (1979b)
Rat Leydig cells	Colchicine stimulates testosterone secretion.	Saltarelli et al. (1984)
Mouse Leydig cells	LH, cytochalasin, and vinblastine cause increase in androgen secretion and changes of cell shape toward regular and rounded; they produce an increase in 5,3β-HSD activity.	Bilinska (1989, 1992) Bilinska et al. (1989)
Sertoli cell of immature pig testis	FSH causes disorganization of microfilaments, while testosterone promotes their organization.	Chevalier and Dufaure (1981)
Rat Sertoli cells	FSH plus testosterone and FSH alone, but not testosterone alone, resulted in the peripheral distribution of actin and vinculin, which otherwise remained in stress-fiber-like structures.	Cameron and Muffly (1991)
Rat Sertoli cells in culture	FSH transforms cells from epithelial type to an astrocytic or fibroblast cell type. This is in part due to diminished F-actin stress fibers.	Tung et al. (1993)
Rat Graafian follicles	Addition of antiactin antiserum to the medium inhibits effects of LH.	Zor et al. (1978)
Rat granulosa cells	Gonadotropic regulation of steroidogenesis requires microtubules and microfilaments (cholesterol → mitochondria).	Carnegie et al. (1987) Carnegie and Tsang (1988)
Rat granulosa cells	Effects of gonadotropins on differentiation are associated with the coordinated regulation of cytoskeletal proteins involved in cell contact formation.	Ben-Ze'ev et al. (1987)
Rat granulosa cells	Affects expression of cytoskeletal proteins.	Kranen et al. (1993)
Rat granulosa cells	FSH causes breakdown of actin bundle filaments, GnRH counteracts this.	Amsterdam et al. (1994)
Granulosa cells (human, porcine, rat)	Down-regulation of the actin network seems to be a prerequisite for up-regulation of the steroidogenic enzymes.	Amsterdam et al. (1992)
Human granulosa cells	Gonadotropins alter morphology.	Soto et al. (1986)
Avian granulosa cells	Colchicine suppresses conversion of pregnenolone to progesterone.	Iczkowski and Hertelendy (1991)
Bovine cumulus oophorus	Microfilaments, microtubules, and intermediate filaments fulfill differential roles during gonadotropin-induced expansion.	Sutkovsky et al. (1994)

26

Estrogens

Rat hepatocytes	Dexamethasone or dexamethasone plus insulin reorganizes the cytoskeleton.	Marceau et al. (1983)
Microtubule proteins from porcine brain	Diethylstilbesterol inhibits in vitro polymerization.	Sato et al. (1984, 1987)
Human prostatic tumor cell line	Estrogen and derivatives (diethylstilbesterol) inhibit microtubule assembly and induce metaphase arrest.	Hartley-Asp et al. (1985)
Hamster V79 cells	Diethylstilbesterol and estradiol disrupt normal microtubule network.	Sakakibara et al. (1991); Sato et al. (1992)
Rat hypothalamic cells	Estradiol modifies astrocyte shape and glial fibrillary acidic protein (GFAP) immunoreactivity.	Garcia-Segura et al. (1989)
Rat brain	Estradiol induces redistribution of GFAP immunoreactivity.	Tranque et al. (1987)
Rat astrocytes	Estradiol promotes cell shape changes and redistribution of GFAP.	Torres-Aleman et al. (1992)
Rat neurite in vitro	Estrogen synergizes with insulin in enhancing neurite growth in vitro.	Toran-Allerand et al. (1988)
Rat uterus	17-β-estradiol activates transcription of β- and γ-cytoskeletal actin genes.	Cicatiello et al. (1992)
Rat uterus	17-β-estradiol induces accumulation of β-actin mRNAs.	Escalante et al. (1993)
Rat uterus	Estradiol and progesterone: various steroid hormone regimes produce characteristically different microfilament configurations.	Luxford and Murphy (1992)

Androgens

Embryonic mouse brain cells	Dehydroepiandrosterone enhances astrocytic differentiation.	Bologa et al. (1987)
Peritubular cells of monkey testis	Androgens induced appearance of α–smooth muscle actin.	Schlatt et al. (1993)
Specific motoneurons in rat brain.	Testosterone prevents the reduction of the expression levels of both gap junction and β-actin mRNAs induced by castration and contributes to neuronal plasticity.	Matsumoto et al. (1992, 1993)
Rat brain	Neonatal levels of sex steroids influence distribution of GFAP.	Garcia-Segura et al. (1988)
Rat brain	Estradiol, testosterone, and dihydrotestosterone can regulate the expression of the astrocyte-specific GFAP (role in synaptic plasticity).	Day et al. (1993)
Hippocampus dentate gyrus	Gonadal steroids influence axon sprouting in a sexually dimorphic manner.	Morse et al. (1986)

Progesterone

Rat granulosa cells	Modulation of actin skeleton is related to acute progesterone secretion.	Aharoni et al. (1993)

TABLE V

Interrelationships between Hormones of the Brain–Thyroid–Parathyroid Axis and the Cytoskeleton

Hormone and target	Functions	References
TRH (thyrotropin releasing hormone)		
Rat pituitary GH_4C_1 cells	Mediates reorganization of vinculin and fragmentation of actin filaments.	Kiley et al. (1992)
Anterior pituitary cells from estradiol-treated rats	Causes fast microtubule-associated protein phosphorylation.	Martinez de la Escalera and Weiner (1992)
TSH (thyrotropin)		
Mouse thyroid	Promotes the endocytotic accumulation of colloid in the thyroid cells; colchicine prevents this change.	Williams and Wolff (1970); Reaven and Reaven (1975)
Rat thyroid	Stimulates the release of ^{131}I-labeled hormones. Colchicine inhibits the response to TSH but does not alter the basal or unstimulated release.	Williams and Wolff (1970)
Rat thyroid	Induces changes in the morphology and organization of microfilament structures.	Tramontano et al. 1982
Rat thyroid	Colchicine, vinblastine, and cytochalasin B inhibit or completely block in vitro TSH-induced folliculogenesis of the thyroid gland. TSH affects the synthesis of several cytoskeletal proteins.	Pic et al. (1984); Passareiro et al. (1984)
Dog thyroid	Stimulates formation of villous processes that engulf stored colloid; stimulates thyroid secretion and colloid droplet formation. All are disturbed by colchicine and cytochalasin B. The inhibition of thyroid secretion bears on the first step of secretion: the endocytosis of colloid.	Nève et al. (1972)
Dog thyroid	Induces reorganization of microfilaments, disrupts actin-containing stress fibers, probably by dephosphorylating the actin-binding proteins cofilin and destrin.	Saito et al. (1994)
Bovine thyroid	Increases the amount of actin associated with lysosomes isolated from thyroid cells by centrifugation, and increases level of membrane-associated actin.	Dickson et al. (1979)
Human thyroid adenoma	Causes drastic change in morphology, including a total cytoplasmic arborization; induces disintegration of microfilament bundles.	Westermark and Porter (1982)
Porcine thyroid	Polarity of thyroid cells is important for iodine uptake, but the follicle structure is required for thyroid hormone synthesis.	Takasu et al. (1992)
Porcine thyroid	Removal of TSH causes redistribution of microfilaments and microtubule and the accumulation of myosin.	Yap and Manley (1994)
Thyroxine (T4)		
Rat astrocytes	Synergizes with insulin and hydrocortisone on differentiation.	Aizenman and de Vellis (1987)

Rat astrocytes	Promotes actin polymerization, induces binding of the deiodinase to F-actin, and promotes rapid internalization of the enzyme.	Farwell et al. (1990)
Rat astrocytes	Modulates the organization of the actin cytoskeleton. Restores the F-actin stress fibers, promoting the binding of deiodinase to F-actin, and stimulates enzyme internalization. Control of actin polymerization may contribute to thyroid hormone's influence on arborization, axonal transport, and cell–cell contact in developing brain.	Siegrist-Kaiser et al. (1990); Farwell and Leonard (1992)
Rat astrocytes	Targets type II iodothyronine 5'-deiodinase to an endosomal pool by activating specific protein–F-actin interactions involved in microfilament-mediated intracellular protein trafficking.	Farwell et al. (1993)
Rat glial cells	An intact actin cytoskeleton is required for thyroid hormone to modulate the energy-dependent regulation of the half-life of the short-lived, membrane-bound enzyme type II 5'-deiodinase.	Leonard et al. (1990)
Rat glial cells	Induces internalization of type II deiodinase.	Safran et al. (1992, 1993)
Mouse optic nerve	Hypothyroidism reduces the delivery of microtubules and neurofilaments to the distal axon.	Stein et al. (1991)
Rat brain	The genes encoding tubulin isotypes are differentially regulated during development.	Aniello et al. (1991)
Rabbit aortic vascular smooth muscle cells	Causes a switch in the expression of myosin isoforms.	Giuriato et al. (1991)
Rat brain	Thyroid hormone deficiency modifies the developmental pattern of expression of various tubulin isoforms.	Nunez et al. (1992)
Hearts of tight skin mouse	Significantly reduces mRNA for cytoskeletal and sarcomeric actin.	Yao and Eghbali (1992)
Developing rat brain	Hypothyroidism leads to increase in T α1 α-tubulin.	Figueiredo et al. (1993)
Triiodothyronine (T3)		
Rat neuronal and glial cells	Elicits 35–40% increase in the soluble tubulin content of glial cells only.	De et al. (1991, 1994)
Neonatal rat brain cells	Stimulates tubulin and actin synthesis.	De et al. (1994)
Human heart cells	Induces an increase in α-actin mRNA.	Collie and Muscat (1992)
Mevalonate		
FRTL-5 rat thyroid cells	Controls cytoskeleton organization.	Bifulco et al. (1993)
Parathyroid hormone		
Bovine parathyroid slices	Microtubules facilitate transport of newly synthesized prohormone to its intracellular site of cleavage to PTH.	Kemper et al. (1975)
Renal proximal cells	Induces changes in brush border topography and cytoskeleton.	Goligorski et al. (1986)
Rat osteoblasts	Promotes the disassembly of cytoskeletal actin and myosin.	Egan et al. (1991)

TABLE VI

Interrelationship between Some Hormones, the Cytoskeleton, and Chemotaxis

Target	Hormone and functions	References
Rabbit neutrophils	Substance P, a potent vasodilatory and smooth muscle contracting agent, binds to formylpeptide chemotaxis receptor.	Marasco et al. (1981)
Pig neutrophils	Calcitonin, somatostatin, and substance P exert differential effects on chemotaxis and phagocytosis.	Brzezinski and Dodlewski (1991)
Human peripheral blood mononuclear cells	β-endorphin and met-enkephalin stimulate chemotaxis.	Van Epps and Saland (1984)
Human monocytes in vitro	ACTH (1–24) and CRF stimulate migration.	Genedani et al. (1990, 1992)
Human polymorphonuclear leukocytes	Insulin, substance P, and somatostatin enhance chemotactic properties.	Cavalot et al. (1992) Partsch and Mattucci-Cerinic (1992)
Human polymorphonuclear leukocytes	Progesterone enhances chemotaxis and random migration.	Miyagi et al. (1992)
Normal and capsicain-treated rats	Substance P induces chemotaxis of neutrophils.	Helme et al. (1987)
Microvessels in rabbit skeletal muscle	Substance P activates leukocytes and platelets.	Öhlen et al. (1989)
Bovine bronchial epithelial cells	Tachykinins (mammals) induce the release of neutrophil chemotactic activity and recruit neutrophils.	Von Essen et al. (1992)
Rat alveolar macrophage	Vasoactive intestinal polypeptide (VIP) inhibits phagocytosis and chemotaxis (function in modulation of immune response in lung).	Litwin et al. (1992)

the cytoskeleton of myoepithelial cells. The best known transmitters of coelenterates are peptides. In insects and other arthropods, several dozen myotropic peptides have been isolated from nervous tissues. Proctolin was the first identified (Brown, 1977) and most investigated insect neuropeptide with pronounced myotropic functions (Slama *et al.,* 1993). In molluscs, FMRFamide-related peptides have been intensively studied (Price *et al.,* 1987). The number of identified molluscan myotropins continues to grow: -RFamides, APGWamide, myomodulins, etc. (Smit *et al.,* 1992; Li *et al.,* 1992, 1994). Some myotropic peptides are probably not released into the hemolymph and are therefore neurotransmitters or neuromodulators. Others can be immunolocalized in neurohemal organs (e.g., the corpora cardiaca) and then released into the hemolymph. The best-documented model insect in this respect is the migratory locust *Locusta migratoria,* from which no fewer than 44 fractions with myotropic activity were isolated. About half of them were present in large enough concentrations to be sequenced (Schoofs *et al.,* 1993). In recent years it has become clear that some peptides that were originally isolated because of their myotropic activity are also important regulators of physiological functions such as control of juvenile hormone biosynthesis, diapause induction, pigmentation, and stimulation of pheromone biosynthesis (Table VII).

3. Movement at the Organismal Level: Hormones and Behavior

Animal behavior can be defined as the sum of all movements made by an organism. Some hormones have drastic effects on behavior and thus on the cytoskeleton of the muscles involved. Well-known behavioral effects of hormones are related to reproduction, feeding and drinking, grooming, coping with stress, etc. A few examples are listed in Table VIII.

F. Oocyte Maturation

Oocyte maturation means the completion of meiosis. The envelope of the germinal vesicle breaks down during this process. There is also a substantial reorganization of the cytoskeleton that leads to the extrusion of the polar bodies and to a rearrangement of the complex cytoskeletal framework present in the egg cortex. The latter plays a role in the localization of some maternal mRNAs and in anchoring some membrane proteins. Depending on the species, a variety of molecules can act as hormonal inducers (Table IX).

IV. Hormones of Invertebrates and the Cytoskeleton

The best-known hormones of invertebrates are the steroid molting hormone (ecdysone and 20-OH-ecdysone) and the terpenoid juvenile hormones of

TABLE VII
Myotropic Peptides and Motility: A Few Examples

Target	Hormone and functions	References
Vertebrates		
Mammals	Cholecystokinin (CCK) stimulates colonic motility, gall bladder contraction, exocrine pancreatic secretion, etc.	Fioramonti and Buéno (1994)
Rat	Tachykinin induces bronchoconstriction by an indirect mechanism.	Joos et al. (1988)
Cultured vascular smooth muscle cells	Angiotensin II stimulates two microtubule-associated protein kinases.	Tsuda et al. (1992)
Human spermatozoa	Oxytocin suppresses motility in vitro.	Ratnasooriya and Premakumara (1993)
Molluscs		
	FMRFamide and related peptides	
Fifty species of bivalved molluscs	Predominantly, but not always, cardioexcitatory.	Painter and Greenberg (1982)
Muscle cells and neurons of Helix	Exhibits range of actions on motility.	Cottrell et al. (1983a,b)
		Lehman and Greenberg (1987)
		Price et al. (1990)
Tentacle retractor muscle of Helix	Evokes contraction.	Falconer et al. (1993)
Heart ventricle of mussel Geukensia	Causes positive chronotropy and inotropy within 5–15s.	Bayakly and Deaton (1992)
Insects: Locusta migratoria as a model		
	Myostimulating peptides	
Gut, oviduct, heart, skeletal muscle	Locustamyotropins (four variants; same active core as in the diapause hormone of Bombyx, the pheromone biosynthesis stimulating neuropeptide (PBAN) of Heliothis and Bombyx, and the red pigment concentrating hormone of Pseudaletia)	See review of Schoofs et al. (1993)

Locustatachykinins (five variants)
Locustapyrokinins (two variants)
Locustasulfakinin (related to gastrin-CCK)
Locustakinin
Locusta AKH I4-10
Locusta Accessory gland myotropin I and II (Lom-AG-MT I and Lom-AG-MT II)
Proctolin
Myoinhibiting peptides
Locustamyoinhibiting peptide
Locustamyosuppressin
Locutamyoinhibin

Gut, oviduct, heart, skeletal muscle

Allatostatin 3 of the blowfly *Calliphora* is the most potent inhibitor (active in *Calliphora* at 10^{-16} to 10^{-13} M) of peristaltic movements known   Duve and Thorpe (1994)

Allatostatins of the cockroach *Diploptera* inhibit contractions of the hindgut of cockroaches Lange et al. (1993); Veelaert et al. (1995)

Other insects   See reviews of Holman et al. (1990, 1991); Schoofs et al. (1993)

33

TABLE VIII

Hormones and Behavior: Some Selected Examples

Target	Hormone and functions	References
Rats	CRF affects grooming and ingestive behaviors.	Morley and Levine (1982)
Rats and/or mice	Neuropeptide Y is powerful stimulant of feeding behavior.	Clark *et al.* (1984) Levine and Morley (1984) Morley *et al.* (1987a,b) Stanley and Leibowitz *et al.* (1985) Corp *et al.* (1990) Steinman *et al.* (1994) Lynch *et al.* (1994)
Several mammalian species	Insulin and IGFs inhibit feeding.	Baskin and Wilcox (1988)
Several mammalian species	ACTH and MSH produce variety of behavioral effects (stretching, yawning syndrome, sexual stimulation, grooming, etc.)	Bertolini and Gessa (1981)
Mollusc *Aplysia*	Oviposition hormone controls oviposition behavior.	Chin *et al.* (1979) Newcomb and Scheller (1987)
Mollusc *Lymnaea*	Oviposition hormone controls oviposition behavior.	Ebberink *et al.* (1985)
Mollusc *Lymnaea*	Neuropeptides control mating behavior.	Li *et al.* (1992)
Moth *Manduca*	Eclosion hormone regulates shape of motoneurons.	Truman and Reiss (1988)

insects and other arthropods. The molting hormone is active at times of active growth (molts, metamorphosis, ovarian development) and can therefore be considered as a major growth factor. Relatively few studies have been devoted to its effects on the cytoskeleton (Table X). The mode of action of juvenile hormone is, despite many years of research, still largely unknown. The prevailing idea is that it acts in the same way as steroid hormones, namely, mainly through a nuclear receptor. In 1983 Van Mellaert *et al.* discovered that benzodioxoles are antagonists of juvenile hormone. Since then it has been shown that these compounds disturb the normal assembly of tubulin both *in vivo* and *in vitro* and that juvenile hormone antagonizes this effect (Chang *et al.*, 1994). This means that a major function of juvenile hormone is to stabilize the cytoskeleton, somewhat like the plant substance taxol, which is being intensively studied because of its potential in the treatment of some types of cancer. Taxol has no juvenile hormone properties when tested in the *Galleria* bioassay (De Loof, unpub-

TABLE IX

Hormones, the Cytoskeleton, and Maturation of Oocytes

Target	Hormone and functions	References
Echinoderms	Radial nerve growth factor causes ovaries to secrete 1-methyladenine, which causes germinal vesicle breakdown and completion of meiosis.	Grant (1978)
Xenopus oocytes	Maturation promoting factor causes hyperphosphorylation of the lamins of the nuclear envelope, followed by depolymerization and breakdown of the envelope.	Arion *et al.* (1988) Miake-Lye and Kirschner (1985) Peter *et al.* (1990) Ward and Kirschner (1990)
Xenopus oocytes	Progesterone-induced changes in the functional organization of the signal transduction pathway require the tubulin component of the cytoskeleton.	Matus-Leibovitch *et al.* (1993)
Xenopus oocytes	Progesterone causes microtubule reorganization and spindle assembly.	Gard (1992)
Locusta migratoria oocytes	Ecdysone causes induction of meiosis.	Goltzené *et al.* (1978)
Drosophila oocytes	A functionally specialized α-tubulin is required for oocyte meiosis.	Matthews *et al.* (1993)
Sheep oocytes	FSH and LH stimulate oocyte maturation.	De Smedt and Szollosi (1991)
Bovine oocytes	FSH, LH, and E2 affect the actin and tubulin cytoskeleton.	Allworth and Albertini (1993)

lished results). Perhaps the absence of juvenile hormone during metamorphosis causes a destabilization of the cytoskeleton of larval cells that leads, either directly or indirectly, to their programmed cell death.

Since 1985 there has been a boom in the isolation and identification of peptide hormones from invertebrates belonging to all major phyla, from coelenterates to arthropods and invertebrate chordates. The majority of these peptides are myotropins. Some of these myotropins, which were at first thought to be relatively unimportant, control important physiological processes such as induction of diapause, synthesis of pheromones, coloration, and biosynthesis of juvenile hormone (Schoofs *et al.*, 1993; Table VII). Well-known myotropins of vertebrates like cholecystokinin and tachykinins have family members in invertebrates (De Loof and Schoofs, 1990; Schoofs *et al.*, 1993). FMRFamide and related peptides, which are widely distributed in the animal kingdom, have been intensively studied in molluscs (Table VII).

TABLE X

Interrelationship between Some Invertebrate Hormones and the Cytoskeleton (see also Table VII)

Hormone and target	Functions	References
20-OH-ecdysone		
Persistent sensory neurons of *Manduca sexta*	Causes expansion of central arborizations during insect metamorphosis.	Levine (1989)
Drosophila Kc cells	Controls β-3-tubulin gene. There are enhancer and silencer elements in an intron.	de Cock *et al.* (1992) Bruhat *et al.* (1993) Tourmente *et al.* (1993)
Transgenic *Drosophila* cells	Influences expression of *Bombyx* actin genes	Abraham *et al.* (1993)
Intersegmental muscle cells of *Manduca*	Represses actin and myosin.	Schwartz (1992)
Drosophila	Induces the stubble–stubbloid gene that affects the organization of microfilament bundles during bristle morphogenesis.	Appel *et al.* (1993)
Fleshfly *Sarcophaga*	Causes changes in cytoskeletal actin pattern in Malpighian tubules during oogenesis.	Meulemans and De Loof (1992)
Juvenile hormones and analogues		
Ovarian follicle cells of *Rhodnius and Locusta*	Reduces their volume (patency).	Abu-Hakima and Davey (1977) Davey *et al.* (1993)
Blatella germanica	F-actin pattern in follicular epithelium changes during oogenesis.	Zhang and Kunkel (1992)
Drosophila ovaries	Stabilizes cytoskeleton and reverses disrupting effects of benzodioxoles on patency and microtuble assembly and coated vesicle numbers in oolemma.	Song *et al.* (1990) Ballarino *et al.* (1991)
Drosophila capitata	Interferes with assembly by microtubules during spermatogenesis and *in vitro*.	Chang *et al.* (1994)
Antigonadotropin		
Locusta migratoria (thoracic ganglia)	Antagonizes reduction in cell volume of ovarian follicle cells by JH III.	Davey *et al.* (1993)
Eclosion hormone		
Moth *Manduca sexta*	Regulates shape of identified motoneurons.	Truman and Reiss (1988)
ACTH/MSH (4–9)		
Central neurons of *Lymnaea*	Stimulates microtubule formation.	Müller *et al.* (1992)
Insulin and neuroparsin		
Locusta central nervous system	Promotes neurite outgrowth.	Vanhems *et al.* (1990)
Myotropins (see Table VII)		

V. Plant Hormones, the Cytoskeleton, and Cell Elongation

The cytoskeleton of plants, reviewed by Seagull (1989), Staiger and Lloyd (1991), Hepler *et al.* (1993), and Shibaoka and Nagai (1994), is more difficult to study than that of animals, and relatively few investigators are active in this domain. One should not assume exact homology between cytoskeletal arrays of plant and animal systems. During interphase the mainly cortical skeleton wraps around the inner face of the plasma membrane, where it helps to align nascent cellulose microfibrils (CMFs) produced by membrane synthases. The microfibrils, in turn, control cell shape by controlling the direction in which the cell can swell, which is influenced by plant growth substances and environmental signals (Staiger and Lloyd, 1991). Most studies on the effects of plant hormones on the cytoskeleton describe effects on the orientation of microtubules (MTs) and cellulose microfibrils (CMFs) in elongating cell systems (root and shoot tissue, seedlings, mature plants) (Table XI).

Plant cells having walls with CMFs deposited transversely to the direction of cell elongation can easily expand longitudinally, while parallel orientation of the CMFs to this axis is an obstacle. In zones with a high capability for longitudinal expansion—thus, in zones where plant hormones that affect elongation act—there are more cells with CMFs that are oriented transversely to the axis of cell elongation than there are in zones with reduced elongation. It has been hypothesized that the direction of the deposition of CMFs is controlled by the interaction of membrane-bound cellulose synthesizing complexes with cortical microtubules (Hepler and Palevitz, 1974; Staiger and Lloyd, 1991; Hepler *et al.*, 1993). Thus, the orientation of CMFs seems to follow from the orientation of the MTs according to this view. Factors that can change the orientation of microtubules will thus affect cell elongation. Plant hormones are known to affect the orientation of MTs in association with their effects on cell elongation. Gibberellin and auxin both induce alteration of MT orientation, which is accompanied by an alteration of the orientation of newly deposited CMFs on the inner surface of the cell wall to the same direction as that of MTs. In contrast with gibberellins and auxin, kinetin, abscisic acid, and ethylene, which usually inhibit cell elongation but promote lateral cell expansion, have been reported to have the opposite effect on MT orientation. They all increase the number of cells with longitudinally oriented MTs (Ishida and Katsumi, 1992). However, there are also data about cells in which MTs and CMFs are not in parallel alignment (Seagull, 1989). According to Emons *et al.* (1992), a third factor might be involved whose effects could be overruled by MTs where, for instance, they are more tightly attached to the plasma membrane than in other cells.

TABLE XI
Interrelationships between Plant Hormones and the Cytoskeleton

Hormone and target	Functions	References
Abscisic acid		
Epicotyl cells of dwarf pea	Abscisic acid and gibberellin A$_3$ cause the reorientation of cortical microtubules.	Sakiyama and Shibaoka (1990) Sakiyama-Sogo and Shibaoka (1993)
Hypocotyl cells of cucumber seedlings	Counteracts gibberellin (GA$_4$)-induced effects on microtubule orientation and cell elongation.	Ishida and Katsumi (1992)
Auxin		
Azuki bean epicotyls	Induces elongation, microtubule reorientation.	Shibaoka and Hogetsu (1977) Bergfeld *et al.* (1988)
Maize coleoptiles	Induces transverse microtubule arrangement.	Zandomeni and Schopfer (1993)
Ethylene		
Pea cells	Mediates the orientation of microtubules and cellulose microfibrils.	Steen and Chadwich (1981) Lang *et al.* (1982)
Vigna (mung bean)	Induces axially oriented microtubules.	Roberts *et al.* (1985)
Gibberellin(s)		
Azuki bean epicotyls and various plants	Regulate direction of cell expansion by altering the orientation of cortical microtubules, which, in turn, control the orientation of cellulose microfibrils. They promote stem elongation and stabilize cortical microtubules by affecting their association with the plasma membrane. Colchicine antagonizes cell elongation induced by GA$_3$ and cold.	Shibaoka (1972, 1974, 1991, 1993) Hogetsu *et al.* (1974) Katsumi and Ishida (1991)

Lettuce hypocotyls	Colchicine reduces or eliminates cell elongation.	Sawhney and Srivastava (1975)
Vigna angularis	Increases number of microfibrils in transverse orientation. Colchicine antagonizes GA_3.	Takeda and Shibaoka (1981)
Pea cells	GA_3 increases resistance of microtubules to colchicine disruption; gibberellin affects the arrangement and cold stability of cortical microtubules and causes reorientation of cortical microtubules without changes in the rate of cell elongation.	Lang *et al.* (1982) Akashi and Shibaoka (1987) Sakiyama-Sogo and Shibaoka (1993)
Onion leaf sheat cells	Stabilizes microtubules. Colchicine-induced cell swelling is reduced by GA_3 treatment.	Mita and Shibaoka (1984)
Cotton	Affects orientation of microtubules and wall microfibrils.	Seagull (1986)
Zea mays mesocotyl cells	Induces parallel microtubule orientation.	Mita and Katsumi (1986) Ishida and Katsumi (1991)
Zea mays root cortex	Several hormones may intervene in orientation of cortical microtubules and in regulation of cell growth polarity.	Blancaflor and Hasenstein (1993)
Mesophyll of wheat	Cytoskeleton plays role in cell shaping.	Wernicke and Jung (1992)
Cucumber hypocotyl cells	Orients cortical microtubules in parallel arrays, transverse to the cell long axis.	Ishida and Katsumi (1992)
Kinetin		
Azukia angularis	Induces microtubules to orient parallel to the long axis of cells. Prevented by colchicine.	Shibaoka (1974)

39

VI. Concluding Remarks

In recent decades endocrinologists studying the mode of action of hormones at the cellular and subcellular level have mainly focused on the interactions between hormones and their membrane and/or nuclear receptor(s), and on the cascades of molecular events that are triggered by these interactions (secondary messengers production and release, interactions with DNA, etc.). Despite all the efforts, the problem of tissue and developmental stage specificity of hormone action is not yet entirely solved. A variety of signaling molecules use the same secondary messengers but nevertheless make their target cells engage in different functions. This problem is related to the mechanisms underlying differentiation during embryonic development (De Loof, 1985). In our opinion the crucial role of the cytoskeleton herein has to date been largely overlooked or underestimated. The basic and universal principle underlying differentiation is likely to be not much more complex than this: "Keep the genome constant, but change again and again its macromolecular and/or ionic environment. Initiate this process by introducing asymmetry in the plasma membrane–cytoskeletal complex first; all the rest will follow from necessity" (De Loof *et al.,* 1992; De Loof, 1993b).

Another unsolved problem in endocrinology is that of "priming." Some cells or tissues do not immediately respond to a hormone when they are first exposed to it or respond only after a lag period or after having been exposed to another hormone(s). Perhaps a partial explanation for priming is that cells have to acquire their proper membrane, cytoplasmic, and nuclear framework before they can engage in certain functions: the "function follows form" principle (De Loof, 1993b).

The "cell as a miniature electrophoresis chamber" concept, in which all known dimensions of cells can simultaneously be taken into account, was proposed (De Loof, 1986). The discovery of the cablelike properties (conductance of electric signals) of actin (Lin and Cantiello, 1993) added strength to the view that in order to completely understand the mechanisms controlling gene expression, the electrical phenomena, which are such an important aspect of cellular communication (De Loof and Vanden Broeck, 1995), should be taken into account more than they are today.

In comparison with other domains of endocrinological research only relatively few studies have as yet been devoted to the interrelationship between hormones and the cytoskeleton. Nevertheless, it is becoming evident that effects of hormones on the cytoskeleton are likely to be the rule rather than the exception. It is unlikely that any hormone can exert its action(s) without using some elements of the cytoskeleton. As the methods for observing changes in the cytoskeleton become more and more refined [e.g., confocal laser microscopy, methods for measuring micromotion of individual cells (Giaever and Keese, 1991), and patch clamp methods of the type used by Lin and Cantiello (1993)], the more important the role of the cytoskeleton in signal transduction will turn out to be.

Acknowledgments

We thank all those who provided reprints and additional information, the Onderzoeksfonds of the K.U.Leuven, and the N.F.W.O. of Belgium for support of our research.

References

Abraham, E. G., Mounier, N., and Bosquet, G. (1993). Expression of a *Bombyx* cytoplasmic actin gene in cultured *Drosophila* cells: Influence of 20–hydroxyecdysone and interference with expression of endogenous cytoplasmic actin genes. *Insect Biochem. Mol. Biol.* **23**, 905–912.

Abu-Hakima, R., and Davey, K. G. (1977). The action of juvenile hormone on the follicle cells of *Rhodnius:* The importance of volume changes. *J. Exp. Biol.* **69**, 33–34.

Ackland, J. F., Schwartz, N. B., and Mayo K. E. (1992). Nonsteroidal signals originating in the gonads. *Physiol. Rev.* **72**, 731–787.

Agard, D. A., and Sedat, J. W. (1983). Three dimensional architecture of a polytene nucleus. *Nature (London)* **302**, 676–681.

Aharoni, D., Dantes, A., and Amsterdam, A. (1993). Cross-talk between adenylate cyclase activation and tyrosine activation leads to modulation of the actin cytoskeleton and to acute progesterone secretion in ovarian granulosa cells. *Endocrinology* **133**, 1426–1436.

Aizenman, Y., and de Vellis, J. (1987). Synergistic action of thyroid hormone, insulin and hydrocortisone on astrocyte differentiation. *Brain Res.* **414**, 301–308.

Akashi, T., and Shibaoka, H. (1987). Effects of gibberellin on the arrangement and cold stability of cortical microtubules in epidermal cells of pea internodes. *Plant Cell Physiol.* **28**, 339–348.

Akner, G., Mossberg, K., Wikstrom, A. C., Sundqvist K. G., and Gustafsson, J. A. (1991a). Evidence for colocalisation of glucocorticoid receptor with cytoplasmic microtubules in human gingival fibroblasts, using two different monoclonal anti-GR antibodies, confocal laser scanning microscopy, and image analysis. *J. Steroid Biochem. Mol. Biol.* **39**, 419–432.

Akner, G., Sundqvist, K. G., Denis, M., Wikstrom, A. C., and Gustafsson, J. A. (1991b). Immunocytochemical localisation of glucocorticoid receptor in human gingival fibroblasts and evidence for a colocalisation of glucocorticoid receptor with cytoplasmic microtubules. *Eur. J. Cell Biol.* **53**, 390–401.

Al-Habori, M. (1993). Mechanism of insulin action, role of ions and the cytoskeleton. *Int. J. Biochem.* **25**, 1087–1099.

Allworth, A. E., and Albertini, D. F. (1993). Meiotic maturation in cultured bovine oocytes is accompanied by remodelling of the cumulus cell cytoskeleton. *Develop. Biol.* **158**, 101–112.

Almahbobi, G., Williams, L. J., and Hall, P. F. (1992a). Attachment of mitochondria to intermediate filaments in adrenal cells: Relevance to the regulation of steroid synthesis. *Exp. Cell Res.* **200**, 361–369.

Almahbobi, G., Williams, L. J., and Hall, P. F. (1992b). Attachment of steroidogenic lipid droplets to intermediate filaments in adrenal cells. *J. Cell Sci.* **101**, 383–393.

Almås, B., Pryme, I. F., Vedeler, A., and Hesketh J. E. (1992a). Insulin: Signal transmission and short-term effects on the cytoskeleton and protein synthesis. *Int. J. Biochem.* **24,** 183–191.

Almås, B., Vedeler, A., and Pryme, I. F. (1992b). The effects of insulin, cycloheximide and phalloidin on the content of actin and p35 in extracts prepared from the nuclear fraction of Krebs II ascites cells. *Mol. Cell. Biochem.* **115,** 187–194.

Amsterdam, A., Dantes, A., and Liscovitch, M. (1994). Role of phospholipase-D and phosphatidic acid in mediating gonadotropin-releasing hormone-induced inhibition of preantral granulosa cell differentiation. *Endocrinology* **135,** 1205–1211.

Amsterdam, A., Plehn-Dujowich, D., and Suh B. S. (1992). Structure–function relationships during differentiation of normal and oncogene-transformed granulosa cells. *Biol. Reprod.* **46,** 513–522.

Aniello, F., Couchie, D., Gripois, D., and Nunez, J. (1991). Regulation of five tubulin isotypes by thyroid hormone during brain development. *J. Neurochem.* **57,** 1781–1786.

Antakly, T., Pelletier, G., Zeytinoglu, F., and Labrie, F. (1979). Effects of colchicine on the morphology and prolactin secretion of rat anterior pituitary cells in monolayer culture. *Am. J. Anat.* **156,** 353–372.

Appel, L. F., Prout, M., Abu-Shumays, R., Hammonds, A., Garbe, J. C., Fristrom, D., and Fristrom, J. (1993). The *Drososphila* stubble–stubbloid gene encodes an apparent transmembrane serine protease required for epithelial morphogenesis. *Proc. Natl. Acad. Sci. U.S.A.* **90,** 4937–4941.

Arion, D., Meijer, L., Brizuela, L., and Beach, D. (1988). cdc2 is a component of the M-phase specific histone H1 kinase: Evidence for identity with MPF. *Cell* **55,** 371–378.

Armato, U., Nussdorfer, F. F., Neri, G., Draghi, E., Andries, P. G., Mazzochi, G., and Mantero, F. (1978). Effects of ACTH and 3′,5′-cyclic purine nucleotides on the morphology and metabolism of normal adult human adrenocortical cells in primary culture. *Cell Tissue Res.* **190,** 187–205.

Armelin, M. C. S., and Armelin, H. A. (1983). Glucocorticoid hormone modulation of both cell surface and cytoskeleton related to growth control of rat glioma cells. *J. Cell Biol.* **97,** 459–465.

Avila, J. (1992). Microtubule functions. *Life Sci.* **50,** 327–334.

Ballarino, J., Song, Q., Ding, T., Chang, F., and Ma, M. (1991). Effects of benzodioxole on microtubule formation in the ovaries of *Drosophila melanogaster*. *Pest. Biochem. Physiol.* **41,** 258–264.

Baskin, D. G., and Wilcox, B. J. (1988). Insulin and insulin-like growth factors in the CNS. *TINS* **11,** 107–111.

Bassell, G. J. (1993). High resolution distribution of mRNA within the cytoskeleton. *J. Cell Biochem.* **52,** 127–133.

Bayakly, N. A., and Deaton, L. E. (1992). The effects of FMRFamide, 5-hydroxytryptamine, and phorbol esters on the heart of the mussel *Geukensia demissa*. *J. Comp. Physiol. B* **162,** 463–468.

Bendayan, M. (1992). Association of secreted insulin with particular domains of the pancreatic B-cell plasma membrane: The actin-rich microvilli. *J. Histochem. Cytochem.* **40,** 327–331.

Ben-Ze'ev, A. (1992). Cytoarchitecture and signal transduction. *Crit. Rev. Eukaryotic Gene Expression* **2,** 265–281.

Ben-Ze'ev, A., Kohen, F., and Amsterdam, A., (1987). Gonadotropin-induced differentiation of granulosa cells is associated with the coordinated regulation of cytoskeletal proteins involved in cell-contact formation. *Differentiation* **34,** 222–235.

Bergfeld, R., Speth, V., and Schopfer, D. (1988). Reorientation of microfibrils and microtubules at the outer epidermal wall of maize coleoptiles during auxin-mediated growth. *Bot. Acta* **101,** 133–138.

Bertolini, A., and Gessa, G. L. (1981). Behavioral effects of ACTH and MSH peptides. *J. Endocrinol. Invest.* **4,** 241–251.

Bestetti, G. E., Spycher, R., Brandli, P., and Rossi, G. L. (1992). Reduced number and LH content of marginated secretory granules in pituitary gonadotropes of streptozocin-induced diabetic male rats. *Horm. Res.* **38,** 177–183.

Bhat, N. R. (1983). Insulin-dependent neurite outgrowth in cultured embryonic mouse brain cells. *Dev. Brain Res.* **11,** 315–318.

Bifulco, M., Laezza, C., Aloj, S. M., and Garbi C. (1993). Mevalonate controls cytoskeleton organization and cell morphology in thyroid epithelial cells. *J. Cell Physiol.* **155,** 340–348.

Bilinska, B. (1989). Visualisation of the cytoskeleton in Leydig cells: *In vitro* effect of luteinizing hormone and cytoskeletal disrupting drugs. *Histochemistry* **93,** 105–110.

Bilinska, B. (1992). LH as a modulator of cytoskeleton arrangement and steroidogenic function in Leydig cells *in vitro. Folia Histochem. Cytobiol.* **30,** 35–37.

Bilinska, B., Viklicky, V., Draber, P., and Wojtusiak A. (1989). Luteinizing hormone–induced modifications of the cytoskeleton in cultured mouse Leydig cells. *Cytobios* **58,** 25–34.

Blancaflor, E. B., and Hasenstein, K. H. (1993). Organization of cortical microtubules in graviresponding maize roots. *Planta* **191,** 231–237.

Blaustein, J. D., and Olster, D. H. (1993). Colchicine-induced accumulation of estrogen receptor and progestin receptor immunoreactivity in atypical areas in guinea-pig brain. *J. Neuroendocrinol.* **5,** 63–70.

Bockus, B. J., and Stiles, C. D. (1984). Regulation of cytoskeletal architecture by platelet-derived growth factor, insulin, and epidermal growth factor. *Exp. Cell Res.* **153,** 186–197.

Bologa, L., Sharma, J., and Roberts, E. (1987). Dehydroepiandrosterone and its sulfated derivative reduce neuronal death and enhance astrocytic differentiation in brain cell cultures. *J. Neurosci. Res.* **17,** 225–234.

Borovsky, D., Carlson, D. A., Griffin, P. R., Shabanowitz, J., and Hunt, D. F. (1990). Mosquito oostatic factor: A novel decapeptide modulating modulating trypsin-like enzyme biosynthesis in the midgut. *FASEB J.* **4,** 3015–3020.

Bourquet, J., Chevalier, J., and Hugon, J. S. (1976). Alterations in membrane-associated particle distribution during antidiuretic challenge in frog urinary bladder epithelium. *Biophys. J.* **16,** 627–639.

Brackman, D., Trydal, T., Lillehaug, J. R., and Aarskog, D. (1992). Reorganization of the cytoskeleton and morphological changes induced by 1,25-dihydroxyvitamin D3 in C3H/10T1/2 mouse embryo fibroblasts: Relation to inhibition of proliferation. *Exp. Cell Res.* **201,** 485–493.

Bretscher, A. (1991). Microfilament structure and function in the cortical cytoskeleton. *Annu. Rev. Cell Biol.* **7,** 337–374.

Brown, B. E. (1977). Occurrence of proctolin in six orders of insects. *J. Insect Physiol.* **3,** 861–864.

Brown, D., and Orci, L. (1983). Vasopressin stimulates formation of coated pits in rat kidney collecting ducts. *Nature (London)* **302,** 253–255.

Bruhat, A., Dreau, D., Drake, M. E., Tourmente, S., Chapel, S., Couderc, J. L., and Dastugue, B. (1993). Intronic and 5' flanking sequences of the *Drosophila* beta 3 tubulin gene are essential to confer ecdysone responsiveness. *Mol. Cell. Endocrinol.* **94,** 61–71.

Brzezinski, P. M., and Dodlewski, A. (1991). The effects of calcitonin, somatostatin and analogue (Glp6) of substance P (6–11) on chemotaxis and phagocytosis of guinea pig neutrophils. *Folia Histochem. Cytobiol.* **29,** 153.

Bylemans, D., Borovsky, D., Hunt, D., Shabanowitz, J., Grauwels, L., and De Loof, A. (1994). Sequencing and characterization of trypsin modulating oostatic factor (TMOF) from the ovaries of the grey fleshfly, *Neobellieria (Sarcophaga) bullata. Regul. Pept.* **50,** 61–72.

Bylemans, D., Proost, P., Samyn, B., Borovsky, D., Grauwels, L., Huybrechts, R., Van Damme, J., Van Beeumen, J., and De Loof, A. (1995). Neb-colloostatin, a second folliculostatin of the grey fleshfly, *Neobellieria bullata. Eur. J. Biochem.* **228,** 45–49.

Cahill, A. L., and Perlman, R. L. (1991). Activation of a microtubule-associated protein-2 kinase by insulin-like growth factor-I in bovine chromaffin cells. *J. Neurochem.* **57**, 1832–1839.

Cameron, D. F., and Muffly, K. E. (1991). Hormonal regulation of spermatid binding. *J. Cell Sci.* **100**, 623–633.

Cantiello, H. F., Stow, J. L., Pratt, A. G., and Ausiello D. A. (1991). Actin filaments regulate epithelial Na^+ channel activity. *Amer. J. Physiol.* **261**, Pt. 1, C882–C888.

Carnegie, J. A., Dardick, I., and Tsang, B. K. (1987). Microtubules and the gonadotropic regulation of granulosa cell steroidogenesis. *Endocrinology* **120**, 819–828.

Carnegie, J. A., and Tsang, B. K. (1988). The cytoskeleton of rat granulosa cell steroidogenesis: Possible involvement of microtubules and microfilaments. *Biol. Reprod.* **38**, 100–108.

Caron, J. M. (1989). Alteration in microtubule physiology in hepatocytes by insulin. *J. Cell Physiol.* **138**, 603–610.

Castellino, F., Heuser, J., Marchetti, S., Bruno, B., and Luini, A. (1992). Glucocorticoid stabilization of actin filaments: A possible mechanism for inhibition of corticotropin release. *Proc. Nat. Acad. Sci. U.S.A.* **89**, 3775–3379.

Cavalot, F., Anfossi, G., Russo, I., Mularoni, E., Massucco, P., Burzacca, S., Mattiello, L., and Trovati M. (1992). Insulin, at physiological concentrations, enhances the polymorphonuclear leukocyte chemotactic properties. *Horm. Metab. Res.* **24**, 225–228.

Chang, F., Jang E. B., Hsu, C. L., Ma, M., and Jurd, L. (1994). Benzodioxole-1,3-benzodioxole derivatives and their effects on the reproductive physiology of insects. *Arch. Insect Biochem. Physiol.* **27**, 39–51.

Chen-Zion, M., Livnat, T., and Beitner, R. (1992). Insulin rapidly stimulates binding of phosphofructokinase and aldolase to muscle cytoskeleton. *Int. J. Biochem.* **24**, 821–826.

Chevalier, M., and Dufaure, J. P. (1981). Effect of FSH, testosterone and Ca^{2+} on Sertoli cell microfilaments in the immature pig testis. *Biol. Cell* **41**, 105–112.

Chin, A. Y., Hunkapillar, M. W., Heller, E., Stuart, D. K., Hood, L. E., and Strumwasser F. (1979). Purification and primary structure of the neuropeptide egg-laying hormone of *Aplysia californica. Proc. Natl. Acad. Sci. U.S.A.* **76**, 6656–6660.

Cicatiello, L., Ambrosino, C., Coletta, B., Scalona, M., Sica, V., Bresciani, F., and Weisz, A. (1992). Transcriptional activation of jun and actin genes by estrogen during mitogenic stimulation of rat uterine cells. *J. Steroid. Biochem. Mol. Biol.* **41**, 523–528.

Clark, J. T., Kalra, P. S., Crowley, W. R., and Kalra, S. P. (1984). Neuropeptide Y and human pancreatic polypeptide stimulate feeding behavior in rats. *Endocrinology* **115**, 427–429.

Clottens, F. L., Holman, G. M., Coast, G. M., Totty, N. F., Hayes, T. K., Kay, I., Mallet, A. I., Wright, M. S., Chung, J.-M., Truong, O., and Bull, D. L. (1994). Isolation and characterization of a diuretic peptide common to the house fly and stable fly. *Peptides* **15**, 971–979.

Collie, E. S., and Muscat, G. E. (1992). The human skeletal alpha-actin promoter is regulated by thyroid hormone: Identification of a thyroid hormone response element. *Cell Growth Differ.* **3**, 31–42.

Condorelli, G., Formisano, P., Villone, G., Smith, R. J., and Beguinot, F. (1989). Insulin and insulin-like growth factor I (IGF I) stimulate phosphorylation of a Mr 175,000 cytoskeleton-associated protein in intact FRTL5 cells. *J. Biol. Chem.* **264**, 12,633–12,638.

Corp, E. S., Melville, L. D., Greenberg, D., Gibbs, J., and Smith, G. P. (1990). Effects of fourth venticular neuropeptide Y and peptide YY on ingestive and other behaviors. *Am. J. Physiol.* **259**, R317–R323.

Cottrell, G. A., Greenberg, M. J., and Price, D. A. (1983a). Differential effects of the molluscan neuropeptide FMRFamide and the related met-enkephalin derivative YGGFMRFamide on the *Helix* tentacle retractor muscle. *Comp. Biochem. Physiol.* **C75**, 373–375.

Cottrell, G. A., Schott, P. C., and Dockray, G. J. (1983b). Identification and probable role of a single neurone containing the neuropeptide *Helix* FMRFamide. *Nature (London)* **304**, 636–640.

Coutavas, E., Ren, M., Oppenheim, J. D., D' Eustachio, P., and Rush, M. (1993). Characterization of proteins that interact with the cell-cycle regulatory protein Ran/TC4. *Nature* **366,** 585–587.

Crivello, J. F., and Jefcoate, C. R. (1978). Mechanism of corticotropin action in rat adrenal cells: I. The effects of inhibitors of protein synthesis and of microfilament formation on corticosterone synthesis. *Biochim. Biophys. Acta* **542,** 315–329.

Cronshaw, J., Reese, B. K., Collie, M. A., and Holmes, W. N. (1992). Cytoskeletal changes accompanying ACTH-induced steroidogenesis in cultured embryonic adrenal gland cells from the Pekin duck. *Cell Tissue Res.* **268,** 157–165.

Cuttita, F., Carney, D. M., Mulshine, J., Moody, T. W., Fedorko, J., Fisher, A., and Minna, J. D. (1985). Bombesin-like peptides can function as autocrine growth factors in human small-cell lung cancer. *Nature (London)* **316,** 823–826.

Davey, K. G., Sevala, V. L., and Gordon, D. R. B. (1993). The action of juvenile hormone and antigonadotropin on the follicle cells of *Locusta migratoria. Invert. Rep. Develop.* **24,** 39–46.

Davies, E., Corner, E. C., Lionberger, J. M., Stankovic, B., and Abe, S. (1993). Cytoskeleton-bound polysomes in plants. III. Polysome–cytoskeletal–membrane interactions in corn endosperm. *Cell Biol. Int.* **17,** 331–340.

Davies, W. L., Jones, R. G., Richemont, P. C., and Goodman, D. P. P. (1985). Activation of actin-containing microfilaments by vasopressin in the amphibian urinary bladder epithelium: A fluorescent study using NBD phallacidin. *Anat. Rec.* **211,** 239–245.

Day, J. R., Laping, N. J., Lampert-Etchells, M., Brown, S. A., O'Callaghan, J. P., McNeill, T. H., and Finch, C. E. (1993). Gonadal steroids regulate the expression of glial fibrillary acidic protein in the adult male rat hippocampus. *Neuroscience* **55,** 435–443.

De, A., Chaudhury, S., and Sarkar, P. K. (1991). Thyroidal stimulation of tubulin and actin in primary cultures of neuronal and glial cells of rat brain. *Int. J. Dev. Neurosci.* **9,** 381–390.

De, A., Das, S., Chaudhury, S., and Sarkar, P. K. (1994). Thyroidal stimulation of tubulin and actin in rat brain cytoskeleton. *Int. J. Dev. Neurosci.* **12,** 49–56.

de Cock, J. G., Klink, E. C., Ferro, W., Lohman, P. H., and Eeken, J. C. (1992). Neither enhanced removal of cyclobutane pyrimidine dimers nor strand-specific repair is found after transcription induction of the beta 3-tubulin gene in a *Drosophila* embryonic cell line Kc. *Mutat. Res.* **293,** 11–20.

De Loof, A. (1985). A possible solution for the problem of tissue specificity of hormone action as related to the process of differentiation. *Annls. Soc. R. Zool. Belg.* **115,** 121–136.

De Loof, A. (1986). The electrical dimension of cells: The cell as a miniature electrophoresis chamber. *Int. Rev. Cytol.* **104,** 251–352.

De Loof, A. (1992). All animals develop from a blastula: Consequences of an undervalued definition for thinking on animal development. *BioEssays* **14,** 373–375.

De Loof, A. (1993a). Schrödinger 50 years ago: "What is life?" The ability to communicate, a plausible reply? *Int. J. Biochem.* **25,** 1715–1721.

De Loof, A. (1993b). Differentiation: "Keep the genome constant but change over and over again its ionic and/or macromolecular environment"? A conceptual synthesis. *Belg. J. Zool.* **123,** 77–91.

De Loof, A., Bylemans, D., Schoofs, L., Janssen, I., Spittaels, K., Vanden Broeck, J., Huybrechts, R., Borovsky, D., Hua, Y.-J., Koolman, J., and Sower, S. (1995). Folliculostatins, gonadotropins, and a model for control of growth in the grey fleshfly, *Neobellieria bullata. Insect Biochem. Molec. Biol.* **25,** 661–667.

De Loof, A., Callaerts, P., and Vanden Broeck, J. (1992). The pivotal role of the plasma membrane–cytoskeletal complex and of epithelium formation in differentiation in animals. *Comp. Biochem. Physiol.* **101A,** 639–651.

De Loof, A., and Schoofs, L. (1990). Homologies between the amino acid sequences of some vertebrate peptide hormones and peptides isolated from invertebrate sources. *Comp. Biochem. Physiol.* **95 B,** 459–468.

De Loof, A., and Vanden Broeck, J. (1995). Communication: The key to defining "Life," "Death" and the force driving evolution. "Organic chemistry–based" versus "artificial" life. *Belg. J. Zool.,* **125,** 5–28.

De Smedt, V., and Szollosi, D. (1991). Cytochalasin D treatment induces meiotic resumption in follicular sheep oocytes. *Mol. Reprod. Dev.* **29,** 163–171.

den Hartig, J. C., van Bergen en Henegouwen, P. M. P., Verkley, A. J., and Boonstra, J. (1992). The EGF–receptor is an actin-binding protein. *J. Cell Biol.* **119,** 349–355.

Dickson, J. G., Malon, P. G., and Ekins, R. P. (1979). The association of actin with a thyroid lysosomal fraction. *Eur. J. Biochem.* **97,** 471–479.

Ding, G., Franki, N., Condeelis, J., and Hayes, R. M. (1991). Vasopressin depolymerizes F-actin in the toad bladder epithelial cell. *Am. J. Physiol.* **260** (*Cell Physiol.* **29**), C9–C16.

Donath, M. Y., Zapf, J., Eppenberger-Eberhardt, M., Froesch, E. R., and Eppenberger, H. M. (1994). Insulin-like growth factor I stimulates myofibril development and decreases smooth muscle alpha-actin of adult cardiomyocytes. *Proc. Natl. Acad. Sci. U.S.A.* **91,** 1686–1690.

Drenckhahn, D., Jons, T., Kollertjons, A., Koob, R., Kraemer, D., and Wagner, S. (1993). Cytoskeleton and epithelial polarity. *Renal Physiol. Biochem.* **16,** 6–14.

Duve, H., and Thorpe, A. (1994). Distribution and functional significance of Leu-callatostatins in the blowfly, *Calliphora vomitoria. Cell Tissue Res.* **276,** 367–379.

Ebberink, R. H. M., van Loenhout, H., Geraerts, W. P. M., and Joosse, J. (1985). Purification and amino acid sequence of the ovulation hormone of *Lymnaea stagnalis. Proc. Natl. Acad. Sci. U.S.A.* **82,** 7767–7771.

Egan, J. J., Gronowicz, G., and Rodan, G. A. (1991). Parathyroid hormone promotes the disassembly of cytoskeletal actin and myosin in cultured osteoblastic cells: mediation by cyclic AMP. *J. Cell Biochem.* **45,** 101–111.

Emons, A. M. C., Derksen, J., and Sassen, M. M. A. (1992). Do microtubules orient plant cell wall microfibrils? *Physiol. Plant.* **84,** 486–493.

Escalante, R., Houdebine, L. M., and Pamblanco, M. (1993). Transferrin gene expression in the mammary gland of the rat. The enhancing effect of 17-beta-oestradiol on the level of RNA is tissue-specific. *J. Mol. Endocrinol.* **11,** 151–159.

Falconer, S. P. W., Carter, A. N., Downes, C. P., and Cottrell, G. A. (1993). The neuropeptide Phe-Met-Arg-Phe-NH$_2$ (FMRFamide) increases levels of inositol 1,4,5-triphosphate in the tentacle retractor muscle of *Helix aspersa. Exp. Physiol.* **78,** 757–766.

Farwell, A. P., DiBenedetto, D. J., and Leonard, J. L. (1993). Thyroxine targets different pathways of internalisation of type II iodothyronine 5' deiodinase in astrocytes. *J. Biol. Chem.* **268,** 5055–5062.

Farwell, A. P., and Leonard, J. L. (1992). Dissociation of actin polymerisation and enzyme inactivation in the hormonal regulation of type II iodothyronine 5'-deiodinase activity in astrocytes. *Endocrinology* **131,** 721–728.

Farwell, A. P., Lynch, R. M., Okulicz, W. C., Comi, A. M., and Leonard, J. L. (1990). The actin cytoskeleton mediates the hormonally regulated translocation of type II iodothyronine 5'-deiodinase in astrocytes. *J. Biol. Chem.* **265,** 18546–18553.

Feuilloley, M., Contesse, V., Lefèbvre, H., Delarue, C., and Vaudry, H. (1994). Effects of selective disruption of cytoskeletal elements on steroid secretion by human adrenocortical slices. *Am. J. Physiol.* (*Endocrinol. Metab.* **29**) **266,** E202–E210.

Feuilloley, M., Netchitailo, P., Delarue, C., Leboulanger, F., Benyamina, M., Pelletier, G., and Vaudry, H. (1988). Involvement of the cytoskeleton in the steroidogenic response of frog adrenal glands to angiotensin II, acetylcholine, and serotonin. *J. Endocrinol.* **118,** 365–374.

Feuilloley, M., Netchitailo, P., Lihrman, I., and Vaudry, H. (1986). Effect of vinblastine, a potent antimicrotubular agent on steroid secretion by perifused frog adrenal cells. *J. Steroid Biochem.* **25,** 143–147.

Figueiredo, B. C., Almazan, G., Ma, Y., Tetzlaff, W., Miller, F. D., and Cuello, A. C. (1993). Gene expression in the developing cerebellum during perinatal hypo- and hyperthyroidism. *Mol. Brain Res.* **17,** 258–268.

Fioramonti, J., and Buéno, L. (1994). Cholecystokinin: A pleiotropic role in the control of gut motility. *Eur. J. Gastroenterol. Hepatol.* **6,** 377–379.

Garcia-Segura, L. M., Suarez, I., Segovia, S., Tranque, P. A., Cales, J. M., Aguilera, P., Olmos, G., and Guillamon, A. (1988). The distribution of glial fibrillary acidic protein in the adult rat brain is influenced by the neonatal levels of sex steroids. *Brain Res.* **456,** 357–363.

Garcia-Segura, L. M., Torres-Aleman, I., and Naftolin, F. (1989). Astrocyte shape and glial fibrillary acidic protein immunoreactivity are modified by estradiol in primary rat hypothalamic cultures. *Dev. Brain Res.* **47,** 298–302.

Gard, D. L. (1992). Microtubule organization during maturation of *Xenopus* oocytes assembly and rotation of the meiotic spindles. *Develop. Biol.* **151,** 516–530.

Genedani, S., Bernardi, M., Baldini, M. G., and Bertolini, A. (1990). ACTH (1–24) stimulates the migration of human monocytes *in vitro*. *Peptides* **11,** 1305–1307.

Genedani, S., Bernardi, M., Baldini, M. G., and Bertolini, A. (1992). Influence of CRF and alpha-MSH on the migration of human monocytes *in vitro*. *Neuropeptides* **23,** 99–102.

Giaever, I., and Keese, C. R. (1991). Micromotion of mammalian cells measured electrically. *Proc. Natl. Acad. Sci. U.S.A.* **88,** 7896–7900.

Giuriato, L., Borrione, A. C., Zanellato, A. M., Tonello, M., Scatena, M., Scannapieco, G., Pauletto, P., and Sartore, S. (1991). Aortic intimal thickening and myosin isoform expression in hyperthyroid rabbits. *Arterioscler. Thromb.* **11,** 1376–1389.

Goligorsky, M. S., Menton, D. N., and Hruska, K. A. (1986). Parathyroid hormone–induced changes of the brush border topography and cytoskeleton in cultured renal proximal tubular cells. *J. Membr. Biol.* **92,** 151–162.

Goltzené, F., Lagueux, M., Charlet, M., and Hoffmann, J. A. (1978). The follicle cell epithelium of maturing ovaries of *Locusta migratoria:* A new biosynthetic tissue for ecdysone. *Z. Phys. Chem.* **359,** 1427–1434.

Goshima, K., Masuda, A., and Owaribe, K. (1984). Insulin-induced formation of ruffling membranes of KB cells and its correlation with enhancement of amino acid transport. *J. Cell Biol.* **98,** 801–809.

Grant, P. (1978). "Biology of Developmental Systems." Holt, Rinehart and Winston, New York.

Guller, S., Corin, R. E., Yuan-Wu, K., and Sonenberg, M. (1991). Up-regulation of vinculin expression in 3T3 preadipose cells by growth hormone. *Endocrinology* **129,** 527–533.

Hall, P. F. (1984). The role of the cytoskeleton in hormone action. *Can. J. Biochem. Cell Biol.* **62,** 653–665.

Hall, P. F., and Almahbobi, G. (1992). The role of the cytoskeleton in the regulation of steroidogenesis. *J. Steroid Biochem. Mol. Biol.* **43,** 769–777.

Hall, P. F., Charponnier, C., Nakamura, M., and Gabbiani, G. (1979a). The role of microfilaments in the response of adrenal tumor cells to adrenocorticotropic hormone. *J. Biol. Chem.* **254,** 9080–9084.

Hall, P. F., Charponnier, C., Nakamura, M., and Gabbiani, G. (1979b). The role of microfilaments in the response of Leydig cells to luteinizing hormone. *J. Steroid Biochem.* **11,** 1361–1366.

Hardy, M. A., and DiBona, D. R. (1982). Microfilaments and the hydrosmotic action of vasopressin in toad urinary bladder. *Am. J. Physiol.* **243,** C200–C204.

Hartley-Asp, B., Deinum, J., and Wallin, M. (1985). Diethylstilbesterol induces metaphase arrest and inhibits microtubule assembly. *Mutation Res.* **143,** 231–235.

Hartwig, J. H., Ausiello, D. H., and Brown, D. (1987). Vasopressin induced changes in the three-dimensional structure of toad bladder apical surface: *Am. J. Physiol.* **253** (*Cell Physiol.* **22**), C707–C720.

Hartwig, J. H., and Kwiatkowski, D. J. (1991). Actin-binding proteins. *Curr. Opin. Cell Biol.* **3**, 87–97.

Helme, R. D., Eglezos, A., and Hosking, C. S. (1987). Substance P induces chemotaxis of neutrophils in normal and capsicain-treated rats. *Immunol. Cell Biol.* **65**, 267–269.

Hepler, P. K., Cleary, A. L., Gunning, B. E. S., Wadsworth, P., Wasteneys, G. O., and Zhang, D. H. (1993). Cytoskeletal dynamics in plant cells. *Cell Biol. Int.* **17**, 127–142.

Hepler, P. K., and Palevitz, B. A. (1974). Microtubules and microfilaments. *Annu. Rev. Plant Physiol.* **25**, 309–362.

Hesketh, J. E. (1986). Insulin inhibits the phosphorylation of the membrane cytoskeletal protein spectrin in pig erythrocytes. *Cell Biol. Int. Rep.* **10**, 623–629.

Hesketh, J. E., and Pryme, I. F. (1988). Evidence that insulin increases the proportion of polysomes that are bound to the cytoskeleton in 3T3 fibroblasts. *FEBS Lett.* **231**, 62–66.

Hinshaw, J. E., and Schmid, S. L. (1995). Dynamin self-assembles into rings suggesting a mechanism for coated vesicle budding. *Nature* **374**, 190–192.

Hitt, A. L., and Luna, E. J. (1994). Membrane interactions with the actin cytoskeleton. *Curr. Opin. Cell Biol.* **6**, 120–130.

Hogetsu, T., Shibaoka, H., and Shimokoriyama, M. (1974). Involvement of cellulose synthesis in actions of gibberellin and kinetin on cell expansion: Gibberellin–coumarin and kinetin–coumarin interactions on stem elongation. *Plant Cell Physiol.* **15**, 265–272.

Holman, G. M., Nachman, R. J., Schoofs, L., Hayes, T. K., Wright, M. S., and De Loof, A. (1991). The *Leucophaea maderae* hindgut preparation: A rapid and sensitive bioassay tool for the isolation of insect myotropins of other insect species. *Insect Biochem.* **21**, 107–112.

Holman, G. M., Nachman, R. J., and Wright, M. S. (1990). Insect neuropeptides *Annu. Rev. Entomol.* **35**, 201–217.

Holmgren, K., Magnusson, K. E., Franki, N., and Hays, R. M. (1992). DH-induced depolymerization of F-actin in the toad bladder granular cell: A confocal microscope study. *Am. J. Physiol.* **262**, C672–C677.

Hooghe-Peeters, E. L., Meda, P., and Orci, L. (1981). Coculture of nerve cells and pancreatic islets. *Dev. Brain Res.* **1**, 287–292.

Hua, Y-J., Bylemans, D., De Loof, A., and Koolman, J. (1994). Inhibition of ecdysone biosynthesis in flies by a hexapeptide isolated from vitellogenic ovaries. *Mol. Cell. Endocrinol.* **104**, R1–R4.

Iczkowski, K. A., and Hertelendy, F. (1991). Participation of the cytoskeleton in avian granulosa cell steroidogenesis. *Gen. Comp. Endocrinol.* **82**, 355–363.

Imai, T., Seo, H., Murata, Y., Ohno, M., Satoh, Y., Funahashi, H., Takagi, H., and Matsui, N. (1990). Adrenocorticotropin increases expression of *c*-fos and beta-actin genes in the rat adrenals. *Endocrinology* **127**, 1742–1747.

Ishida, K., and Katsumi, M. (1991). Immunofluorescence microscopical observation of cortical microtubule arrangement as affected by gibberellin in d5 mutant of *Zea mays* L. *Plant and Cell Physiol.* **32**, 409–417.

Ishida, K., and Katsumi, M. (1992). Effects of gibberellin and abscisic acid on the cortical microtubule organisation in hypocotyl cells of light grown cucumber seedlings. *Int. J. Plant Sci.* **153**, 155–163.

Ito, H., Hiroe, M., Hirata, Y., Tsujino, M., Adachi, S., Shichiri, M., Koike, A., Nogami, A., and Marumo, F. (1993). Insulin-like growth factor-I induces hypertrophy with enhanced expression of muscle specific genes in cultured rat cardiomyocytes. *Circulation* **87**, 1715–1721.

Joos, G. F., Pauwels, R. A., and Van der Straeten, M. E. (1988). The mechanism of tachykinin-induced bronchioconstriction in the rat. *Am. Rev. Respir. Dis.*, **137**, 1038–1044.

Kachadorian, W. A., Ellis, S. J., and Muller, J. (1979). Possible roles for microtubules and microfilaments in ADH action on toad urinary bladder. *Am. J. Physiol.* **236**, (*Renal Fluid Electrolyte Physiol.* 5) F14–F20.

Kadowaki, T., Nishida, E., Kasuga, M., Akiyama, T., Takaku, F., Ishikawa, M., Sakai, H., Kathuria, S., and Fujita-Yamaguchi, Y. (1985). Phosphorylation of fodrin (nonerythroid spectrin) by the purified insulin receptor kinase. *Biochem. Biophys. Res. Commun.* **127**, 493–500.

Karecla, P. I., and Kreis, T. E. (1992). Interaction of membranes of the Golgi complex with microtubules *in vitro. Eur. J. Cell Biol.* **57**, 139–146.

Katsumi, M., and Ishida, K. (1991). The gibberellin control of cell elongation. *In* "Gibberellins" (N. Takahashi, B. O. Phinney, and J. MacMillan, eds.), pp. 211–219. Springer Verlag, New York.

Kehlenbach, R. H., Matthey, J., and Huttner, W. B. (1994). XLαS is a new type of G protein. *Nature* **372**, 804–809.

Kemper, B., Habener, J. F., Rich, A., and Potts, J. (1975). Microtubule and the intracellular conversion of proparathyroid hormone to parathyroid hormone. *Endocrinology* **96**, 903–912.

Kiley, S. C., Parker, P. J., Fabro, D., and Jaken, S. (1992). Hormone and phorbol ester-activated protein kinase C isozymes mediate a reorganization of the actin cytoskeleton associated with prolactin secretion in GH-4c-1 cells. *Mol. Endocrinol.* **6**, 120–131.

Kirkeeide, E. K., Pryme, I. F., and Vedeler, A. (1992). Morphological changes in Krebs II ascites tumour cells induced by insulin are associated with differences in protein composition and altered amounts of free, cytoskeletal-bound, and membrane-bound polysomes. *Mol. Cell. Biochem.* **118**, 131–140.

Kirkeeide, E. K., Pryme, J. F., and Vedeler, A. (1993). Microfilaments and protein synthesis: Effects of insulin. *Int. J. Biochem.* **25**, 853–864.

Koelle, M. R., Talbot, W. S., Segraves, W. A., Bender, M. T., Cherbas, P., and Hogness, D. S. (1991). The *Drosophila* EcR gene encodes an ecdysone receptor. A new member of the steroid receptor superfamily. *Cell* **67**, 59–77.

Kranen, R. W., Overes, H. W., Kloosterboer, H. J., and Poels, L. G. (1993). The expression of cytoskeletal proteins during the differentiation of rat granulosa cells. *Hum. Reprod.* **8**, 24–29.

Kroeger, H. (1963). Experiments on the extranuclear control of gene activity in dipteran polytene chromosomes. *J. Cell Comp. Physiol.* **62** (*Suppl.* 1), 45–59.

Lang, J. M., Eisinger, W. R., and Green, P. B. (1982). Effects of ethylene on the orientation of microtubules and cellulose microfibrils of pea epicotyl cells with polylamellate cell walls. *Protoplasma* **110**, 5–14.

Lange, A. B., Chan, K. K., and Stay, B. (1993). Effects of allatostatin and proctolin on antennal pulsatile organ and hindgut muscle in the cockroach *Diploptera punctata. Arch. Insect Biochem. Physiol.* **24**, 79–92.

Larson, D. E., Berger, L. C., Sienna, N., and Sells, B. H. (1991). Cytoskeletal association of messenger RNAs during growth stimulation. *Biochem. Soc. Trans.* **19**, 1099–1103.

Lehman, H. K., and Greenberg, M. J. (1987). The actions of FMRFamide-like peptides on visceral and somatic muscles of the snail *Helix aspersa. J. Exp. Biol.* **131**, 55–68.

Leonard, J. L., Siegrist-Kaiser, C. A., and Zuckerman, C. J. (1990). Regulation of type II iodothyronine 5′ deiodinase by thyroid hormone. *J. Biol. Chem.* **265**, 940–946.

Leroux-Nicollet, I., Noel, M., Baribault, H., Goyette, R., and Marceau, N. (1983). Selective increase in cytokeratin synthesis in cultured rat hepatocytes in response to hormonal stimulation. *Biochem. Biophys. Res. Commun.* **114**, 556–563.

Levine, A. S., and Morley, J. E. (1984). Neuropeptide Y: A potent inducer of consummatory behavior in rats. *Peptides* **5**, 1025–1029.

Levine, R. B. (1989). Expansion of the central arborisations of persistent sensory neurons during insect metamorphosis: The role of the steroid hormone 20-hydroxyecdysone. *J. Neurosci.* **9**, 1045–1054.

Li, K. W., Smit, A. B., and Geraerts, W. P. M. (1992). Structural and functional characterization of neuropeptides involved in the control of male mating behaviour of *Lymnaea stagnalis. Peptides* **13**, 633–638.

Li, K. W., Van Golen, F. A., Van Minnen, J., Van Veelen, P. A., Van der Greef, J., and Geraerts, W. P. M. (1994). Structural identification, neuronal synthesis, and role in male copulation of myomodulin-A of *Lymnaea:* A study involving direct peptide profiling of nervous tissue by mass spectrometry. *Mol. Brain Res.* **25,** 355–358.

Lin, E. C., and Cantiello, H. F. (1993). A novel method to study the electrodynamic behavior of actin filaments. Evidence for cable-like properties of actin. *Biophys. J.* **65,** 1371–1378.

Litwin, D. K., Wilson, A. K., and Said, S. I. (1992). Vasoactive intestinal polypeptide (VIP) inhibits rat alveolar macrophage phagocytosis and chemotaxis *in vitro. Reg. Pept.* **40,** 63–74.

Lobie, P. E., Mertani, H., Morel, G., Morales-Bustos, O., Norstedt, G., and Waters, M. J. (1994). Receptor-mediated nuclear translocation of growth hormone. *J. Biol. Chem.* **269,** 21330–21339.

Loesser, K. E., Cain, L. D., and Malamed, S. (1994). The peripheral cytoplasm of adrenocortical cells: Zone-specific responses to ACTH. *Anat. Rec.* **239,** 95–102.

Luna, E. J., and Hitt, A. L. (1992). Cytoskeleton–plasma membrane interactions. *Science* **258,** 955–963.

Luxford, K. A., and Murphy, C. R. (1992). Changes in the apical microfilaments of rat uterine epithelial cells in response to estradiol and progesterone. *Anat. Rec.* **233,** 521–526.

Lynch, W. C., Hart, P., and Babcock, A. M. (1994). Neuropeptide Y attenuates satiety: Evidence from a detailed analysis of patterns of ingestion. *Brain Res.* **636,** 28–34.

Marasco, W. A., Showel, H. J., and Becker E. L. (1981). Substance P binds to the formylpeptide chemotaxis receptor on the rabbit neutrophil. *Biochem. Biophys. Res. Commun.* **99,** 1065–1072.

Marceau N., Goyette R., Pelletier G., and Antakly T. (1983). Hormonally induced changes in the cytoskeleton organisation of adult and newborn rat hepatocytes cultured on fibronectin precoated substratum: Effect of dexamethasone and insulin. *Cell. Mol. Biol.* **29,** 421–436.

Martinez de la Escalera, G. M., and Weiner, R. I. (1992). Hypothalamic regulation of microtubule-associated protein phosphorylation in lactrotrophs. *Neuroendocrinology* **55,** 327–335.

Matsumoto, A., Arai, Y., and Hyodo, S. (1993). Androgenic regulation of expression of beta-tubulin messenger ribonucleic acid in motoneurons of the spinal nucleus of the bulbocavernosus. *J. Neuroendocrinol.* **5,** 357–363.

Matsumoto, A., Arai, Y., Urano, A., and Hyodo, S. (1992). Effect of androgen on the expression of gap junction and beta-actin RNAs in adult rat motoneurons. *Neurosci. Res.* **14,** 133–144.

Matthews, K. A., Rees, D., and Kaufman, T. C. (1993). A functionally specialized alpha-tubulin is required for oocyte meiosis and cleavage mitosis in *Drosophila. Development* **117,** 977–991.

Mattson, P., and Kowal, J. (1982). Effects of cytochalasin B on unstimulated and adrenocorticotropin-stimulated adrenocorticotropin tumor cells *in vitro. Endocrinology* **111,** 1632–1647.

Matus-Leibovitch, N., Gershengorn, M. C., and Oron, Y. (1993). Differential effects of cytoskeletal agents on hemispheric functional expression of cell membrane receptors in *Xenopus* oocytes. *Cell. Mol. Neurobiol.* **13,** 625–637.

Mays, R. W., Beck, K. A., and Nelson, W. J. (1994). Organization and function of the cytoskeleton in polarized epithelial cells: A component of the protein sorting machinery. *Curr. Opin. Cell Biol.* **6,** 16–24.

Mercer, J. G., Lawrence, C. B., and Copeland, P. A. (1992). Corticotropin-releasing factor binding sites undergo axonal transport in the rat vagus nerve. *J. Neuroendocrinol.* **4,** 281–285.

Messina, J. L. (1992). Induction of cytoskeletal gene expression by insulin. *Mol. Endocrinol.* **6,** 112–119.

Meulemans, W., and De Loof, A. (1990a). Cytoskeletal F-actin patterns in Malpighian tubules of insects. *Tissue and Cell* **22,** 283–290.

Meulemans, W., and De Loof, A. (1990b). The dynamics of thick actin bundles localized at the basal membrane of whole-mounted insect excretory epithelia during metamorphosis. *Eur. J. Cell Biol.* **53,** *Suppl.* **31,** 23.

Meulemans, W., and De Loof, A. (1992). Changes in cytoskeletal actin pattern in the Malpighian tubules of the fleshfly *Sarcophaga bullata* (Parker) (Diptera, Calliphoridae), during metamorphosis. *Int. J. Insect Morphol. Embryol.* **21,** 1–16.

Miake-Lye, R., and Kirschner, M. W. (1985). Induction of early mitotic events in a cell-free system. *Cell* 41, 165–175.

Michiels, F., Habets, G. G. M., Stam, J. C., van der Kammen, R. A. and Collard, J. G. (1995). A role for Rac in Tiam1-induced membrane ruffling and invasion. *Nature* **375,** 338–340.

Mill, J. E., Chao, M. V., and Ishii, D. N. (1985). Insulin, insulin-like growth factor II and nerve growth factor effects on tubulin mRNA levels and neurite formation. *Proc. Natl. Acad. Sci. U.S.A.* **82,** 7126–7130.

Mita, T., and Shibaoka, H. (1984). Gibberellin stabilizes microtubules in onion leaf sheat cells. *Protoplasma* **119,** 100–109.

Mita, T., and Katsumi, M. (1986). Gibberellin control of microtubule arrangement in the mesocotyl epidermal cells of the d5 mutant of *Zea mays* L. *Plant Cell Physiol.* **27,** 651–659.

Miyagi, M., Aoyama, H., Morishita, M., and Iwamoto, Y. (1992). Effects of sex hormones on chemotaxis of human peripheral polymorphonuclear leukocytes and monocytes. *J. Periodontol.* **63,** 28–32.

Miyasaka, E., Miyasaka, T., and Hayashi, K. (1992). Bombesin/GRP stimulates a protein kinase in Swiss 3T3 cells that phosphorylates microtubule-associated protein kinase. *J. Clin. Biochem. Nutr.* **12,** 11–17.

Miyata Y., and Yahara, I. (1991). Cytoplasmic 8S glucocorticoid receptor binds to actin filaments through the 90-kDa heat shock protein moiety. *J. Biol. Chem.* **226,** 8779–8783.

Morley, J. E., Hernandez, E. N., and Flood, J. F. (1987a). Neuropeptide Y increases food intake in mice. *Am. J. Physiol.* **253,** (*Regulatory Integrative Comp. Physiol.* **22**), R516–R522.

Morley, J. E., and Levine, A. S. (1982). Corticotrophin releasing factor, grooming and ingestive behavior. *Life Sci.* **31,** 1459–1464.

Morley, J. E., Levine, A. S., Gosnell, B. A., Kneip, J., and Grace, M. (1987b). Effect of neuropeptide Y on ingestive behaviors in the rat. *Am. J. Physiol.* **252,** R599–R609.

Morrison, R. S., De Vellis, J., Lee, Y. I., Bradshaw, R. A., and Eng, L. F. (1985). Hormones and growth factors induce the synthesis of glial fibrillary acidic protein in rat brain astrocytes. *J. Neurosci. Res.* **14,** 167–176.

Morse, K. J., Scheff, S. W., and Dekosky, S. T. (1986). Gonadal steroids influence axon sprouting in the hippocampus dentate gyrus: A sexually dimorphic response. *Expl. Neurol.* **94,** 649–658.

Moss, R., Pryme, I. F., and Vedeler, A. (1991). Difference in patterns of proteins associated from polysomes in free, cytoskeleton-bound, and membrane-bound fractions in MCP-II cells incubated with insulin. *Biochem. Soc. Trans.* **19,** 1138–1139.

Moss, R., Pryme, I. F., and Vedeler, A. (1993). The effect of insulin on proteins associated with free, cytokeletal-bound, and membrane-bound polysome populations. *Cell Biol. Int.* **17,** 1065–1073.

Mrotek, J. J., and Hall, P. F. (1975). The influence of cytochalasin B on the response of adrenal tumor cells to ACTH and cyclic AMP. *Biochem. Biophys. Res. Commun.* **64,** 891–896.

Mrotek, J. J., and Hall, P. F. (1977). Response of adrenal tumor cells to adrenocorticotropin: Site of inhibition by cytochalasin B. *Biochemistry* **16,** 3177–3181.

Müller, L. J., Moorer van Delft, C. M., and Boer, H. H. (1992). The ACTH/MSH (4–9) analogue ORG-2766 stimulates microtubule formation in axons of central neurons of the snail *Lymnaea stagnalis. Peptides* **13,** 769–774.

Muller, J., Katchadorian, W. A., and DiScala, V. A. (1980). Evidence that ADH-stimulated intramembrane particle aggregates are transferred from cytoplasmic to luminal membranes in toad bladder epithelial cells. *J. Cell Biol.* **85,** 83–95.

Nakayasu, H., and Berezney, R. (1991). Nuclear matrins: Identification of the major nuclear matrix proteins. *Proc. Nat. Acad. Sci. U.S.A.* **88,** 10312–10316.

Nanhua, C., and Masters, C. (1988). The influence of insulin and glucagon on the interactions between glycolytic enzymes and cellular structure. *Biochem. Int.* **16**, 903–911.

Nemere, I., Feld, C., and Norman, A. W. (1991). 1,25-Dihydroxyvitamin D3–mediated alterations in microtubule proteins isolated from chick intestinal epithelium: Analyses by isoelectric focusing. *J. Cell Biochem.* **47**, 369–379.

Netchitailo, P., Perroteau, I., Feuilloley, M., Pelletier, G., and Vaudry, H. (1985). *In vitro* effect of cytochalasin B on adrenal steroidogenesis in frog. *Mol. Cell. Endocrinol.* **43**, 205–213.

Nève, P., Ketelbant-Balasse, P., Williams, C., and Dumont, J. E. (1972). Effect of inhibitors of microtubules and microfilaments on dog thyroid slices *in vitro*. *Exp. Cell Res.* **74**, 227–244.

Newcomb, R. W., and Scheller, R. A. (1987). Proteolytic processing of the *Aplysia* egg-laying hormone and R3–14 neuropeptide precursors. *J. Neurosci.* **7**, 853–863.

Nichols, N. R., Osterburg, H. H., Masters, J. N., Millar, S. L., and Finch, C. E. (1990). Messenger RNA for glial fibrillary acidic protein is decreased in rat brain following acute and chronic corticosterone treatment. *Molec. Brain Res.* **7**, 1–7.

Norman, J. C., Price, L. S., Ridley, A. J., Hall, A. and Koffer, A. (1994) Actin filament organization in activated mast cells is regulated by heterotrimeric and small GTP-binding proteins. *J. Cell Biol.* **146**, 1005–1015.

Nunez, J., Couchie, D., Aniello, F., and Bridoux, A. M. (1992). Thyroid hormone effects on neuronal differentiation during brain development. *Acta Med. Austriaca.* **19**, *Suppl.* **1P**, 36–39.

O'Callaghan, J. P., Brinton, R. E., and McEwen, B. S. (1991). Glucocorticoids regulate the synthesis of glial fibrillary acidic protein in intact and adrenalectomized rats but do not affect its expression following brain injury. *J. Neurochem.* **57**, 860–869.

Öhlen, A., Thureson-Klein, Å., Lindbom, L., Persson, M. G., and Hedqvist, P. (1989). Substance P activates leukocytes and platelets in rabbit microvessels. *Blood Vessels* **26**, 84–94.

Ohno, M., Seo, H., Imai, T., Murata, Y., Miyamoto, N., Satoh, Y., Funahashi, H., Takagi, H., and Matsui, N. (1992). ACTH increases expression of c-fos, c-jun, and beta-actin genes in the dexamethasone-treated rat adrenals. *Endocrinol. Jpn.* **39**, 377–383.

Önfelt, A. (1992). Bombesin impairs spindle function in mitotic V 79 Chinese hamster cells by a receptor-dependent mechanism. *Mutat. Res.* **270**, 97–102

Orci, L., Gabbay, K. H., and Malaise, W. J. (1972). Pancreatic beta-cell web: Its possible role in insulin secretion. *Science* (*Washington, DC*) **175**, 1128–1129.

Osawa, S., Betz, G., and Hall, P. F. (1984). Role of actin in the responses of adrenal cells to ACTH and cyclic AMP: Inhibition by DNase I. *J. Cell Biol.* **99**, 1335–1342.

O'Shea Alvarez, M. S. (1991). 3T3 cells in adipocytic conversion. *Arch. Invest. Med. (Mexico)* **22**, 235–244.

Painter, S. D., and Greenberg, M. J. (1982). A survey of the responses of bivalve hearts to the molluscan neuropeptide FMRFamide and to 5-hydroxytryptamine. *Biol. Bull.* **162**, 311–332.

Pappas, T. C., Gametchu, B., and Watson, C. S. (1995). Membrane estrogen receptors identified by multiple antibody labeling and impeded-ligand binding. *FASEB J.*, **9**, 404–410.

Partsch, G., and Matucci-Cerinic, M. (1992). Effect of substance P and somatostatin on migration of polymorphonuclear (PMN) cells *in vitro*. *Inflammation* **16**, 539–547.

Passareiro, H., Lamy, F., Lecocq, R., Dumont, J. E., and Nunez, J. (1984). Effects of TSH on the synthesis of several proteins of the cytoskeleton during differentiation of thyroid cells in culture. *Ann. Endocrinol.* **45**, 53.

Pearl, M., and Taylor, A. (1983). Actin filaments and vasopressin-stimulated water flow in toad urinary bladder. *Am. J. Physiol.* **245** (*Cell Physiol.* **14**), C28–C39.

Perrot-Applanat, M., Lescop, P., and Milgrom, E. (1992). The cytoskeleton and the cellular traffic of the progesterone receptor. *J. Cell Biol.* **119**, 337–348.

Peter, M., Nakagawa J., Dorée, M., Labbé, J. C., and Nigg, E. A. (1990). *In vitro* disassembly of the nuclear lamina and M phase-specific phosphorylation of lamins by cdc2 kinase. *Cell* **61**, 591–602.

Pic, P., Rémy, L., Athouel-Haon, A.-M., and Mazzella, E. (1984). Evidence for a role of the cytoskeleton in *in vitro* folliculogenesis of the thyroid gland of the fetal rat. *Cell Tissue Res.* **237,** 499–508.

Pienta, K. J., and Coffey, D. S. (1992). Nuclear–cytoskeletal interactions: Evidence for physical connections between the nucleus and cell periphery and their alteration by transformation. *J. Cell. Biochem.* **49,** 357–365.

Pipeleers, D. G., Pipeleers-Marichal, M. A., and Kinnis, D. M. (1976). Regulation of tubulin synthesis in islets of Langerhans. *Proc. Natl. Acad. Sci. U.S.A.* **73,** 3188–3191.

Pipeleers, D. G., Pipeleers-Marichal, M. A., and Kinnis, D. M. (1977). Physiological regulation of total tubulin and polymerized tubulin in tissues. *J. Cell Biol.* **77,** 351–357.

Pollard, T. D., Almo, S., Quirk, S., Vinson, V., and Lattman, E. E. (1994). Structure of actin binding proteins: Insights about function at atomic resolution. *Annu. Rev. Cell Biol.* **10,** 207–249.

Prat, A. G., Bertorello, A. M., Ausiello, D. A., and Cantiello, H. F. (1993). Activation of epithelial Na^+ channels by protein kinase A requires actin filaments. *Am. J. Physiol.* **265,** C224–C233.

Pratt, W. B., Sanchez, E. R., Bresnick, E. H., Meshinchi, S., Scherrer, L. C., Dalman, F. C., and Welsh, M. J. (1989). Interaction of the glucocorticoid receptor with the Mr 90,000 heat shock protein: an evolving model of ligand-mediated transformation and translocation. *Cancer Res.* **49,** 2222s–2229s.

Price, D. A., Davies, N. W., Doble, K. E., and Greenberg, M. J. (1987). The variety and distribution of the FMRFamide-related peptides in molluscs. *Zool. Sci.* **4,** 395–410.

Price, D. A., Lesser, W., Lee, T. D., Doble, K. E., and Greenberg, M. J. (1990). Seven FMRFamide-related and SCP-related cardioactive peptides from *Helix. J. Exp. Biol.* **154,** 421–437.

Proux, J. P., Miller, C. A., Li, J. P., Carney, R. L., Girardie, A., Delaage, M., and Schooley, D. A. (1987) Identification of an arginine vasopressin-like diuretic hormone from *Locusta migratoria. Biochem. Biophys. Res. Commun.* **149,** 1890–1896.

Pryme, I. F., and Hesketh, J. E. (1990). Insulin induces cell adhesion and normal flattened morphology in Krebs II ascites cells. *Cell. Biol. Int. Rep.* **14,** 447–445.

Raikhinstein, M., and Hanukoglu, I. (1994). Cloning of ACTH-regulated genes in the adrenal cortex. *J. Steroid Biochem. Mol. Biol.* **49,** 257–260.

Rao, K. M. K., Bretschart, J. M., and Virgi, M. A. (1985). Hormone-induced actin polymerisation in rat hepatoma cells and human leukocytes. *Biochem. J.* **230,** 709–714.

Ratnasooriya, W. D., and Premakumara, G.A. S. (1993). Oxytocin suppresses motility of human spermatozoa *in vitro. Med. Sci. Res.* **21,** 13–14.

Ravindra, R., and Grosvenor, C. E. (1990). Involvement of cytoskeleton in polypeptide hormonal secretion from the anterior pituitary lobe: A review. *Mol. Cell. Endocrinol.* **71,** 165–176.

Ray, F., and Scrott, C. A. (1978). Stimulation of steroid synthesis by normal rat adrenocortical cells in response to antimicrotubular agents. *Endocrinology* **103,** 1281–1288.

Ray, L. B., and Sturgill, T. W. (1987). Rapid stimulation by insulin of a serine/threonine kinase in 3T3-L1 adipocytes that phosphorylates microtubule-associated protein 2 *in vitro. Proc. Natl. Acad. Sci. U.S.A.* **84,** 1502–1506.

Reaven, E. P., and Reaven, G. M. (1975). A qualitative ultrastructural study of microtubule content and secretory granule accumulation in parathyroid glands of phosphate- and colchicine-treated rats. *J. Clin. Invest.* **56,** 49–55.

Recio-Pinto, E., and Ishii, D. N. (1988). Insulin and insulin-like growth factor receptors regulating neurite formation in cultured neuroblastoma cells. *J. Neurosci. Res.* **19,** 312–320.

Remacle, C. (1992) The cytoskeleton. *Belg. J. Zool.* **122,** 3–16.

Remy-Jouet, I., Delarue, C., Feuilloley, M., and Vaudry, H. (1994). Involvement of the cytoskeleton in the mechanism of action of endothelin on frog adrenocortical cells. *J. Steroid Biochem. Mol. Biol.* **50,** 55–59.

Roberts, I. N., Lloyd, C. W., and Roberts, K. (1985). Ethylene-induced microtubule reorientations: Mediation by helical arrays. *Planta* **164**, 439–447.

Rothman, J. E. (1994). Mechanisms of intracellular protein transport. *Nature* **372**, 55–63.

Safran, M., Farwell, A. P., and Leonard, J. L. (1992). Thyroid hormone-dependent redistribution of the 55-kilodalton monomer of protein disulfide isomerase in cultured glial cells. *Endocrinology* **131**, 2413–2418.

Safran, M., Farwell, A. P., Rokos, H., and Leonard, J. L. (1993). Structural requirements of iodothyronines for the rapid inactivation and internalisation of type II iodothyronine 5' deiodinase in glial cells. *J. Biol. Chem.* **268**, 14224–14229.

Saito, T., Lamy, F., Roger, P. P., Lecocq, R., and Dumont, J. E. (1994). Characterization and identification as cofilin and destrin of two thyrotropin- and phorbol ester–regulated phosphoproteins in thyroid cells. *Exp. Cell Res.* **212**, 49–61.

Sakakibara, Y., Saito, I., Ichinoseki, K., Oda, T., Kaneko, M., Saito, H., Kodama, M., and Sato, Y. (1991). Effects of diethylstilbesterol and its methylesters on aneuploidy induction and microtubule distribution in Chinese hamster V79 cells. *Mutation Res.* **263**, 269–276.

Sakiyama, M., and Shibaoka, H. (1990). Effects of abscisic acid on the orientation and cold stability of cortical microtubules in epicotyl cells of the dwarf pea. *Protoplasma* **157**, 165–171.

Sakiyama-Sogo, M., and Shibaoka, H. (1993). Gibberellin A3 and abscisic acid cause the reorientation of cortical microtubules in epicotyl cells of the decapitated dwarf pea. *Plant Cell Physiol.* **34**, 431–437.

Saltarelli, D., Llosa-Hermier, P., Tertein-Clary, C., and Hermier, C. (1984). Effects of antimicrotubular agents in cAMP production and in steroidogenic response of isolated rat Leydig cells. *Biol. Cell* **52**, 259–266.

Sato, Y. P., Murai, T., Oda, H., Saito, H., Kodama, M., and Hirato, A. (1987). Inhibition of microtubule polymerisation by synthetic estrogens: Formation of a ribbon structure. *J. Biochem.* **101**, 1247–1252.

Sato, Y., Murai, M., Tsumuraya, H., Saito, H., Kodama, M. (1984). Disruptive effect of diethylstilbesterol on microtubules. *Jpn. J. Cancer Res.* **75**, 1046–1048.

Sato, Y., Sakakibara, Y., Oda, T., Aizu-Yokota, E., and Ichinoseki, K. (1992). Effect of estradiol and ethynylestradiol on microtubule distribution in Chinese hamster V79 cells. *Chem. Pharm. Bull.* **40**, 182–184.

Sauman, I., and Berry, S. J. (1994). An actin infrastructure is associated with eukaryotic chromosomes: Structural and functional significance. *Eur. J. Cell Biol.* **64**, 348–356.

Sawhney, K. V., and Srivasatava, L. M. (1975). Wall fibrils and microtubules in normal and gibberellic acid–induced growth of lettuce hypocotyl cells. *Can. J. Bot.* **53**, 824–835.

Sawin, K. E., and Endow S. A. (1993). Meiosis, mitosis, and microtubule motors. *BioEssays* **15**, 399–407.

Scherrer, L. C., and Pratt, W. B. (1992). Association of the transformed glucocorticoid receptor with the cytoskeletal protein complex. *J. Steroid Biochem. Mol. Biol.* **41**, 719–721.

Schlatt, S., Weinbauer, G. F., Arslan, M., and Nieschlag, E. (1993). Appearance of alpha–smooth muscle actin in peritubular cells of monkey testes is induced by androgens, modulated by follicle-stimulating hormone, and maintained after hormonal withdrawal. *J. Androl.* **14**, 340–350.

Schoofs, L., Vanden Broeck, J., and De Loof, A. (1993). Mini-review: The myotropic peptides of *Locusta migratoria*: Structures, distribution, functions, and receptors. *Insect Biochem. Mol. Biol.* **23**, 859–881.

Schwartz, L. M. (1992). Insect muscle as a model for programmed cell death. *J. Neurobiol.* **23**, 1312–1326.

Seagull, R. W. (1986). Changes in microtubule orientation and wall microfibril orientation during *in vitro* cotton development: An immunofluorescent study. *Can. J. Bot.* **64**, 1373–1381.

Seagull, R. W. (1989). The plant cytoskeleton. *CRC Crit. Rev. Plant Sci.* **8**, 473–497.

Segawa, A., and Yamashina, S. (1989). Role of microfilaments in exocytosis. *Cell Struc. Func.* **14,** 531–544.

Shibaoka, H. (1972). Gibberellin–colchicine interaction in elongation of azuki bean epicotyl sections. *Plant Cell Physiol.* **13,** 461–469.

Shibaoka, H. (1974). Involvement of wall microtubules in gibberellin promotion and kinetin inhibition of stem elongation. *Plant Cell Physiol.* **15,** 255–263.

Shibaoka, H. (1991). Microtubules and the regulation of cell morphogenesis by plant hormones. *In* "The Cytokeletal Basis of Plant Growth and Form" (C. W. Lloyd, ed.), pp. 159–168. Academic Press, London.

Shibaoka, H. (1993). Regulation by gibberellins of the orientation of cortical microtubules in plant cells. *Aust. J. Plant Physiol.* **20,** 461–470.

Shibaoka, H., and Hogetsu, T. (1977). Effects of ethyl-N-phenylcarbamate on wall microtubules and on gibberellin- and kinetin-controlled cell expansion. *The Botanical Magazine (Tokyo)* **90,** 317–321.

Shibaoka, H., and Nagai, R. (1994). The plant cytoskeleton. *Curr. Opin. Cell. Biol.* **6,** 10–15.

Siegrist-Kaiser, C. A., Juge-Aubry, C., Tranter, M. P., Eckenbarger, D. M., and Leonard, J. J. (1990). Thyroxine-dependent modulation of actin polymerisation in cultured astrocytes. *J. Biol. Chem.* **265,** 5296–5302.

Simantov, R., Shkolnik, T., and Sachs, L. (1980). Desensitisation of enucleated cells to hormones and role of cytoskeleton in control of normal hormonal response. *Proc. Natl. Acad. Sci. U.S.A.* **77,** 4798–4802.

Singer, R. H. (1992). The cytoskeleton and mRNA localization. *Curr. Top. Dev. Biol.* **4,** 15–19.

Sinnett-Smith, J., Zachary, I., Valverde, A. M., and Rozengurt, E. (1993). Bombesin stimulation of P125 focal adhesion kinase tyrosine phosphorylation: Role of protein kinase-C, Ca^{2+} mobilization, and the actin cytoskeleton. *J. Biol. Chem.* **268,** 14261–14268.

Slama, K. (1982). Hormonal regulation of morphogenesis in invertebrates, evolutionary aspects. *Zhurnal Obhchej Bilogii* **43,** 805–822.

Slama, K., Konopinska, D., and Sobotka, W. (1993). Effects of proctolin on autonomic physiological functions in insects. *Eur. J. Entomol.* **90,** 23–35.

Small, J. V., Furst, D. O., and Thornell, L. E. (1992). The cytoskeletal lattice of muscle cells. *Eur. J. Biochem.* **208,** 559–572.

Smit, A. B., Jiménez, C. R., Dirks, R. W., Croll, R. P., and Geraerts, W. P. M. (1992). Characterization of a cDNA clone encoding multiple copies of the neuropeptide APGWamide in the mollusc *Lymnaea stagnalis. J. Neurosci.* **12,** 1709–1715.

Soifer, D., Braun, T., and Hechter, O. (1971). Insulin and microtubules in rat adipocytes. *Science (Washington, D.C.)* **172,** 269–271.

Song, Q., Ma, M., Ding, T., Ballarino, J., and Wu, S.-J. (1990). Effects of benzodioxole, J 2581 (5-ethoxy-6-(4'-methoxyphenyl)methyl-1,3-benzodioxole) on vitellogenesis and ovarian development of *Drosophila melanogaster. Pest. Biochem. Physiol.* **37,** 12–23.

Soto, E. A., Kliman, H. J., and Paavola, L. G. (1986). Gonadotropin and cyclic adenosine 3′,5′-monophosphate (cAMP) alter the morphology of cultured human granulosa cells. *Biol. Reprod.* **34,** 559–569.

Staiger, C. J., and Lloyd, C. W. (1991). The plant cytoskeleton. *Curr. Opin. Cell Biol.* **3,** 33–42.

Stanley, B. G., and Leibowitz, S. F. (1985). Neuropeptide Y injected in the paraventricular hypothalamus: A powerful stimulant of feeding behavior. *Proc. Natl. Acad. Sci. U.S.A.* **82,** 3940–3943.

Steen, D. A., and Chadwick, A. V. (1981). Ethylene effects in pea stem tissue: Evidence of microtubule mediation. *Plant Physiol.* **67,** 460–466.

Stein, S. A., McIntire, D. D., Kirkpatrick, L. L., Adams, P. M., and Brady, S. T. (1991). Hypothyroidism selectively reduces the rate and amount of transport for specific SCb proteins in the hyt/hyt mouse optic nerve. *J. Neurosci. Res.* **30,** 28–41.

Steinman, J. L., Gunion, M. W., and Morley, J. E. (1994). Forebrain and hindbrain involvement of neuropeptide Y in ingestive behaviors of rats. *Pharmacol. Biochem. Behav.* **47,** 207–214.

Südhof, T. C. (1995). The synaptic vesicle cycle: A cascade of protein-protein interactions. *Nature* **375,** 645–653.

Surprenant, K. (1993). Microtubules, ribosomes, and RNA: Evidence for cytoplasmic localization and translational regulation. *Cell Motil. Cytoskel.* **25,** 1–9.

Sutovsky, P., Flechon, J. E., and Pavlok, A. (1994). Microfilaments, microtubules and intermediate filaments fulfill differential roles during gonadotropin-induced expansion of bovine cumulus oophorus. *Reprod. Nutr. Dev.* **34,** 415–426.

Takasu, N., Ohno, S., Komiya, I., and Yamada, T. (1992). Requirements of follicle structure for thyroid hormone synthesis cytoskeletons and iodine metabolism in polarized monolayer cells on collagen gel and in double layered follicle-forming cells. *Endocrinology* **131,** 1143–1148.

Takeda, K., and Shibaoka, H. (1981). Effects of gibberellins and colchicine on microfilament arrangement in epidermal cell walls of *Vigna angularis* Ohwi and Ohashi epicotyls. *Planta* **151,** 393–398.

Taylor, A., Martinez, M., Reaven, E., and Maffly, R. (1973). Vasopressin: Possible role of microtubules and microfilaments in its action. *Science (Washington, DC)* **181,** 347–350.

Thomas, S. M., Soriano, P., and Imamoto, A. (1995). Specific and redundant roles of Src and Fyn in organizing the cytoskeleton. *Nature* **376,** 267–271.

Toran-Allerand, C. D., Bentham, W., Miranda, R. C., and Anderson, J. P. (1991). Insulin influences astroglial fibrillary acidic protein (GFAP) expression in organotypic cultures. *Brain Res.* **558,** 296–304.

Toran-Allerand, C. D., Ellis, L., and Pfenninger, K. A. (1988). Estrogen and insulin synergism in neurite growth enhancement *in vitro:* Mediation of steriod effects by interactions with growth factors? *Dev. Brain Res.* **4,** 87–100.

Torres-Aleman, I., Rejas, M. T., Pons, S., and Garcia-Segura, L. M. (1992). Estradiol promotes cell shape changes and glial fibrillary acidic protein redistribution in hypothalamic astrocytes *in vitro:* A neuronal-mediated effect. *Glia* **6,** 180–187.

Tourmente, S., Chapel, S., Dreau, D., Drake, M. E., Bruhat, A., Couderc, J. L., and Dastugue, B. (1993). Enhancer and silencer elements within the first intron mediate the transcriptional regulation of the beta 3 tubulin gene by 20-hydroxyecdysone in *Drosophila* Kc cells. *Insect Biochem. Mol. Biol.* **23,** 137–143.

Tramontano, D., Avivi, A., Ambesi-Impiobato, F. S., Barak, L., Geiger, B., and Schlessinger, J. (1982). Thyrotropin induces changes in the morphology and the organisation of microfilament structures in cultured thyroid cells. *Exp. Cell Res.* **137,** 269–275.

Tranque, P. A., Suarez, I., Olmos, G., Fernandez, B., and Garcia-Segura, L. M. (1987). Estradiol-induced redistribution of glial fibrillary acid protein immunoreactivity in the rat brain. *Brain Res.* **406,** 348–351.

Trifaro, J. M., Vitale, M. L., and Rodriguez Del Castillo, A. (1992). Cytoskeleton and molecular mechanisms in neurotransmitter release by neurosecretory cells. *Eur. J. Pharmacol.* **225,** 83–104.

Truman, J. W., and Reiss, S. E. (1988). Hormonal regulation of the shape of identified motoneurons in the moth *Manduca sexta. J. Neurosci.* **8,** 765–775.

Tsuda, T., Kawahara, Y., Ishida, Y., Koide, M., Shii, K., and Yokoyama, M. (1992). Angiotensin II stimulates two myelin basic protein microtubule-associated protein-2 kinases in cultured vascular smooth muscle cells. *Circ. Res.* **71,** 620–630.

Tung, P. S., Burdzy, K., and Fritz, I. B. (1993). Proteases are implicated in the changes in the Sertoli cell cytoskeleton elicited by follicle-stimulating hormone or by dibutyryl cyclic AMP. *J. Cell Physiol.* **155,** 139–148.

Vanden Broeck, J., De Loof, A., and Callaerts, P. (1992). Electrical-ionic control of gene expression. *Int. J. Biochem.* **24,** 1907–1916.

Van Epps, D. E., and Saland, L. (1984). Beta-endorphin and met-enkephalin stimulate human peripheral blood mononuclear cell chemotaxis. *J. Immunol.* **132**, 3046–3053.

Vanhems, E., Delbos, M., and Girardie, J. (1990). Insulin and neuroparsin promote neurite outgrowth in cultured locust CNS. *Eur. J. Neurosci.* **2**, 776–782.

Van Mellaert, H., De Loof, A., and Jurd, L. (1983). Anti–juvenile hormone effects of newly described chemosterilants: Benzyl-1,3-benzodioxoles and benzylphenols. *Entomol. Exp. Appl.* **33**, 83–89.

Vedeler, A., Pryme, I. F., and Hesketh, J. E. (1990). Insulin and step-up conditions cause a redistribution of polysomes among free, cytoskeletal, and membrane-bound fractions in Krebs II ascites cells. *Cell Biol. Int. Rep.* **14**, 211–218.

Vedeler, A., Pryme, I. F., and Hesketh, J. E. (1991). Insulin induces changes in the subcellular distribution of actin and 5'-nucleotidase. *Mol. Cell. Biochem.* **108**, 67–74.

Veelaert, D., Schoofs, L., Tobe, S. S., Yu, C. G., Vullings, H. G. B., Couillaud, F., and De Loof, A. (1995). Immunological evidence for an allatostatin-like neuropeptide in the central nervous system of *Schistocerca gregaria, Locusta migratoria,* and *Neobellieria bullata. Cell Tissue Res.* **279**, 601–611.

Vemuri, M. C., Raju, M. N., and Malkota, S. K. (1993). Recent advances in nuclear matrix function. *Cytobios* **76**, 117–128.

Von Essen, S. G., Rennard, S. I., O'Neill, D., Ertl, R. F., Robbins, R. A., Koyama, S., and Rubinstein, I. (1992). Bronchial epithelial cells release neutrophil chemotactic activity in response to tachykinins. *Amer. J. Physiol. (Lung Cell. Mol. Physiol.* 7) **263**, L226–L231.

Ward, G. E., and Kirschner, M. W. (1990). Identification of cell cycle-regulated phosphorylation sites on nuclear lamin C. *Cell* **61**, 561–577.

Wernicke, W., and Jung, G. (1992). Role of the cytoskeleton in cell shaping of developing mesophyll of wheat (*Triticum aestivum* L.). *Eur. J. Cell Biol.* **57**, 88–94.

Westermark, B., and Porter, K. R. (1982). Hormonally induced changes in the cytoskeleton of human thyroid cells in culture. *J. Cell Biol.* **94**, 42–50.

Willey, J. C., Lechner, J. F., and Harris, C.C, (1984). Bombesin and the C terminal tetradecapeptide of gastrin-releasing peptide are growth factors for normal human bronchial epithelial cells. *Exp. Cell Res.* **153**, 245–248.

Williams, J. A., and Wolff, J. (1970). Possible roles of microtubules in thyroid secretion. *Proc. Natl. Acad. Sci. U.S.A.* **67**, 1901–1908.

Wünsch, S., Schneider, S., Schwab, A., and Oberleiter, H. (1993). 20-OH-ecdysone swells nuclear volume by alkalinization in salivary glands of *Drososphila melanogaster. Cell Tissue Res.* **274**, 145–151.

Yahara, I., Miyata, Y., Minami, Y., Rimura, Y., Matsumoto, S., Koyasu, S., Yonezawa, N., Nishida, E., and Sakai, H. (1991). HSP90, a carrier of key proteins that regulates cell function. *In* "Heat Shock" (B. Maresca and S. Lundquist, eds.), pp. 119–122, Springer Verlag, Berlin.

Yamaguchi, Y., Dalle-Molle, E., and Hardison, W. G. (1991). Vasopressin and A23187 stimulate phosphorylation of myosin light chain-1 in isolated rat hepatocytes. *Am. J. Physiol.* **261**, G312–G319.

Yao, J., and Eghbali, M. (1992). Decreased collagen mRNA and regression of cardiac fibrosis in the ventricular myocardium of the tight skin mouse following thyroid hormone treatment. *Cardiovasc. Res.* **26**, 603–607.

Yap, A. S., and Manley, S. W. (1994). Thyrotropin inhibits the intrinsic locomotility of thyroid cells organized as follicles in primary culture. *Exp. Cell. Res.* **214**, 408–417.

Zandomeni, K., and Schopfer, P. (1993). Reorientation of microtubules at the outer epidermal wall of maize coleoptiles by phytochrome, blue light photoreceptor, and auxin. *Protoplasma* **173**, 103–112.

Zettl, K. S., Sjaastad, M. D., Riskin, P. M., Parry, G., Machen, T. E., and Firestone, G. L. (1992). Glucocorticoid-induced formation of tight junctions in mouse mammary epithelial cells *in vitro. Proc. Natl. Acad. Sci. U.S.A.* **89**, 9069–9073.

Zhang, Y. J., and Kunkel, J. G. (1992). Program of F-actin in the follicular epithelium during oogenesis of the German cockroach, *Blatella germanica. Tissue Cell* **24,** 905–917.

Zor, U., Strulovici, B., and Lindner, H. P. (1978). Implication of microtubules and microfilaments in the response of ovarian adenylate–cyclase AMP system to gonadotropins and prostaglandin E_2. *Biochem. Biophys. Res. Commun.* **80,** 983–992.

Twisted Liquid Crystalline Supramolecular Arrangements in Morphogenesis

Marie-Madeleine Giraud-Guille
Histophysique et Cytophysique—Ecole Pratique des Hautes Etudes
Observatoire Océanologique, Université P. et M. Curie and CNRS, 66650
Banyuls-sur-Mer, France

Supramolecular assemblies following liquid crystalline cholesteric geometries have been described in biological systems from optical properties observed in polarized-light microscopy and structural data obtained in electron microscopy. Major biological macromolecules are disccused, including structural polymers of the extracellular matrix, genetic material in nuclei and chromosomes, and proteins of the cytoplasm. The liquid crystalline assembly properties of biological polymers have been demonstrated by experiments *in vitro* with molecules at basic structural levels, such as molecular chains of cellulose and chitin, triple helices of collagen, and double helices of DNA, and also with entities at higher states of organization as they appear in cells and tissues, such as cellulose and chitin crystallites, and collagen fibrils. It appears that the building of cellular and extracellular edifices implies self-ordering processes of the liquid crystalline type and that the study of these mesomorphic states will help resolve basic questions about the structure and morphogenesis of densely packed biological structures.

KEY WORDS: Cholesteric mesophases, Extracellular matrix, Nuclei and chromosomes, Collagen, Chitin, Cellulose, DNA, Polarized-light microscopy, Electron microscopy.

I. Introduction

A. The Question of Biological Organization

The basic questions of cellular biology are related to growth and form: How do macromolecules interact and assemble in three-dimensional structures?

Copyright © 1996 by Academic Press, Inc.
All rights of reproduction in any form reserved.

How do cells differentiate and grow? How do tissues organize and form patterns? Since the advent of molecular biology, immense consequence of Watson and Crick's model, such questions have been approached mainly through analysis at the molecular level. These efforts have resulted in enormous contributions: to the identification of thousands of molecules involved in morphogenetic regulation (e.g., the class of polymers forming the cytoskeleton), and to highlighting the role of genetic controls and chemical regulators in the cell machinery, (e.g., homeobox genes and soluble growth factors). Nevertheless, molecular biologists have become aware of the limits of such a reductionist approach, recalling that the inventory of molecules is insufficient to deduce the behavior of cells or groups of cells, and suggesting that new concepts are necessary to understand the dynamic organization of macromolecules at a three-dimensional level (Wilkins, 1985).

Increasing experimental data support the idea that, in addition to molecular recognition and signals, physical forces play a role in the organization of living matter. Results appear at different levels of the hierarchy: macromolecules, supramolecular complexes, subcellular structures, and cells. At the cellular level the extracellular matrix has been shown to offer mechanical resistance to cells and to modulate their shape by mechanical stresses; moreover, the stability of tissue morphogenesis requires a balance between cytoskeleton tension and the extracellular matrix scaffold (Ingber *et al.*, 1994). On the question of the passage from the molecular to the supramolecular level, new concepts are emerging at the frontier between condensed matter physics, topochemistry, and molecular biology (Rigny *et al.*, 1994). This review utilizes this new approach and focuses on intermediate states of matter known as *liquid crystals* that are recognized at the cellular and tissue level.

B. Fluidity and Order in Cells and Tissues

Compacted assemblies, partially ordered structures, and more or less fluid states are properties shared in common by biological structures and liquid crystals. Indeed, models developed in condensed matter physics can help interpret the structure and dynamic organization of molecular assemblies in biology. An example of a liquid crystal analogue already familiar to molecular and cellular biologists, even if not expressed in these terms, is the cell membrane, where fluidity and order coexist; the fluid mosaic model described by Singer and Nicolson (1972) corresponds to a two-dimensional solution of oriented globular proteins and phospholipids and behaves like a liquid crystal of the smectic type (Luzzati and Husson, 1962).

The presence in cell and tissue compartments of biopolymers behaving like liquid crystals of a different type, referred to as *cholesteric,* is much less well known and is the subject of this chapter. This theory was first postulated by Bouligand (1972), who showed the similarity of the three-dimensional arrangements between biological polymers in living structures and molecules in liquid crystals. General reviews on liquid crystalline order in biological materials were then published in works aimed at physicists (Bouligand, 1978a,b) or at researchers working in specialized fields, such as collagen (Bouligand and Giraud-Guille, 1985b), DNA (Livolant, 1991), cellulose (Reis *et al.,* 1994), or fibrous composites (Neville, 1993).

This review aims to interest cellular biologists in general and focuses on the major macromolecules produced by cells, proteins, polysaccharides, and nucleic acids, when they form liquid crystalline molecular arrangements of the cholesteric type. We introduce this subject, unfamiliar to many, by giving basic definitions of true liquid crystals and their biological analogues and describing the properties and consequences of the cholesteric geometry. We then discuss the many situations encountered *in vivo* of supramolecular assemblies showing cholesteric geometries. Finally, we report on recent developments obtained at the molecular level *in vitro* with major biopolymers —cellulose, chitin, collagen, and DNA—demonstrating that structural analogies with cholesteric liquid crystals observed *in vivo* can directly be related to self-assembly properties described in condensed matter physics. This experimental approach thus becomes an interesting way to attack many problems concerning morphogenesis: the dynamic assembly and disassembly of the matrix during growth or repair, the condensation and decondensation of genetic material during the cell cycle, and more generally the multiple assembly steps leading from molecules to final structures.

II. Properties of Liquid Crystals and Their Analogues

A. Intermediate States of Matter

Liquid crystals are defined as ordered liquids formed by molecules presenting a well-defined shape. They were discovered at the end of the nineteenth century by a botanist who observed a turbid appearance in cholesterol benzoate heated just above 145°C (Reinitzer, 1989; English trans.). The physicist Lehman (1908) and his colleagues then studied the strong optical anisotrope properties of this substance and of many other derivatives. A history of liquid crystals was published by Kelker (1973), and several books review their physical and chemical properties (Gray, 1962; De Gennes, 1974). In short, liquid crystals have an organizational arrangement intermediate between the disorder of ordinary liquids and the three-dimensional periodicity of crystals. In a defined range of temperature and concentration, molecules are mobile and oriented in one, two, or three dimensions; the fluid character varies inversely with the three-dimensional order. Liquid crystals are most often made of rodlike organic molecules (a typical length is 20 Å), but much longer molecules, namely, within biological systems, also form similar arrangements. The molecular configurations often consist of stiff central parts and flexible extremities, as is the case with methoxybenzylidenebutylaniline (MBBA), which is used in watches and pocket computers.

Liquid crystals that appear within pure chemical systems and whose order is a function of the temprerature are called *thermotropics*. Liquid crystals that emerge in mixed systems and whose degree of order is directly related to the concentration of the sample are called *lyotropics*. These are always the type found in biological structures.

Liquid crystal structures are studied by analyzing X-ray diffraction patterns when the order parameter is between a few angstroms and 1000 Å or by analyzing the textures obtained in polarized-light microscopy when the order parameter is between 100 nm and 100 μm or more.

B. Smectics, Nematics, and Cholesterics

Using microscopic observations, Friedel (1922) was the first to classify liquid crystals into what he called *mesomorphic states* or *mesophases*. He described three principal organizations with elongated rods corresponding to *smectics, nematics,* and *cholesterics* (Fig. 1). Another type, *discotics,* has more recently been described in the case of disklike molecules (Bouligand, 1980).

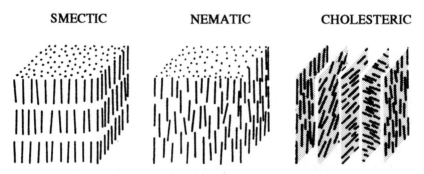

| SMECTIC | NEMATIC | CHOLESTERIC |

FIG. 1   Structure of liquid crystals made of rodlike molecules. In smectics, molecules lie normal to layers of equal thickness; they diffuse freely within the layers but not between them. In nematics, molecules are all parallel, the centers of gravity of the elongated molecules are randomly oriented, and the order is purely orientational. In cholesterics, molecules lie parallel in each plane and rotate by a small and constant angle from one plane to the next. This structure derives from a nematic by introducing a twist between adjacent parallel molecules. (Redrawn from Bouligand, 1978d).

The smectic state, characterized by X-ray studies, shows the highest degree of order: molecules stand parallel in stacked layers of equal thickness and they diffuse freely within each layer but cannot move from one layer to the next. Molecules are disposed either normal to the layers, in smectic A, or oblique to them, in smectic C. In a chiral form, the molecular directions rotate regularly in the successive layers.

Nematic phases have the lowest degree of order: molecules are all parallel and equidistant but do not form any layers, as demonstrated by X-ray studies, and the molecules are free to move in the direction of their length.

A chiral form of the nematic is the cholesteric phase. This name comes from the fact that these particular liquid crystals were discovered in several cholesterol esters. Cholesteric mesophases have a high degree of optical activity as a consequence of their particular structure. Molecules align in parallel form in successive planes, with the molecular directions rotating regularly from one plane to the next.

C. Biological Analogues of Liquid Crystals

As this review shows, many biological molecules (e.g., polysaccharides, proteins, and nucleic acids) form supramolecular organizations very close to that of liquid crystals. In biological materials these molecules do not give true solutions but rather form colloids—that is, molecules are assembled in micelles large enough to scatter light strongly. Colloids can be in either a sol state or a gel state due to the presence of links between the micelles (Frey-Wissling, 1953). The gels can appear in more or less dense states or even be consolidated to form solid structures. The terms "biological analogues of liquid crystals" and "pseudomorphoses" were introduced by Bouligand (1972) to describe those systems that obey liquid crystalline geometries but have lost their liquid character. Examples of birefringent biological materials that can be genuine liquids are biological membranes, nucleic material, and certain gland secretions. Nonfluid analogues are cell walls and skeletal structures of plants and animals. The fibrillar material is aligned in a way geometrically similar to the arrangement of molecules in cholesteric liquid crystals; however, the length of the polymer, its crystallization, and the presence of cross-links destroy the liquid character that must exist during the first steps of secretion of the organic components by the cells.

D. Properties and Consequences of Cholesteric Geometry

1. Twisted Plywood Model

A classic model introduced by Bouligand (1965) represents the way molecules are organized in a cholesteric structure and explains how typical series of arced patterns observed in sections of cells and tissues result not from authentic curved filaments but originate from the successive molecular orientations found in the twisted plywood arrangement. The model is constructed as follows. The molecular directions are represented by parallel and equidistant straight lines on a series of rectangles, with the orientation of the lines rotating from one rectangle to the next by a small and constant angle (Fig. 2). A periodicity is visible each 180° rotation of the molecular directions corresponds to the half cholesteric pitch $P/2$. A cholesteric axis is defined normal to the stratification. The rotation is chosen to be left-handed, as has been found in all biological twisted materials studied so far.

Directly visible on the oblique sides of the pyramid are what appear to be superposed series of parallel nested arcs. The concavities of the arcs are reversed on opposite sides of the model. In biological systems this particular geometry has often been described as a *twisted plywood,* but depending on its appearance in optical or electron microscopy, or on the research field of the authors, a variety of descriptions of the cholesteric structure are found in the literature (Table I).

Two major types of twists are found in liquid crystals and their biological analogues and are defined by the disposition of the fibrillar elements either in parallel planes (*planar twist*) or coaxial cylinders (*cylindrical twist*) (Fig. 3). The *toroidal twist* is a third, less common, configuration (Bouligand *et al.,* 1985a).

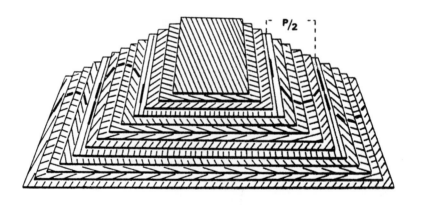

TWISTED PLYWOOD MODEL

FIG. 2   The model is drawn as a set of discrete planes represented by stacked sheets of cardboard forming a pyramid. On each card straight parallel lines represent the molecular directions. Superposed series of nested arcs are visible on the sides of the pyramid due to the regular variations of the molecular directions.

TABLE I

Descriptions of Cholesteric Geometries

Terms used by physicists or chemists (liquid crystal and polymer field)
  Cholesteric mesophases
  Cholesteric texture
  Chiral nematic structure

Terms used by microscopists (aspect in optical or electron microscopy)
  Fingerprint-like patterns
  Arced patterns

Terminology introduced by Bouligand and colleagues in France
  Twisted plywood
  Multidirectional plywood
  Twisted fibrillar phase

Terminology introduced by Neville and colleagues in the United Kingdom
  Helicoids
  Helicoidal architecture
  Helicoidal arrays

A convenient representation of the cholesteric geometry, proposed by Friedel and Kléman (1970), follows drawing conventions in which molecules parallel to the visualization plane are represented as lines, molecules normal to this plane are dots, and molecules oblique to it are nails. In the last case the tip of the nail is pointed toward the observer (Fig. 3).

The cholesteric periodicity is responsible for the characteristic properties that are observed when helicoidal materials are studied in polarized-light microscopy (periodic extinctions) and in transmission electron microscopy (arced patterns). The presence of deformations, defects, and artifacts causes a departure from the ideal model.

2. Periodic Extinctions in Polarized Light

The most distinctive characteristic of liquid crystals, while in a fluid state, is their ability to restore light when viewed between crossed polars in optical microscopy; the intensity of birefringence depends on the degree of alignment, the state of compaction, and the preparation thickness.

Considering the cholesteric model, maximum light is transmitted by molecules parallel to the slide and coverslip plane, minimum light by molecules normal to it, and intermediate amounts of light in between. Furthermore, molecules lying in the direction of the two polars produce extinctions. The results of these optics laws are shown in Fig. 4 for two positions of the crossed polars shifted by 45°. In both cases light and dark bands alternate, the only difference being in the position of the bands. Thus in oblique

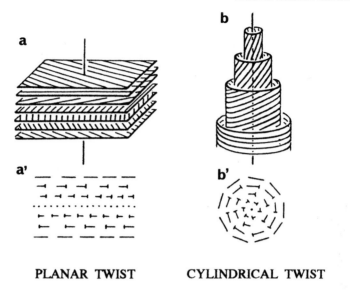

PLANAR TWIST   CYLINDRICAL TWIST

FIG. 3   (a) In a planar twist equidistant straight lines are drawn on horizontal planes, and the direction of the lines rotates regularly from plane to plane. (a') In the conventional notation for a cholesteric geometry applied to a planar twist, lines represent molecules longitudinal to the drawing plane, and dots represent molecules perpendicular to it; molecules in oblique position are represented by nails whose points are directed toward the observer. (b) In a cylindrical twist equidistant helices are drawn on a series of coaxial cylinders, and the angle of the helices rotates regularly from one cylinder to the next. (b') Conventional representation of a cholesteric geometry applied to a cylindrical twist. (Adapted from Bouligand *et al.*, 1985a.)

sections, when the microscope stage is rotated, the position of the extinction bands will regularly vary, producing what appear to be mobile fringes. This phenomenon can be clearly observed in semithin sections of biological materials analyzed in polarized-light microscopy using crossed polars, for example, in decalcified cuticles or compact bones (Section III,A).

3. Arced Patterns in Transmission Electron Microscopy

A major consequence of the helicoidal geometry is that materials having this spatial configuration produce superposed series of arced patterns when viewed in sections oblique to the cholesteric axis, which is the most frequently encountered situation. As explained in the discussion of the twisted plywood model, this visualization is due not to curved filaments but to the superposition of the successive molecular orientations within the section plane. The elucidation of the origin of the arced patterns is what initially allowed the twisted plywood model to be proposed and analogies to be

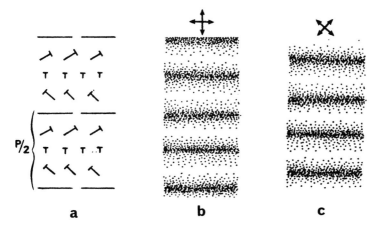

MOBILE FRINGES IN POLARIZED LIGHT

FIG. 4   (a) Conventional representation of a cholesteric structure in oblique section planes (as defined in figure caption 3b). A periodicity is visible for each 180° rotation of the molecular directions, defined as the half cholesteric pitch *P*/2. (b) and (c) Extinction positions observed in polarized-light microscopy correspond to molecules parallel to the directions of the crossed polars, which creates alternating extinguished and illuminated bands. The dark bands are shifted every 45° rotation of the crossed polars, which generates mobile fringes as the microscope stage is rotated.

made between the arrangement of macromolecules in biological structures and that of molecules in cholesteric liquid crystals (Bouligand, 1972). The arcs usually appear in electron microscopy at magnifications ranging from 10,000X to 80,000X, depending on the tissue and the concerned polymer, but sometimes they are easily visualized in optical microscopy, as in ascidian tunics or fish dermal plates. Arced patterns can be disposed on regular parallel lamellae and extend over hundreds of micrometers (e.g. in crab cuticles). They can also appear on concentric lamellae, as in compact bone osteons.

As the obliqueness of the angle section changes, the number of lamellae remains constant, but the apparent thickness and shape of the arcs vary. For section planes close to the vertical, arcs appear narrower; in vertical sections they disappear. Fibrils in longitudinal sections then alternate with fibrils in transverse sections. Vertical sections, devoid of arcs, can be tilted in the electron microscope by means of a goniometric stage. When a section is tilted to either side of a rotation axis parallel to the layers, the series of nested arcs then reappears with opposite concavities (Livolant *et al.*, 1978). This goniometric effect constitutes a validation of the twisted plywood model and is best seen on semithin sections observed in high-voltage electron microscopy (Fig. 5).

4. Deviations, Defects, and Artifacts

The twisted plywood model represents an ideal situation, whereas biological materials exhibit deviations of the fibrillar directions around an average position in a small volume of the material. Liquid crystals and their analogues show different kinds of deviations or defects recognized by their appearance in microscopy and resulting from the displacement, insertion, or removal of material. Double spiral patterns arise when distorted lamellae are viewed in sections cut normal to the deformation axis (Fig. 6a,b) (Bouligand, 1978a). *Dislocations* are translation defects that result from the translation of one part of the crystal with respect to another, and they occur in both solid and liquid crystals. They appear in polarized-light microscopy as bifurcating or interrupted lines and correspond to either edge or screw-type dislocations (Fig. 6c,d). *Disclinations* are rotation defects that occur only in liquid crystals. They can be recognized in polarized-light microscopy as triple points corresponding to the junction of three layers ($-\pi$ type) or as cofocal parabolas ($+\pi$ type) (Fig. 6e,f) (Bouligand, 1981). Molecules lie either parallel or normal to the defect line, introducing further λ or τ distinctions and appearing, respectively, as light or dark cores in polarized light.

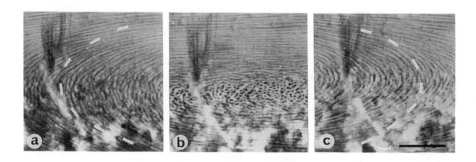

GONIOMETRIC EFFECT

FIG. 5 Goniometric effect observed in the organic matrix of the crab cuticle. The semithin section is observed in high-voltage electron microscopy with a goniometric stage whose tilt axis is parallel to the row of arcs. (a) and (c) In oblique views ($+10°$ and $-10°$ tilt) arcs appear, their direction depending on the sense of the tilt. (b) In horizontal view ($0°$ tilt) arcs disappear. The organic fibrils appear in either transverse or longitudinal position. Bar = 1 μm.

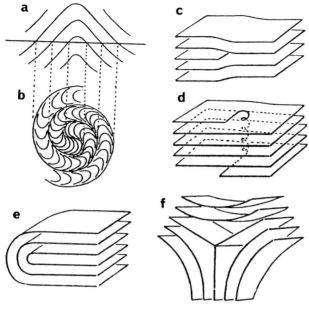

DEVIATIONS AND DEFECTS

FIG. 6   Schematic drawing of deviations or defects in cholesteric liquid crystals or their analogues. The planes represent levels where molecules have the same direction. They are
distant of $P/2$ (as defined in Fig. 2). (a) and (b) Double spiral patterns arising in sections cut
perpendicular to distorted lamellae. (c) and (d) Edge and screw dislocations corresponding
to translation of material. (e) and (f) $+\pi$ and $-\pi$ disclinations corresponding to rotated
lamellar directions after insertion or removal of material (Redrawn from Bouligand 1978d).

Microtomy artifacts have been shown to result from the twisted plywood
geometry. In the course of sectioning a twisted fibrous material, the knife
either passes between fibrils or cuts them and slightly curves their ends.
This process leads to periodic waves at both surfaces of the sections that
have been visualized by high-voltage electron microscopy, shadowing of
thin sections, and stereoscopic views (Giraud-Guille, 1986). This microtomy
artifact induces thickness variations. Consequently, as photons or electrons
are transmitted, irregular image contrast leads to alternating clear and dark
bands (Fig. 7) (Bouligand, 1986; Bunning *et al.,* 1994).

In the case of fractures through cholesteric geometries, different aspects
are obtained in scanning electron microscopy or on replicas obtained by
freeze-fractures. Modulations of the arc shapes occur for different positions
of the steps relative to the cholesteric structure (Livolant and Bouligand,
1989). Simply stated, when the cleavage plane is perpendicular to the choles-

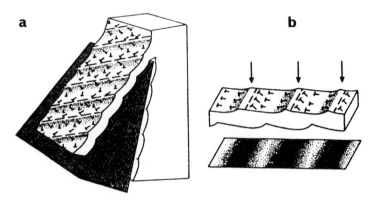

MICROTOMY ARTIFACTS

FIG. 7 Schematic of microtomy of a twisted fibrous material. (a) The knife either passes along the fibrils or cuts them perpendicularly, tilting up the sectioned ends. This cut results in regular steps on both sides of the section. (b) The thickness variations lead to alternating clear and dark bands when the sections are observed in microscopy.

teric stratification, a lamellar aspect is seen, and each lamella has homogeneous structure. When the cleavage plane is oblique to the cholesteric stratification, a lamellar aspect is still seen, but the whole structure presents an inhomogeneous staircase aspect. The surfaces of the steps then correspond to fibrillar directions parallel to the cleavage plan, and the countermarches to all other directions (Giraud-Guille, 1994).

III. Cholesteric Supramolecular Order in Cells and Tissues

A. Extracellular Matrices

The structural components of biological skeletal support systems are, respectively, cellulose in plant cell walls, chitin in many invertebrates and fungi, and collagen in vertebrates and invertebrates. These structural components appear as associations of unit molecular chains, which are called microfibrils, fibrils, or crystallites depending on the particular polymer or tissue. A common characteristic of these biopolymers is to be present in a very condensed state in extracellular matrices and to have regular spatial distributions. They are quite often ordered in a chiral nematic fashion *sensu stricto* or present strong relationships with the cholesteric model. We review all these situations.

1. Invertebrate Integument

a. Arthropod Cuticle Arthropods, with the three main groups—insects, crustaceans, and arachnids—represent about 80% of all known animal species. They are adapted to all types of ecological niches in water, land, and air. A major reason for the evolutionary success of members of this phylum is their cuticle, which is produced by epidermal cells and covers the animal surface, forming an interface between the organism and its environment. The basic chemical composition and structure of arthropod cuticles have been repeatedly investigated. Considerable progress was made in elucidating the architecture of the arthropod cuticle when it was theorized that its organic matrix has same spatial geometry as that described in certain liquid crystals (Bouligand, 1972). Although many other biological structures have since been shown to exhibit this state of matter, the arthropod cuticle remains the best example of a cholesteric liquid crystal analogue.

Much of the cuticle's properties can be attributed to the complex arrangement of chitin and protein, which has been resolved by electron microscopy and X-ray diffraction studies. The organic phase of the cuticle is essentially composed of chitin associated with proteins and is secondarily impregnated by a mineral phase of calcite. Chitin is a long-chain polysaccharide composed of repeating units of N-acetyl-D-glucosamine. About 18 to 25 individual molecular chains are arranged in two or three rows to form single crystalline rods. Individual chitin crystals can be identified in electron microscopy using a diffraction contrast method (Giraud-Guille *et al.*, 1990). Proteins are bound to the periphery of the chitin crystals, forming a fibril (Hadley, 1986).

The chitin–protein microfibrils are positioned horizontal and parallel in successive planes; however, progressing through the thickness of the cuticle, the fibrous direction rotates continuously and counterclockwise (Bouligand, 1965). When viewed in section in optical microscopy, this structure creates in all chitin-containing layers horizontal laminae whose thickness varies from 0.1 μm to several microns. In polarized-light microscopy periodic extinction patterns appear at every 90° rotation of the fibrillar directions, yielding mobile fringes when the microscope stage is rotated (Fig. 8a). In scanning electron microscopy irregular fracture planes lead to the illusion of an inhomogeneous structure with alternating dense and loose lamellae (Fig. 8b). In transmission electron microscopy, in sections oblique to the cuticle surface, the superposed laminae appear as parallel series of nested arcs (Fig. 8c); in vertical sections arced patterns are absent. The density of the twisted architecture can vary as a function of vertical gradients in the

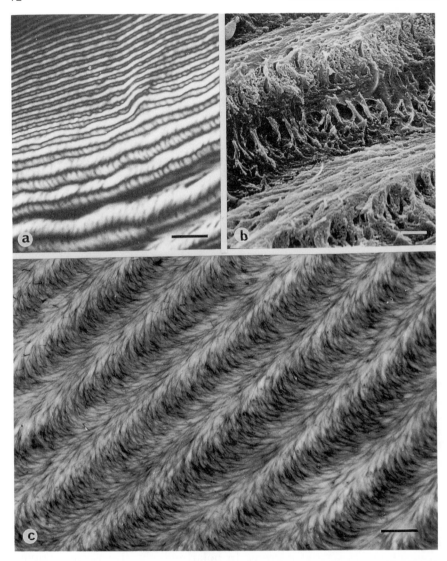

FIG. 8 Organic matrix of the decalcified crab cuticle of *Carcinus maenas* L. (a) Semithin section viewed in polarized-light microscopy. The twisted plywood geometry yields periodic extinctions every 90° rotation of the molecular orientations. Bar = 20 μm. (b) Fracture observed in scanning electron microscopy. A periodicity is still visible but with an inhomogeneous aspect due to a different behavior of the fracture in regard to the cholesteric geometry. Bar = 2 μm (c) Ultrathin oblique sections observed in transmission electron microscopy. The rows of nested arcs are due to the successive rotation of the fibrillar directions. Bar = 2 μm.

density and lamina thickness. Other vertical or oblique components of the cuticle such as projections of the epicuticle, pore canals, and muscle insertions do not participate to the twisted architecture.

b. Annelid Cuticle The cuticle of annelids is an extracellular matrix produced by epidermal cells, but in this group the structural polymer, instead of being formed by chitin rods, as in arthropods, is composed of nonstriated collagen fibrils. Electron micrographs of certain nematode cuticles, as in *Mermis,* reveal quite regular series of nested arcs, attesting to a continuous cholesteric twist of the fibrils in this material (Lee, 1970).

In contrast, other annelid cuticles present discontinuous plywood structures. In this case a discrete number of fibrillar orientations induce angular discontinuities (Bouligand and Giraud, 1980). Cases of only two to three fibrillar directions are described in the cuticle of deep-sea marine worms. In the abyssal polychaete species *Alvinella pompejana,* the three-dimensional organization of collagen shows interesting characteristics: the fibrils are helical and form a quasi-orthogonal network (Gaill and Bouligand, 1987a). Moreover, within each fibril the presence of a mutual twist between individual microfibrils was demonstrated in the specie *Paralvinella grasslei,* by goniometric studies and described as a cylindrical twist (Fig. 9) (Lepe-

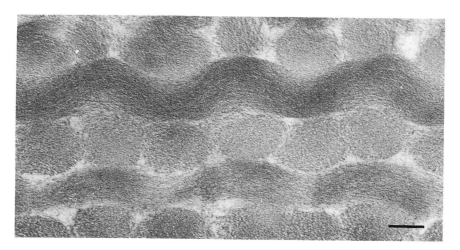

FIG. 9 Ultrastructure of the annelid cuticle *Paralvinella* after paraformaldehyde fixation. Fibrils appear composed of thin microfibrils. In longitudinal view their helical structure is verified by tilting the section. In transverse view the twisted packing of microfibrils is demonstrated by goniometric studies. Bar = 0.1 μm. (Reproduced by kind permission of L. Lepescheux (1988).)

scheux, 1988). The coexistence of orthogonality and twist at these two different levels recalls a model described in some liquid crystals referred to as *blue phases* by physicists, appearing at the transition between an isotropic and a cholesteric order (Meiboom *et al.*, 1981). The organic matrix of the protective tube of this same species has also been related to some deviation from the cholesteric model. The fibrous material made of polysaccharides differs from true cholesterics by the twist handedness, which is inverted each 180° rotation, and which induces, for certain section planes, sinusoidal or sigmoidal patterns (Gaill and Bouligand, 1987b). In another abyssal species *Riftia pachyptilla,* a mixed situation is present: in the anterior part of the animal three sets of fibrils produce, in sections tangential to the cuticle surface, a more or less regular hexagonal pattern, whereas in the posterior part of the animal two sets of fibrils form a regular orthogonal plywood (Gaill *et al.*, 1991).

c. Other Invertebrates   Many other invertebrate groups also yield examples of cholesteric geometries in tissues situated at the interface between the animal and its environment that require a protection role and possess performant mechanical properties.

Certain species of sponges form gemmules, small groups of cells protected by a shell that later develop into new individuals. Gemmule shells made of collagen fibrils show stacked series of nested arcs in *Ephydatia fluvatilis* (De Vos, 1972) and *Suberites domunculata* (Carrière *et al.*, 1974).

Among ascidians, the tunic of *Halocynthis papillosa,* composed of cellulose fibrils associated with proteins, was shown by Bouligand (1965) and Gubb (1975) to form a twisted plywood. In the bottom laminae situated near the epidermal cells regular arced patterns appear, but in the superficial laminae the cholesteric geometry is disrupted, possibly by mechanical tractions due to the growth of the body wall matrix.

Among echinoderms, the class Holothuroidea exhibits examples of cholesteric packing of collagen fibrils, as was shown, for example, in *Holothuria forskäli* in cuvierian tubules that are appendages of the alimentary canal (Dlugosz *et al.*, 1979). A micrograph showing the helicoidal packing of collagen fibrils in the dermis of *Havelockia inermis* was published by Bouligand (1978c), but neither that material nor the extracellular matrix of other groups cited by the author, such as Thyone's, were looked at in detail.

In molluscans, Hunt and Oates (1984) demonstrated that, in the periostracum of the marine gastropod *Buccinum undatum,* both the protein and the chitin fibrillar constituents form helical arrangements.

2. Vertebrate Skeletal Structures

In the years following the introduction of the electron microscope, photomicrographs of decalcified human compact bones published by Frank *et al.*,

(1955) showed quite regular series of arced patterns, but the origin of the arcs was not interpreted at that time as a regular twist of the collagen fibrillar orientations. Other patterns, corresponding to zigzags, were also described in the same Haversian systems, thus demonstrating a greater complexity of vertebrate skeletal structures in comparison with those observed in invertebrate integuments. This fact was further confirmed in studies on fish scales, where the collagen network presents a complex geometry described as a double twisted system.

a. **Human Bone**   Samples of compact bones, fractured transversely to the longitudinal bone axis and etched by nitric acid, reveal at low magnification in scanning electron microscopy adjacent osteons; each osteon is a cylindrical structure about 0.2 mm in diameter comprising concentric lamellae around a Haversian canal. A model proposed by Gebhardt (1906), generally regarded as the basic structure of osteons, consists of collagen fibrils running helically around the Haversian canal with a 90° change in the orientation of the fibrils from one lamella to the next. Three osteon types were further described by Ascenzi and Bonucci (1968) as a function of their aspect in cross-section in polarized-light microscopy, respectively called bright, intermediate, or dark osteons. The authors proposed that in bright osteons the collagen fibrils have a marked transversal course, in dark osteons a marked longitudinal course, and in intermediate osteons both transverse and longitudinal directions occur. Morphological descriptions in scanning electron microscopy generated a variety of terms and interpretations. The most recent works refer to the existence of dense and loose lamellae (Marotti and Muglia, 1988) or thin and thick lamellae (Weiner and Traub, 1992).

Theorizing that the arced patterns observed in ultrathin sections of compact bone are generated by the cholesteric model, Giraud-Guille (1988) described the two main spatial distributions of fibrils in osteons as producing either orthogonal or twisted plywoods. This concept unifies the diverse situations encountered in the literature at the optical or electron microscopy level and was rendered possible by studies on isolated osteons belonging to the three bright, dark, and intermediate types. The intermediate type osteons, observed among crossed nicols with polarized light, show a regular alternanation of extinguished and illuminated bands and produce a mobile fringe phenomenon when the microscope stage is rotated (Fig. 10a). Ultrathin sections from the same samples, in planes oblique to the osteon axis, reveal arced patterns formed by collagen fibrils lying either parallel to the section plan, transverse to it, or in intermediate oblique positions (Fig. 10c). In this case the cholesteric twist appears between successive concentric lamellae. In the bright and dark osteon types arced patterns are absent, and only two directions of fibrils are seen on ultrathin sections. Observations in scanning electron microscopy often give the illusion of different types

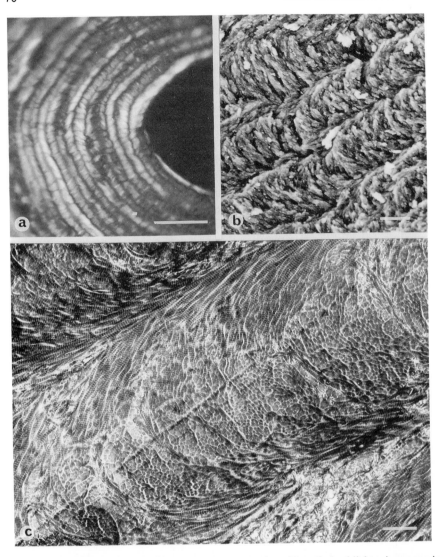

FIG. 10   (a) Decalcified human compact bone osteon viewed in polarized-light microscopy in sections transverse to its longitudinal axis. Periodic extinctions occur concentric to the osteon canal. Bar = 10 μm. (b) Horse radius bone viewed in scanning electron microscopy. The collagen fibrils display parallel series of nested arcs. Bar = 10 μm. (Reproduced by kind permission of M. Raspanti and A. Ruggeri.) (c) Decalcified human compact bone osteon observed in transmission electron microscopy. The cross-striated collagen fibrils create series of nested arcs in oblique section planes. Bar = 1 μm.

of lamellae as a consequence of certain fracture cleavage planes through the cholesteric geometry (Section II,D,4). In samples of heat-deproteinated equine bone observed in scanning electron microscopy, regular series of arced patterns are visualized, corroborating in this material the twisted plywood model (Fig. 10b) (Raspanti et al., 1994).

From data obtained with circularly polarized light and computer image analysis, Portigliatti-Barbos et al. (1984) concluded that a spiral distribution of the osteon types exists along the femoral schaft. This observation supports the hypothesis that mechanical factors such as tension and compression modulate the plywoods in different regions of the bone matrix. Twisted and orthogonal plywoods coexist in vertebrate compact bones and are closely related in structure. Indeed, a transformation was observed, within the same osteon, from twisted to orthogonal lamellae and vice versa, with the cholesteric patterns always corresponding to the most recently secreted levels (Giraud-Guille, 1988).

b. Fish Scales and Scutes

A polymorphism of the collagen three-dimensional network can be seen in the exoskeleton of fishes. The body of most teleosts is covered by thin and flexible elasmoid scales, and almost the whole thickness of the scales is formed by a dense connective tissue, the isopedin. Scales of the coelacanth *Latimeria chalumnae* first showed that this network of collagen bundles forms a complex three-dimensional arrangement. The collagen layers, composed of bundles of parallel fibrils, form successive plies that cross at an angle that differs slightly from 90°; every second layer then rotates by a small angle of about 25° in a given sense along an axis normal to the scale, so that the whole system is twisted (Fig. 11a). Oblique sections show stacked series of nested arcs when either odd or even layers are considered (Giraud et al., 1978) This supramolecular assembly defined as a double twisted plywood was shown to exist in many other scales but never appeared as regular or as extended. The spatial organization of collagen fibrils in elasmoid scales of 24 teleost families was studied, and this double twisted plywood generally appeared among more "primitive" families of fishes, whereas more or less orthogonal plywoods were found among more advanced families (Meunier and Castanet, 1982).

The body of fishes in certain taxons, instead of having classical elasmoid scales, is protected by scutes or dermal plates forming a rigid cuirass. This structure has recently been described in *Ostracion lentiginosum* as two opposite mineralized plates separated by an unmineralized dense collagenous network where a twisted plywood geometry is suggested (Meunier and Francillon-Vieillot, 1995). Indeed, on semithin sections observed in polarized light the collagen network yields particularly large series of arced patterns (Fig. 11b); at long distance the cholesteric laminae are not planar but correspond to a more complex geometry.

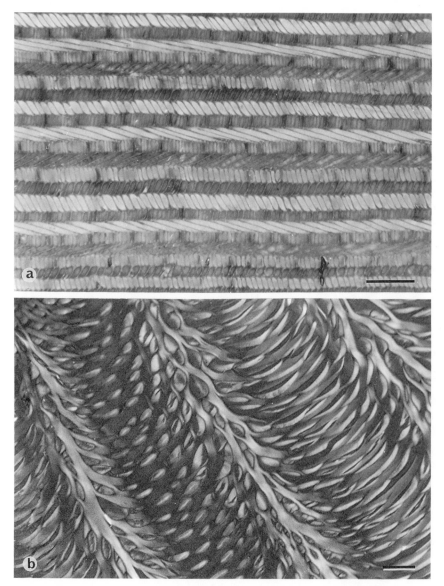

FIG. 11 (a) Coelacanth scale observed in sections oblique to the body surface. A complex distribution of the collagen bundles is observed in optical microscopy between crossed polars, with a periodic pattern appearing every seven layers. Bar = 20 μm. (b) Dermal plate of the fish *Ostracion* observed in polarized light. The collagen fibrils form large series of arced patterns. Bar = 20 μm. (Reproduced by kind permission of L. Besseau.)

3. Membranes of Animal Eggs

Examples of helicoidal architectures have been described in the egg membranes of various animals, such as nematodes, crustaceans, insects, and fishes. For example, fully formed eggshells of the parasitic nematodes *Trichuris suis* (Wharton and Jenkins, 1978) and *Trichuris muris* (Preston and Jenkins, 1984) reveal in their middle layers, composed of chitin microfibrils encased in a protein coat, superposed series of nested arcs, attesting to a twisted plywood architecture.

The eggs of anostracan crustaceans are capable of withstanding long periods of desiccation because the eggs are surrounded by a thick shell. The inner layer of the shell surrounding the plasma membrane of the gastrula in the crustacean *Branchipus stagnalis* shows parabolic lamellae typical of twisted fibrous systems (Gilchrist, 1978).

Mazur *et al.* (1982) performed an exhaustive study of silkmoth chorions. Their model illustrates *in vivo* departures from the ideal cholesteric model. The chorion is composed almost entirely of proteins secreted inward around the oocyte by the apical surface of a monolayer of epithelial cells. The bulk of the lamellar chorion consists of fibrils in lamellar arrangements appearing in oblique sections as stacked rows of parallel arcs. These arcs disappear in transverse sections, leaving alternating bands of lines and dots. Many holes, distortions (bend lamellae), and defects (dislocations and disclinations) are present in the chorion, which give the structure a quite irregular aspect in comparison with the quasi-perfect twisted plywoods of arthropod cuticles (Fig. 12).

The egg envelope of the teleost fish *Cichlasoma nigrofasciata* is composed of two concentric layers. The glycoprotein fibrils forming the inner, thick, laminated *zona pellucida* present regular series of arced patterns, and the cholesteric geometry is suggested to occur in a nonrigid, nonpolymerized state (Busson-Mabillot, 1973).

4. Plant Cell Walls

Neville in England and Roland's group in France have produced evidence that Bouligand's helicoidal model is applicable to a great number of plant cell walls (Fig. 13). The twisting microfibrils are essentially cellulose but can also be chitin, as in fungi. On the basis of tilting observations under the electron microscope Neville *et al.* (1976) established the validity of rotating fiber components for the cell wall of *Chara vulgaris* oospores and extended the model to the architecture of all major algae groups. Micrographs showing regular series of arced patterns were then published in the literature that referred to the helicoidal architecture in the rigid cell wall of the sclereids of poplar (Nanko *et al.*, 1978) and beech bark

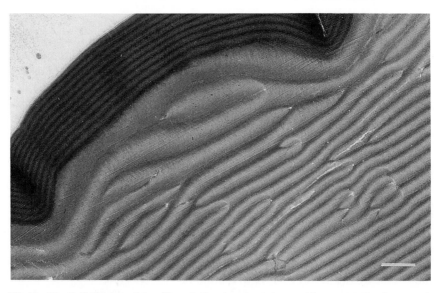

FIG. 12   Eggshell chorion of the silkmoth *Bombyx mori* observed in cross section in transmission electron microscopy. At this low magnification the arced patterns are visible only in the larger lamellae, but the cholesteric periodicity is strongly underlined by the microtomy artifact. The electron-dense outer crust appears as thin, even lamellae; the middle lamellar chorion shows distortions and many translation defects. Bar = 5 μm. (Reprinted from Mazur *et al.* (1982) with permission from Plenum Publishing Corporation.)

(Parameswaran and Sinner, 1979) and in subunits of fungal cell walls (Bonfante-Fasolo and Vian, 1984). In 1984, Neville and Levy published an extensive list of 50 plant species, for which electron micrographs of cell walls showed arced patterns, belonging to an extensive range including algae, fungi, moss, fern, monocotyledons, dicotyledons, and conifers. These helicoids were found in various cell types—spores, collenchyma, sclerenchyma, tracheids, sclereids, and parenchyma.

The plant helicoids just cited arise as liquid crystals but quickly become cross-linked to form stable arrays. In contrast, naturally occurring cholesteric liquid crystals were described in plant tissues within the cellulosic mucilage of mature seeds in the quince *Cydonia oblonga* (Willison and Abeysekera, 1988).

The versatility of the helicoidal expression in plants was considered by Roland *et al.* (1989) who showed the progression from liquid constructions, as in the quince, to hard systems, as in sclerified walls. These authors suggest that the matrix acts as a surfactant around the rigid cellulose crystallites, preventing flocculation and providing mobility. They also showed that various turbulences in the growing plant disturb the regular planar helicoidal

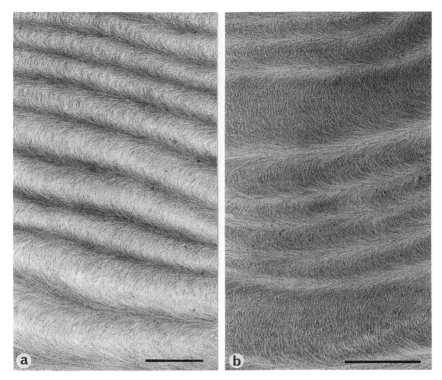

FIG. 13   Helicoidal patterns in plant cell walls observed in transmission electron microscopy after NaClO$_2$ delignification, DMSO extraction, and PATAg test for visualization of polysaccharides. (a) Regular series of arced patterns in stone cell of pear. (b) Alternation of wide and narrow arcs in endocarp of prune. Bar = 0.5 μm. (Reproduced by kind permission of D. Reis *et al.* (1994).)

pattern. In the external laminae, plant surface extension produces shears and stretches, inducing randomization, near the plasmalemna, instabilities cause alterations of the helicoidal pattern (Roland *et al.,* 1992). Cytochemical labeling reveals the close association between cellulose and glucuronoxylans, which form an acidic environment around cellulose microfibrils. Labeling also emphasizes the role of xylan components in helicoidal ordering assembly (Reis *et al.,* 1994).

B. Nuclei and Chromosomes

Just after the helicoidal packing of chitin was elucidated in arthropod cuticles, a regular series of arced patterns formed by DNA was observed in a quite unrelated structure—the chromosome of a primitive unicellular algae, the dinoflagellate *Prorocentrum micans* (Fig. 14). Bouligand *et al.* (1968) first proposed a model referring to the twisted packing of the DNA molecule in this material, where the main part of chromatin, devoid of histones, remains condensed throughout the cell cycle. The cholesteric model was further validated by goniometric studies in classic and high-voltage transmission electron microscopy (Livolant and Bouligand, 1978) and confirmed by cryosubstitution and cryosectioning techniques of vitrified specimens (Gautier *et al.*, 1987).

A cholesteric supramolecular ordering of DNA was further extended to other genetic material as arced patterns were observed, for example, in mitochondrial DNA in the ciliate *Bodo* (Brugerolle and Mignot, 1977) and the bacterial nucleoid of *Rhizobium* (Gourret, 1978); the cholesteric organization is present during the stationary condensed phase and disappears during the growth stage, where DNA synthesis occurs.

Much less evidence of twisted organizations of DNA exists among eukaryotes. It may be that the nucleosomic structure of the chromatin filament

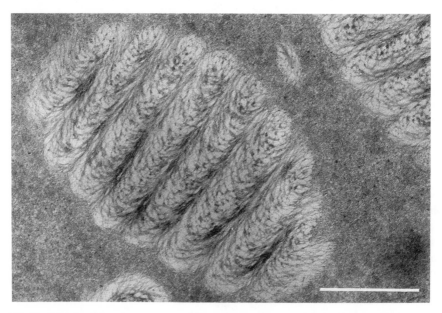

FIG. 14   Dinoflagellate chromosomes of *Prorocentrum micans* observed in transmission electron microscopy. DNA filaments draw regular superposed series of arcs in sections oblique to the chromosome longitudinal axis. Bar = 0.5 μm. (Reproduced by kind permission of M. O. Soyer-Gobillard.)

and the presence of histones either prevent or mask these long-range ordered assemblies. However, there is more or less direct evidence for a
cholesteric organization in sperm nuclei from many species, where the
amount of protein associated with DNA is smaller than in classic eukaryotic
cells. In the stallion sperm head, a lamellar structure with a periodicity of
330 Å is observed on freeze-fracture replicas, which probably corresponds
to the cholesteric stratification, but the half cholesteric pitch is too small
to visualize arcs in ultrathin sections 500 Å thick (Livolant, 1984). Furthermore, circular dichroism spectroscopy of this chromatin after extraction
shows characteristic properties of a cholesteric organization (Sipski and
Wagner, 1977).

In addition to helicoidal organizations, smecticlike structures and hexagonal packing of DNA have been described from X-ray diffractions and
microscopic analysis (Livolant, 1991). These different compacted forms of
DNA *in vivo* are important to consider in regard to the multiple liquid
crystalline phases observed *in vitro* that are directly related to the polymer
concentration (Section IV,C).

C. Cytoplasmic Inclusions, Cytoskeleton, and
Gland Secretions

A review of the literature shows cytoplasmic inclusions form helicoidal
assemblies, for example, proteins in mouse pancreatic acinar cells, hemoglo-
bin in sickle-cell anemia erythrocytes, and viruses in plant cells and insect
hosts (Bouligand, 1978a, 1978b).

Actin filaments have been directly related to liquid transverse order in
the stereocilium of cochlear hair cells. Each stereocilium, responsible for
the transduction of sound, comprises large polar bundles composed of
thousands of actin filaments that show no lateral order in transverse section.
Oblique section show periodic motifs of arced patterns, called *festooning*
by the authors, whose period depends on the angle of the section (De
Rosier *et al.*, 1980). These characteristics are typical of biological analogues
of liquid crystals (Section II,D,3).

The liquid crystalline nature of the cytoskeleton was shown in the plank-
ton chaetognath species *Sagitta setosa* (Duvert *et al.*, 1984), where the
keratin-like filaments are present in a viscous colloidal state in the inner
epidermal cells of certain regions of the body. In the cytoplasm of the red
alga *Alsidium helminthochorton* the fibrillar proteins yield stacked rows
of nested arcs in oblique section planes and alternating longitudinal and
transverse views of fibrils in transverse section planes, thus obeying the
geometric properties of the cholesteric model (Pellegrini and Pellegrini,
1985).

Examples of nematic or twisted nematic order of gland secretions are also described. The protein in the oothecal glands of the praying mantis was shown to exist in a true liquid crystalline state. The cholesteric spherulites, observed with a polarizing microscope, are seen to flow out from freshly punctured collaterial gland tubules. Electron micrographs of the spherulites after fixation show both classic parabolic patterns and double spiral patterns that arise from certain section planes through distorted lamellae (Neville and Luke, 1971). The optical birefringence of natural secretions from major ampullae of the spider *Nephila clavipes* and silk glands of the silkworm *Bombyx mori* suggests their liquid crystalline nature. The secretions become liquid crystalline after leaving the gland but before solidifying into a fiber, ensuring uniform molecular alignments (Kerkam *et al.*, 1991).

IV. Cholesteric Ordering of Biopolymers *in Vitro*

Experimental data demonstrating the physical tendency of biological mac-romolecules to form twisted supramolecular structures is important for understanding factors governing the three-dimensional arrangements of biopolymers in many living tissues. Evidence of liquid crystalline order formed *in vitro* with biological molecules first appeared sporadically in the literature (e.g., fibrillar polysaccharides (Marchessault *et al.*, 1959) and polypeptide solutions (Robinson, 1961)). Important developments were experimental data demonstrating cholesteric order within condensed phases of cellulosic polymers and DNA, and more recently the obtainment of long-range cholesteric assemblies with the three major polymers of extracellular matrices—cellulose, chitin, and collagen (Table II).

A. Cellulose and Chitin

1. Cellulose-Based Polymers

Evidence of a true cholesteric mesophase formed with molecular chains of cellulose-based polymers was first reported with the formation of cholesteric liquid crystalline phases in concentrated aqueous solutions of hydroxy-propyl cellulose (Werbowyj and Gray, 1976). Since then, many other cellu-lose derivatives have been found to exhibit typical textures of cholesteric mesophases either on temperature change (thermotropics) or in solution (lyotropics). Their formation and properties are influenced by the rigidity of the chains (reaching at most 150 Å in length), the occurrence of steric

TABLE II

Microscopic Evidence for *in Vitro* Cholesteric Assemblies of Biological Polymers

Twisting unit	Physical state	Identification level	Reference
DNA	Solution in 0.1 *M* NaCl	Cholesteric textures in PL[a]	Robinson (1961)
tRNA	Concentrated solution	Cholesteric spherulites in PL	Spencer *et al.* (1962)
DNA sonicated	High salt concentration	Banded globules in PL	Lerman (1973)
Hydroxy-propyl cellulose	Aqueous solution	Cholesteric textures in PL	Werbowyj and Gray (1976)
DNA fragments	Tris–HCl solution	Cholesteric textures in PL	Livolant (1983)
Collagen	Aggregates in gels	Twist between fibrils	Bouligand *et al.* (1985a)
Xanthan polymer	Aqueous solution	Cholesteric textures in PL	Livolant (1986)
Cellulose acetate	Solid film	Helicoidal structure in TEM[b]	Giasson *et al.* (1988)
Cellulosic polymers	Variety of solvents	Cholesteric textures in PL	Ambrosino and Sixou (1989)
Collagen monomers	Acetic acid solution	Cholesteric textures in PL	Giraud-Guille (1992)
Cellulose crystallites	Aqueous suspension	PL + TEM	Revol *et al.* (1992)
DNA fragments	Quick-frozen mesophase	Helicoidal structure in TEM	Leforestier and Livolant (1993)
Chitin crystallites	Aqueous suspension	PL + TEM	Revol and Marchessault (1993)
Collagen fibrils	Gel at neutral pH	Helicoidal structure in TEM	Besseau and Giraud-Guille (1995)

[a] PL, fingerprint patterns observed in polarized-light microscopy.
[b] TEM, arced patterns visualized in transmission electron microscopy.

effects, and the flexibility of side groups (Ambrosino and Sixou, 1989). Circular dichroism demonstrates that solid films prepared from cholesteric liquid crystalline phases of cellulose and cellulose acetate retain their supramolecular helicoidal structure (Ritcey and Grey, 1988). These films embedded and sectioned for observation in electron microscopy show a laminated aspect, microscopy artifacts, and defects—all typical of cholesteric structures observed in microscopy (Giasson *et al.*, 1988).

2. Crystallites and Fibrils

The tendency for extracellular polysaccharides to spontaneously form organized structures has also been observed at higher states of their organization, with crystallite suspensions and mixed cellulose–glucuronoxylan fibrils, which correspond to supramolecular assemblies found *in vivo.*

A most significant advance with fibrillar polysaccharides was the observation of long-range cholesteric assemblies obtained with aqueous crystallite suspensions, of both cellulose and chitin, above critical concentrations of 3–5% by weight (Revol *et al.,* 1992; Revol and Marchessault, 1993). The colloidal dispersion of crystallites separates within minutes into an anisotropic phase formed by small tactoids exhibiting birefringent bands (Fig. 15a). After longer times the tactoids fuse and produce continuous textures with fingerprint-like patterns (Fig. 15b). The chiral nematic order is retained when the mesophase is slowly dried to form a solid film; periodic arced fibrillar patterns are visible in ultrathin sections from such films (Fig. 15c). The periodicity corresponding to the half cholesteric pitch measures 30 μm in the initial chitin mesophase and 5 μm in the solid film.

The source of the chirality in these systems is not yet understood. Chitin and cellulose at the molecular level are optically active, but it is not clear how these chiral properties are transferred to the crystallites, which correspond to rigid rods made of ordered parallel arrays of chains. Moreover, at the concentrations at which the cholesteric order appears the crystallites are spaced about 20 nm apart. Revol and Marchessault (1993) attribute the cholesteric order to a helical geometry of the rods that generates a twist of the effective ionic envelope around each crystallite and induces chiral interactions between adjacent rods.

Some authors propose that a "twisting agent" induces the cholesteric assembly of cellulose microfibrils based on the fact that *in vivo* the microfibrils are mixed with other components. Indeed, cytochemical results show that inside the mature seeds of the quince *Cydonia oblonga* L. glucuronoxylans form a charged coat around each cellulose microfibril, with a prevalent helicoidal organization. This water-extracted cellulosic mucilage, concentrated to a viscous state by centrifugation, presents anisotropic domains when observed in optical microscopy with polarized light. After embedding in a hydrosoluble resin some sigmoidal patterns are visible in transmission electron microscopy (Reis *et al.,* 1991).

B. Collagen

Twisted architectures were first observed *in vitro* within collagen-containing systems in ultrastructural studies of frog skin cultured on purified collagen

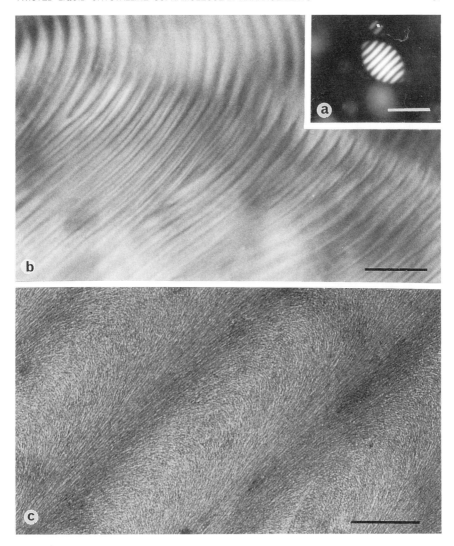

FIG. 15 Cholesteric liquid crystalline phases of cellulose and chitin (a) Small ellipsoidal regions (tactoïds) form spontaneously in a few minutes in aqueous suspensions of cellulose crystallites. The birefringent bands are observed between crossed polars in optical microscopy. Bar = 0.1 mm. (b) Fingerprint-like patterns in an aqueous suspension of chitin crystallites one day after sonication or shearing. Periodic extinction lines, observed in polarized light, correspond to the half cholesteric pitch $P/2$. Bar = 100 μm. (c) Electron micrograph of a chiral nematic mesophase of cellulose crystallites, slowly evaporated to form a film, showing periodic series of arced patterns. Bar = 1 μm. (Reproduced by kind permission of J. F. Revol et al. (1992) and J. F. Revol and R. H. Marchessault (1993).)

gels. Fibrils at the bottom of petri dishes, and consequently at long distances from the cells, showed remarkable local architectures. Examined in the absence of cells, the gels that were prepared from 2.5 mg/ml collagen solutions confirmed the occurrence of cholesteric geometries at two levels of organization: (1) microfibrils gathered into twisted bundles that further condensed in cross-striated fibrils and (2) fibrils assembled in aggregates about 1 μm in diameter that presented a twisted packing (Bouligand et al., 1985a).

1. Monomers at Acid pH

Results obtained in cell-free assembled collagen gels (Bouligand et al., 1985a) presented limitations because the extent of the cholesteric order was restricted to several micrometers and because loss of fluidity within the gel prohibited further evolution of the in vitro system. For these reasons, progress in the search for ordered assemblies of collagen was obtained at the molecular level in acid-soluble solutions of triple helical monomers. At first anisotropic distributions of molecules were monitored in polarized-light microscopy, in 1-μl drops of acid-soluble type I calfskin collagen at 5 mg/ml that were allowed to slowly evaporate on a slide. Concentric arrangements of the molecular aggregates appeared with more or less complex birefringent patterns progressing from the center of the drop to its periphery (Giraud-Guille, 1987). The textures were described as concentric bands, radiating series of arcs, and undulating patterns. These results can be related to the anisotropic behavior of collagen in a condensed state, which gives the appearance of molecular alignments and undulations but does not produce the authentic textures of twisted nematic phases because the displacements of the molecular aggregates are limited within the highly concentrated samples.

This drawback was overcome when the collagen solutions were first subjected to ultrasonic waves, thus inducing mobility in the samples, which had been evaporated to a viscous state. The birefringence of collagen drops placed between the slide and coverslip is progressively enhanced as a function of the concentration. After several hours, small cholesteric globules can spontaneously emerge within the anisotropic phase, showing concentric cholesteric bands (Fig. 16a). After several days typical fingerprint-like patterns of cholesteric liquid crystals are observed in polarized light (Fig. 16b) (Giraud-Guille, 1989). The sequence of events between disordered isotropic solutions and cholesteric domains can also pass through nematic and precholesteric textures corresponding to molecular alignments and undulations. All these situations have been described, geometrically interpreted, and related to in vivo situations (Giraud-Guille, 1992).

It was further shown that the rate of cholesteric liquid crystal formation correlates positively with sonication time. The extent of the cholesteric

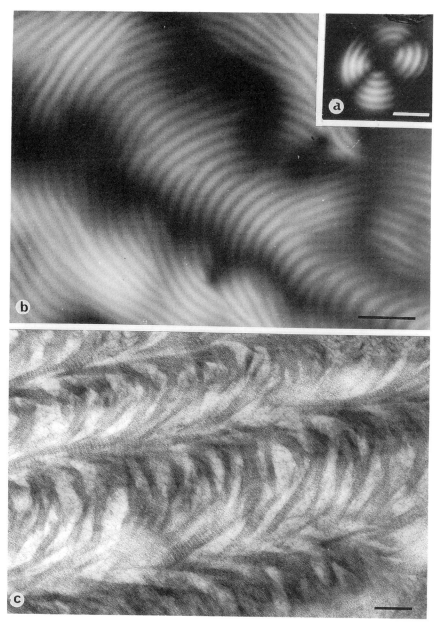

FIG. 16 Cholesteric liquid crystalline phases of collagen (a) Cholesteric spherulite spontaneously emerging in a viscous solution of collagen observed between crossed polars in optical microscopy. Bar = 10 μm. (b) Fingerprint-like textures observed in polarized light in a concentrated solution of acid-soluble collagen after sonication. Bar = 10 μm. (c) Stabilized cholesteric phase at neutral pH showing series of arced patterns in transmission electron microscopy formed by cross-striated collagen fibrils. Bar = 1 μm.

order then significantly improves, increasing from local cholesteric domains restricted to the air interfaces, to large twisted domains forming up to a third of the sample surface. The effect of ultrasonic waves was shown, by electrophoresis and electron microscopy of shadowed molecules, to reduce the molecular aggregates present in the collagen molecular population and to induce the fragmentation of triple helical monomers (Giraud-Guille *et al.*, 1994).

2. Fibrils at Neutral pH

Ultrastructural studies of collagen liquid crystalline assemblies by the usual techniques of fixation for transmission electron microscopy are inadequate because any dilution changes the three-dimensional order in the sample, with consequent loss of the cholesteric order. Such studies were rendered possible after stabilization of the viscous mesophases under ammonia vapor. The passage from a sol state at acid pH to a gel state at neutral pH, without any dilution of the samples, retains the initial three-dimensional order as controlled in polarized light. The resulting gel was dissected for precise textures identified in microscopy and then classically prepared for transmission electron microscopy. The cholesteric domains give images of superposed series of nested arcs formed either with supramolecular collagen aggregates or with typical cross-striated collagen fibrils. Such micrographs (Fig. 16c) allow relevant comparisons with collagen matrices *in vivo* because they mimic patterns observed in thin sections of collagen skeletal structures, for example, in decalcified compact bone osteons (Fig. 10c). These results attest to the involvement of collagen liquid crystalline assemblies in matrix morphogenesis (Besseau and Giraud-Guille, 1995).

C. DNA

Traditionally the condensed states of DNA were considered to be functionally inactive in both replication and transcription. Recent experimental data show that accelerated DNA renaturation occurs in condensed DNA (Sikorav and Church, 1991). Moreover, electron microscopy and biological properties of polyamine-compacted DNA plead in favor of a transcriptional activity associated with condensation (Baeza *et al.*, 1992). Understanding the liquid crystalline properties of DNA at high concentrations is therefore of special interest in biology.

Condensed forms of DNA have been observed by different experimental methods Jordan *et al.* (1972) first studied a mixture of DNA and polyethylene glycol called Psi DNA. Excluded-volume interactions undergo a spontaneous arrangement characterized by the typical circular dichroism spectra of DNA, which is interpreted as long-range ordering in compact states. Lerman *et al.* (1976) described hexagonal packing from sonicated DNA in alcoholic solution.

Significant *in vitro* studies were then conducted either with sonicated calf thymus DNA, which yielded DNA fragments between 50 nm and 7.9 μm long (Livolant, 1983), or with 50 nm long DNA in aqueous solutions prepared by selective digestion of calf thymus chromatin with micrococcal nuclease (Strzelecka *et al.*, 1988).

At concentrations, lower than 50 mg/ml, the DNA molecules are randomly oriented, and the solution is a classic isotropic liquid. As the concentration increases, ranging from 150 to 250 mg/ml, cholesteric phases form (Fig. 17a). In these phases dislocations appear as bifurcating lines, and disclinations appear as points around which the molecular orientations rotate. Even more complex textures appear as double spiral patterns and polygonal patterns due to curvatures and distortions of the layers (Livolant,

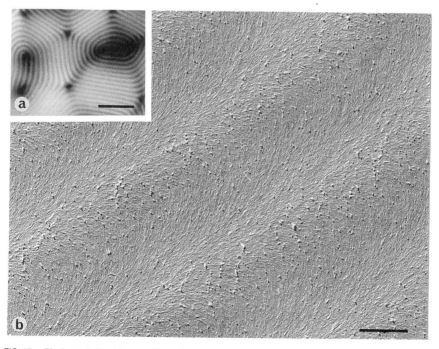

FIG. 17   Cholesteric liquid crystalline phases of DNA. (a) Typical fingerprint patterns observed in polarized-light microscopy between crossed circular polars. Bar = 10 μm. (b) Series of nested arcs visible in electron microscopy after freeze-fracture. Bar = 1 μm. (Reproduced by kind permission of F. Livolant (1991).)

1986). Cholesteric germs can grow directly within an isotropic phase; in the case of long molecules (over 50 nm), successive transition steps described as precholesteric stages, are observed (Livolant, 1987). At higher concentrations, the cholesteric phase transforms and presents a columnar longitudinal order and a hexagonal lateral order (Livolant *et al.,* 1989).

Ultrastructural studies were conducted to determine to what extent liquid crystalline properties of DNA are involved in the condensation of chromatin within the living cell. Because no stable links maintain the three-dimensional structure in liquid crystals, which precludes any chemical fixation, the water-rich DNA mesophases were fixed by quick-freezing against a copper block cooled with liquid helium to prevent the formation of ice crystals. Electron microscopy observation of replicas of fractured specimens, shadowed with Pt and C, reveal periodic patterns that recall the fingerprint patterns observed in polarizing microscopy. At high magnifications arced patterns are recognized (Fig. 17b) (Leforestier and Livolant, 1991).

DNA mesophases can be considered as models of simple chromatin organization. Indeed, arced patterns, dislocations, and disclinations are observed in living systems, namely, in cholesteric chromosomes (Section II,B). More relevant comparisons with the chromatin organization will require addition of other components, such as histones or other proteins and RNA, which may modify the three-dimensional configurations of liquid crystalline DNA.

Transfer ribonucleic acids (tRNA), for example, long ago were shown to form cholesteric assemblies: in concentrated solutions of tRNA the rod-shaped molecules spontaneously form twisted liquid crystalline structures that appear in polarized-light microscopy as regular spherulites showing double spiral contours (Spencer *et al.,* 1962).

V. Concluding Remarks

A. On Twisted Liquid Crystalline Order

1. Reasons of Misknowledge

The first publications on twisted liquid crystalline arrangements in biology are more than 20 years old, yet this subject remains unfamiliar to many cellular and molecular biologists, for two possible reasons. First, the analogy between biological structures and cholesteric assemblies applies to a wide range of materials, which shows the significance of the liquid crystalline hypothesis but also generates dispersion of data on this subject due to the large bibliographic domains concerned. Second, the successful results in obtaining cholesteric assemblies *in vitro* with biological macromolecules have not been directly available to cellular biologists because cholesteric mesophases were first identified under polarized-light microscopy, a deserted tool in cellular biology departments, and because these data were most often published in chemistry or physics journals.

Fortunately, a direct link has been established between *in vivo* and *in vitro* results by publication of data on cholesteric mesophases at the ultrastructural level, which reproduce (at the same scale and with identic macromolecular constituents) the exact geometries identified in biological structures (see Figs. 15c, 16c, 17b, and associated references).

2. Origin of Cholesteric Twist

Theoretical treatments on phase equilibria of rodlike particles in suspension predict that above a critical concentration, which depends primarily on the axial ratio of the rods, phase separation occurs (Flory, 1956a, 1956b).

The spontaneous alignment of elongated molecules at high concentration originates mainly from steric constraints involving the spatial configuration of the molecules: hard, rod-shaped molecules do not interpenetrate, and the excluded volume is reduced when the molecules lie parallel (Flory, 1961). From Flory's calculation it appears that the alignment requires an L/D ratio greater than 10, but displacements of very long molecules are difficult in highly concentrated solutions, which prevents a good alignment.

At high concentrations geometric factors associated with the chirality of the molecules govern the phase separation. A constant characteristic of biological polymers forming cholesteric liquid crystals is that they are helicoidal. A model proposed by Rudall (1956) suggests that twisted assemblies of asymmetric molecules are determined mainly by the contact between helicoidal domains. To minimize steric hindrance, helical polymers first align, thus gaining maximum space. This displacement creates parallel and equidistant grooves, oblique to the helical axis, where a second layer of molecules can be disposed. A twisted system is generated by this oblique packing over successive levels.

At low concentrations the source of chirality is less understandable because long distances can separate the twisting units. For example, Revol (1993) estimated at 20 nm the space between two chitin crystallites within suspensions forming cholesteric phases. This author proposes that the twist originates from interactions between electrical double layers of adjacent rigid rods, which implies that a twisted ionic envelope surrounds each crystallite.

3. Hierarchical Levels of Twist

The building units in liquid crystals and their biological analogues are not finite in size and possibly vary from the molecular to the subcellular level.

This observation thus allows comparisons between systems observed at quite different scales. On the question of the assembly of collagenous matrices, Birk *et al.* (1991) underline the variety of factors that operate from collagen molecules to matrix architecture and recall that physicochemical forces described as self-assembly are involved in the constitution of collagen fibrils. The example of collagen is particularly demonstrative in regard to liquid crystalline behaviors. Indeed, as this review recalls, hierarchical levels of twist have appeared at various scales ranging from angstroms to microns. The smallest entities, observed at the molecular level, correspond to a twist between triple helical rods 1.2 nm wide and about 30 nm long, and the largest entities, observed at the tissue level, correspond to a twist between 5 μm large orthogonal bundles, described in fish scales. At intermediate sizes, a cholesteric twist has been described between monomers in cholesteric phases, between microfibrillar aggregates in gels, and between fibrils in bone and in gels.

B. Implications in Regard to Cell Machinery

1. Fluidity at Time of Secretion

The hypothesis first formulated by Bouligand (1972) was that an intermediate helicoidal fluid phase is involved in the morphogenesis of widely observed structures. The presence in many biological analogues of cholesteric liquid crystals, of distortions, or of certain defects, such as $+\pi$ and $-\pi$ disclinations, which can emerge only in liquid states, confirms this hypothesis.

Neville (1993) judiciously remarked that ultrastructural observations of anthropod cuticles above the epidermal cells reveal uniform series of arced patterns, attesting to fluidity at the time of secretion. Indeed, products of neighboring cells join up neatly if the system is liquid; otherwise, many irregularities and faults would appear along the cell boundaries.

The presence of a constant helicoidal pitch (i.e., a constant layer thickness) can also be an argument in favor of a more liquid character of the system. The appearance of a variable pitch suggests that cholesteric analogues are rapidly stiffened after their deposition. Stabilization of the mesophase occurs through reversible weak links in materials where decondensation of the cholesteric packing occurs, which is the case for nucleic material. Conversely, intermolecular covalent cross-links are progressively established in skeletal structures, for example, the sclerotization process that hardens insect cuticles; extracellular matrices can also be secondarily hardened by the nucleation and growth of mineral crystals, such as calcium carbonate or hydroxyapatite.

2. Cytoskeleton and Helicoids

Hypotheses on the involvement of intracellular cytoskeletal proteins in the positioning of extracellular polymers are regularly published in the literature. These interpretations are based mainly on immunofluorescence or electron microscopy data showing relationships between fibronectin and actin in fibroblasts in culture (Hynes and Destree, 1978), parallel alignment between microtubules and cellulose microfibrils in plant cell walls (Lloyd, 1982), or coalignment of microtubules and actin filaments with collagen in vertebrate skeletal structures (Dane and Tucker, 1986; Zylberberg et al., 1988). According to many authors cytoskeletal proteins within the cell should control the orientation of extracellular polymers beyond the cell membrane. In the case of unidirectional systems this hypothesis is conceivable, but when considering helicoids—where a small and constant change of orientation occurs throughout the structure—this would imply successive polymerizations and depolymerizations of the cytoskeletal proteins with each change of orientation. This situation is quite unrealistic. Experimental data support this doubt. In the green alga *Oocystis solitaria,* parallel orientations of microfibrils occur even after prevention of microtubule formation by colchicine treatment (Grimm et al., 1976). In scleroblasts of fish scales, microtubules were shown to repolymerize parallel to the collagen alignment after their depolymerization at $-2°C$ (Mac Beath and Fujiwara, 1989).

We suggest that the coalignment of cytoskeletal elements in regard to extracellular fibrillar material is initiated by mechanical forces, such as tension or compression, and thus constitutes a response to the orderly spatial arrangements in the extracellular matrix instead of being the cause of them.

C. Future Prospects

Highly condensed packing of macromolecules frequently occurs in living systems showing more or less transitory liquid crystal states, such as the extracellular space of animal tissues, genetic material, and plant cell walls. Considerable progress has been made in the last 10 years in the ability to reproduce *in vitro* these assemblies with all major families of polymers, nucleic acids, proteins, and polysaccharides.

In the future it will become essential to study the detailed mechanisms for chiral ordering and stabilization of solutions or suspensions of rodlike biological polymers in concentrated states. The influence of various parameters such as molecular weight, degree of polymerization and substitution, and dilution in a mixture of solvents will have to be considered. These studies will need to be conducted both at the molecular level—for example, as molecular chains of oligomers or polymers in the case of polysaccharides. or as helical rods of defined length in the case of proteins and nucleic acids—and at the subcellular level, as crystallites or fibrils.

As experimental data on these simplified one-component systems accumulate, other factors will have to be studied, such as the presence of added cellular components (e.g., xylans with cellulose, proteins with chitin, or histones with DNA).

The questions on the first steps of secretion of cholesteric systems in the case of extracellular matrices still need to be investigated and more data are required on the interrelationships between cellular morphology, migration, proliferation, and the extracellular environment. In this context, the behavior of cells confronted by ordered matrices following cholesteric geometries will be of considerable interest.

Acknowledgments

I am very grateful to the following colleagues for providing photographs as well as advice and interesting discussions: Dr. L. Besseau, Dr. Y. Bouligand, Dr. A. Leforestier, Dr. L. Lepescheux, Dr. F. Livolant, Dr. G. D. Mazur, Dr. M. Raspanti, Dr. D. Reis, Dr. J. F. Revol, and Dr. M. O. Soyer-Gobillard.

Thanks are also due to M. Albert and D. Sainte-Hilaire for expert help in microtomy and to B. Riviere, M. J. Bodiou, and J. Lecomte for assistance in preparing the figures.

References

Ambrosino, S., and Sixou, P. (1989). Cellulosic mesophases *In* "Microemulsions and Liquid Crystals" (M. A. El-Nokaly, ed.), Vol. 383, pp. 142–155. ACS Symposium Series, American Chemical Society, Washington, D.C.

Ascenzi, A., and Bonucci, E. (1968). The compressive properties of single osteons. *Anat. Rec.* **161,** 377–392.

Baeza, I., Gariglio, P., Rangel, L. M., Chavez, P., Cervantes, L., Arguello, C., Wong, C., and Montañez, C. (1987). Electron microscopy and biochemical properties of polyamine-compacted DNA. *Biochemistry* **26,** 6387–6392.

Besseau, L., and Giraud-Guille, M. M. (1995). Stabilization of cholesteric phases of collagen to ordered gelated matrices. *J. Mol. Biol.* **251,** 197–202.

Birk, D. E., Silver, F. H., and Trelstad, R. L. (1991). Matrix assembly. *In* "Cell Biology of Extracellular Matrices" (E. D. Hay, ed.), pp. 221–254. Plenum Press, New York.

Bonfante-Fasolo, P., and Vian, B. (1984). Wall texture in the spore of a vesicular-arbuscular mycorrhizal fungus. *Protoplasma* **120,** 51–60.

Bouligand, Y. (1965). Sur une architecture torsadée répandue dans de nombreuses cuticles d'arthropodes. *C. R. Acad. Sci. (Paris)* **261,** 3665–3668.

Bouligand, Y. (1972). Twisted fibrous arrangements in biological materials and cholesteric mesophases. *Tissue and Cell* **4,** 189–217.

Bouligand, Y. (1978a). Liquid crystalline order in biological materials. *In* "Liquid Crystalline Order in Polymers" (A. Blumstein, ed.), pp. 261–297. Academic Press, New York.

Bouligand, Y. (1978b). Liquid crystals and their analogs in biological systems *In* "Liquid Crystals" (L. Liébert, ed.), Solid State Physics, Supplement 14, pp. 259–294. Academic Press, New York.

Bouligand, Y. (1978c). Cholesteric order in biopolymers. *In* "Mesomorphic Order in Polymers and Polymerization in Liquid Crystalline Media" (A. Blumstein, ed.), Vol. 74, pp. 237–247. ACS Symposium Series, American Chemical Society, Washington, D.C.

Bouligand, Y. (1978d). Aspects expérimentaux des défauts dans les structures mésomorphes. *J. Microsc. Spectrosc. Electron.* **3**, 373–386.

Bouligand, Y. (1980). Defects and textures of hexagonal discotics. *J. Physique.* **41**, 1307–1315.

Bouligand, Y. (1981). Geometry and topology of defects in liquid crystals. *In* "Physique des Défauts" (R. Balian *et al.,* eds.). North-Holland, pp. 668–771.

Bouligand, Y. (1986). Theory of microtomy artefacts in arthropod cuticle. *Tissue and Cell* **18**, 621–643.

Bouligand, Y., Denèfle, J. P., Lechaire, J. P., and Maillard, M. (1985a). Twisted architectures in cell-free assembled collagen gels: Study of collagen substrates used for cultures. *Biol. Cell* **54**, 143–162.

Bouligand, Y., and Giraud, M. M. (1980). Structures périodiques continues et discontinues dans les tissus de soutien (principaux exemples: Arthropodes et Prochordés). *Bull. Soc. Zool. Fr.* **105**, 323–329.

Bouligand, Y., and Giraud-Guille, M. M. (1985b). Spatial organization of collagen fibrils in skeletal tissue: Analogies with liquid crystals. *In* "Biology of Invertebrate and Lower Vertebrate Collagens" (A. Bairati and R. Garrone, eds.), pp. 115–134. Plenum Press, New York.

Bouligand, Y., Soyer, M. O., and Puiseux-Dao, S. (1968). La structure fibrillaire et l'orientation des chromosomes chez les dinoflagellés. *Chromosoma* **24**, 251–287.

Brugerolle, G., and Mignot, J. P. (1979). Distribution et organisation de l'ADN dans le complexe kinétoplaste mitochondrie chez un Bodonidé, Protozoaire kinétoplastidé: Variations au cours du cycle cellulaire. *Biol. Cell* **35**, 111–114.

Bunning, T. J., Vezie, D. L., Lloyd, P. F., Haaland, P. D., Thomas, E. L., and Adams, W. W. (1994). Cholesteric liquid crystals: Image contrast in the TEM. *Liquid Crystals* **16**, 769–781.

Busson-Mabillot, S. (1973). Evolution des enveloppes de l'ovocyte et de l'oeuf chez un poisson téléostéen. *J. Microscopie* **18**, 23–44.

Carrière, D., Connes, R., and Paris, J. (1974). Ultrastructure et nature chimique de la coque et du vitellus gemmulaires chez l'éponge marine: *Suberites domunculata. C. R. Acad. Sc. Paris* **278**, 1577–1580.

Dane, P. J., and Tucker, B. (1986). Supracellular microtubule alignments in cell layers associated with the secretion of certain fish scales. *J. Cell Sci.* **5**, 273–291.

De Gennes, P. G. (1974). "The Physics of Liquid Crystals." Cambridge University Press, London.

De Rosier, D. J., Tilney, L. G., and Egelman, E. (1980). Actin in the inner ear: The remarkable structure of the stereocilium. *Nature* **287**, 291–296.

De Vos, L. (1972). Fibres géantes de collagène chez l'éponge *Ephydiata fluviatilis. J. Microscopie* **15**, 247–252.

Dlugosz, J., Gathercole, L. J., and Keller, A. (1979). Cholesteric analogue packing of collagen fibrils in the cuvierian tubules of *Holothuria forskäli* (Holothuroidea, Echinodermata). *Micron* **10**, 81–87.

Duvert, M., Bouligand, Y., and Salat, C. (1984). The liquid crystalline nature of the cytoskeleton in epidermal cells of the chaetognath *Sagitta. Tissue and Cell* **16**, 469–481.

Flory, P. J. (1956a). Statistical thermodynamics of semi-flexible chain molecules. *Proc. Roy. Soc. London, Ser. A,* **234**, 60–73.

Flory, P. J. (1956b). Phase equilibria in solutions of rod-like particles. *Proc. Roy. Soc. London, Ser. A,* **234**, 73–89.

Flory, P. J. (1961). Phase changes in proteins and polypeptides. *J. Polymer Sci.* **59**, 105–128.

Frank, R., Frank, P., Klein, M., and Fontaine, M. (1955). L'os compact humain normal au microscope électronique. *Arch. Anat. Microsc. Morph. Exp.* **44**, 191–206.

Frey-Wissling, A. (1953). "Submicroscopic Morphology of Protoplasm." Elsevier, Amsterdam.

Friedel, G. (1922). Les états mésomorphes de la matière. *Ann. Physique* **19**, 273–474.

Friedel, J., and Kléman, M. (1970). Application of dislocation theory to liquid crystals. In "Fundamental Aspects of Dislocation Theory" (J. A. Simmons, R. de Wit, and R. Bullough, eds.), Vol. 1, pp. 607–635. National Bureau of Standards, US Special Publication.

Gaill, F., and Bouligand, Y. (1987a). Supercoil of collagen fibrils in the integument of *Alvinella*, an abyssal annelid. *Tissue and Cell* **19**, 625–642.

Gaill, F., and Bouligand, Y. (1987b). Alternating positive and negative twist of polymers in an invertebrate integument. *Mol. Cryst. Liq. Cryst.* **153**, 31–41.

Gaill, F., Herbage, D., and Lepescheux, L. (1991). A discrete helicoid of collagenous fibrils: The cuticle of deep-sea hydrothermal vent worms (*Riftia pachyptila*). *Matrix* **11**, 197–205.

Gautier, A., Michel-Salamin, L., Tosi-Couture, E., McDowall, A. W., and Dubochet, J. (1987). Electron microscopy of the chromosomes of dinoflagellates *in situ*: Confirmation of Bouligand's liquid crystalline hypothesis. *J. Ultr. and Mol. Struct. Res.* **97**, 10–30.

Gebhardt, W. (1906) Ueber funktionell wichtige anordnungsweisen der gröberen und feineren bauelemente der wilberltierckknochens. *Roux Arch. Entw. Mech.* **20**: 187–322.

Giasson, J., Revol, J. F., and Ritcey, A. M. (1988). Electron microscopic evidence for cholesteric structure in films of cellulose and cellulose acetate. *Biopolymers* **27**, 1999–2004.

Gilchrist, B. M. (1978). Scanning electron microscope studies of the egg shell in some *Anostraca* (Crustacea: Branchiopoda). *Cell Tiss. Res.* **193**, 337–351.

Giraud, M. M., Castanet, J., Meunier, F. J., and Bouligand, Y. (1978). The fibrous structure of coelacanth scales: A "twisted plywood." *Tissue and Cell* **10**, 671–686.

Giraud-Guille, M. M. (1986). Direct visualisation of microtomy artefacts in sections of twisted fibrous extracellular matrices. *Tissue and Cell* **18**, 603–620.

Giraud-Guille, M. M. (1987). Cholesteric twist of collagen *in vivo* and *in vitro*. *Mol. Cryst. Liq. Cryst.* **153**, 15–30.

Giraud-Guille, M. M. (1988). Twisted plywood architecture of collagen fibrils in human compact bone osteons. *Calcif. Tisue Int.* **42**, 167–180.

Giraud-Guille, M. M. (1989). Liquid crystalline phases of sonicated type I collagen. *Biol. Cell* **67**, 97–101.

Giraud-Guille, M. M. (1992). Liquid crystallinity in condensed type I collagen solutions: A clue to the packing of collagen in extracellular matrices. *J. Mol. Biol.* **224**, 861–873.

Giraud-Guille, M. M. (1994). Liquid crystalline order of biopolymers in cuticles and bones. *Micr. Res. and Techn.* **27**, 420–428.

Giraud-Guille, M. M., Besseau, L., Herbage, D., and Gounon, P. (1994). Optimization of collagen liquid crystalline assemblies: Influence of sonic fragmentation. *J. Struct. Biol.* **113**, 99–106.

Giraud-Guille, M. M., Chanzy, H., and Vuong, R. (1990). Chitin crystals in arthropod cuticles revealed by diffraction contrast transmission electron microscopy. *J. Struct. Biol.* **103**, 232–240.

Gourret, J. P. (1978). Description et interprétation des nucléoides structurés observés dans des bactéroïdes de *Rhizobium*. *Biol. Cell* **32**, 299–306.

Gray, G. W. (1962). "Molecular Structure and Properties of Liquid Crystals." Academic Press, London.

Grimm, I., Sachs, H., and Robinson, D. J. (1976). Structure, synthesis and orientation of microfibrils. II. The effect of colchicine on the wall of *Oocystis solitaria*. *Cytobiologie* **14**, 61–74.

Gubb, D. C. (1975). A direct visualisation of helicoidal architecture in *Halocynthia papillosa* by scanning electron microscopy. *Tissue and Cell* **7**, 19–32.

Hadley, N. F. (1986). The arthropod cuticle. *Scientific American* **255**, 98–106.

Hunt, S., and Oates, K. (1984). Chitin helicoids accompany protein helicoids in the periostracum of a whelk *Buccinum. Tissue and Cell* **16**, 565–575.

Hynes, R. O., and Destree, A. T. (1978). Relationships between fibronectin (LETS protein) and actin. *Cell* **15**, 875–886.

Ingber, D. E., Dike, L., Hansen, L., Karp, S., Maniotis, A., McNamee, H., Mooney, D., Plopper, G., Sims, J., and Wang, N. (1994). Cellular tensegrity: Exploring how mechanical changes in the cytoskeleton regulate cell growth, migration, and tissue pattern during morphogenesis. *Int. Rev. Cyt.* **150**, 173–224.

Jordan, C. F., Lerman, L. S., and Venable, J. H. (1972). Structure and circular dichroism of DNA in concentrated polymer solutions. *Nature New Biol.* **236**, 67–70.

Kelker, H. (1973). History of liquid crystals. *Mol. Cryst. Liq. Cryst.* **21**, 1–48.

Kerkam, K., Viney, C., Kaplan, D., and Lombardi, S. (1991). Liquid crystallinity of natural silk secretions. *Nature* **349**, 596–598.

Lee, D. L. (1970). The ultrastructure of adult female *Mermis nigrescens* (Nematoda). *J. Zool. London* **161**, 513–518.

Leforestier, A., and Livolant, F. (1991). Cholesteric liquid crystalline DNA: A comparative analysis of cryofixation methods. *Biol. Cell* **71**, 115–122.

Leforestier, A., and Livolant, F. (1993). Supramolecular ordering of DNA in the cholesteric liquid crystalline phase: An ultrastructural study. *Biophys. J.* **65**, 56–72.

Lehman, O. (1908). Flüssige Kristalle und die Theorien des Lebens, *J. Ambr. Barth,* Leipzig.

Lepescheux, L. (1988). Spatial organization of collagen in annelid cuticle: Order and defects. *Biol. Cell* **62**, 17–31.

Lerman, L. S. (1973). Chromosomal analogues: Long-range order in ψ condensed DNA. *Cold Spring Harbor Symp. Quant. Biol.* **38**, 59–73.

Lerman, L. S., Wilkerson, J. H., Venable, J. R., and Robinson, B. H. (1976). DNA packing in single crystals inferred from freeze-fracture-etch replicas. *J. Mol. Biol.* **108**, 271–293.

Livolant, F. (1983). Cholesteric organization of DNA *in vivo* and *in vitro. Eur. J. Cell Biol.* **33**, 300–311.

Livolant, F. (1984). Cholesteric organization of DNA in the stallion sperm head. *Tissue and Cell* **16**, 535–555.

Livolant, F. (1986). Cholesteric liquid crystalline phases given by three helical biological polymers: DNA, PBLG and xanthan. A comparative analysis of their textures. *J. Physique* **47**, 1605–1616.

Livolant, F. (1987). Precholesteric liquid crystalline states of DNA. *J. Physique* **48**, 1051–1066.

Livolant, F. (1991). Ordered phases of DNA *in vivo* and *in vitro. Physica* **176**, 117–137.

Livolant, F., and Bouligand, Y. (1978). New observations on the twisted arrangement of dinoflagellate chromosomes. *Chromosoma* **68**, 21–44.

Livolant, F., and Bouligand, Y. (1989). Freeze-fractures in cholesteric mesophases of polymers. *Mol. Cryst. Liq. Cryst.* **166**, 91–100.

Livolant, F., Giraud, M. M. and Bouligand, Y. (1978). A goniometric effect observed in sections of twisted fibrous materials. *Biol. Cell* **31**, 159–168.

Livolant, F., Levelut, A. M., Doucet, J., and Benoit, J. P. (1989). The highly concentrated liquid-crystalline phase of DNA is columnar hexagonal. *Nature* **339**, 724–726.

Lloyd, C. W. 1982. "The Cytoskeleton in Plant Cell Growth and Development." Academic Press, New York.

Luzzati, V., and Husson, F. (1962). The structure of the liquid-crystalline phases of lipid–water systems. *J. Cell Biol.* **12**, 207–219.

Marchessault, R. H., Morehead, F. F., and Walter, N. M. (1959). Liquid crystal systems from fibrillar polysaccharides. *Nature* **184**, 632–634.

Marotti, G., and Muglia, M. A. (1988). A scanning electron microscope study of human bony lamellae. Proposal for a new model of collagen lamellar organization. *Arch. Ital. Anat. Embryol.* **93**, 163–175.

Mazur, G. D., Regier, J. C., and Kafatos, F. C. (1982). Order and defects in the silkmoth chorion, a biological analogue of a cholesteric liquid crystal. *In* "Insect Ultrastructure" (R. C. King and H. Akai, eds.), Vol. 1, pp. 151–185. Plenum Publishing Corp., New York.

Mc Beath, E., and Fujiwara, K. (1989). Coalignment of microtubules, cytokeratin intermediate filaments, and collagen fibrils in a collagen-secreting cell system. *Eur. J. Cell Biol.* **50**, 510–521.

Meiboom, S., Sethna, J. P., Anderson, P. W., and Brinkman, W. F. (1981). Theory of the blue phase of cholesteric liquid crystals. *Phys. Rev. Lett.* **46**, 1216–1219.

Meunier, F. J., and Castanet, J. (1982). Organisation spatiale des fibres de collagène de la plaque basale des écailles des téléostéens. *Zoologica Scripta* **11**, 141–153.

Meunier, F. J., and Francillon-Vieillot, H. (1995). Structure et minéralisation des scutes d'*Ostracion lentiginosum* (Teleostei, Tetraodontiforme, Ostraciidae). *Ann. Sci. Nat.* **16**, 33–47.

Nanko, H., Saiki, H., and Harada, H. (1978). Cell wall structure of the sclereids in the secondary phloem of *Populus euramericana. Mokuzai Gakkaishi* **24**, 362–368.

Neville, A. C. (1993). "Biology of Fibrous Composites: Development beyond the Cell Membrane." University Press, Cambridge.

Neville, A. C., Gubb, D. C., and Crawford, R. M. (1976). A new model for cellulose architecture in some plant cell walls. *Protoplasma* **90**, 307–317.

Neville, A. C., and Levy, S. (1984). Helicoidal orientation of cellulose microfibrils in *Nitella opaca* internode cells: Ultrastructure and computed theoretical effects of strain orientation during wall growth. *Planta* **162**, 370–384.

Neville, A. C., and Luke, B. M. (1971). A biological system producing a self-assembling cholesteric protein liquid crystal. *J. Cell Sci.* **8**, 93–109.

Parameswaran, N., and Sinner, M. (1979). Topochemical studies on the wall of beech bark sclereids by enzymatic and acidic degradation. *Protoplasma* **101**, 197–215.

Pellegrini, M., and Pellegrini, L. (1985). On the occurrence of twisted fibrillar structures in the cytoplasm of the red alga *Alsidium helminthochorton* (la tourette) Kütz.: Ultrastructural and cytochemical observations. *Protoplasma* **126**, 54–61.

Portigliatti-Barbos, M., Bianco, P., Ascenzi, A., and Boyde, A. (1984). Collagen orientation in compact bone. II. Distribution of lamellae in the whole of the human femoral schaft with reference to its mechanical properties. *Metab. Bone Dis. & Rel. Res.* **5**, 309–315.

Preston, C. M., and Jenkins, T. (1984). *Trichuris muris*: Structure and formation of the eggshell. *Parasitology* **89**, 263–273.

Raspanti, M., Guizzardi, S., De Pasquale, V., Martini, D., and Ruggeri, A. (1994). Ultrastructure of heat-deproteinated compact bone. *Biomaterials* **15**, 433–437.

Reinitzer, F. (1989). Contributions to the knowledge of cholesterol. *Liquid Crystals* **5**, 7–18.

Reis, D., Vian, B., Chanzy, H., and Roland, J. C. (1991). Liquid crystal–type assembly of native cellulose–glucuronoxylans extracted from plant cell wall. *Biol. Cell* **73**, 173–178.

Reis, D., Vian, B., and Roland, J. C. (1994). Cellulose-glucuronoxylans and plant cell wall structure. *Micron* **25**, 171–187.

Revol, J. F., Bradford, H., Giasson, J., Marchessault, R. H., and Gray, D. G. (1992). Helicoidal self-ordering of cellulose microfibrils in aqueous suspension. *Int. J. Biol. Macromol.* **14**, 170–171.

Revol, J. F., and Marchessault, R. H. (1993). *In vitro* nematic ordering of chitin crystallites. *Int. J. Biol. Macromol.* **15**, 329–335.

Rigny, P., Tambourin, P., and Thoulouze, D. (1994). De la matière au vivant les systèmes moléculaires organisés. *In* "Images de la Recherche," CNRS-Editions, Paris.

Ritcey, A. M., and Gray, D. G. (1988). Cholesteric order in gels and films of regenerated cellulose. *Biopolymers* **27,** 1363–1374.

Robinson, C. (1961). Liquid-crystalline structures in polypeptide solutions. *Tetrahedron* **13,** 219–234.

Roland, J. C., Reis, D., and Vian, B. (1992). Liquid crystal order and turbulence in the planar twist of the growing plant cell walls. *Tissue and Cell* **24,** 335–345.

Roland, J. C., Reis, D., Vian, B., and Roy, S. (1989). The helicoidal plant cell wall as a performing cellulose-based composite. *Biol. Cell* **67,** 209–220.

Rudall, K. M. (1956). Protein, ribbons, and sheets. *Lectures on the Scientific Basis of Medicine* **5,** 217–230.

Sikorav, J. L., and Church, G. M. (1991). Accelerated DNA renaturation: Complementary recognition in condensed DNA. *J. Mol. Biol.* **222,** 1–23.

Singer, S. J., and Nicolson, G. L. (1972). The fluid mosaic model of the structure of cell membranes. *Science* **175,** 720–731.

Sipski, M. L., and Wagner, T. E. (1977). Probing DNA quaternary ordering with circular dichroism spectroscopy: Studies of equine sperm chromosomal fibers. *Biopolymers* **16,** 573–582.

Spencer, M., Fuller, W., Wilkins, F. R. S., and Brown, G. L. (1962) Determination of the helical configuration of ribonucleic acid molecules by X-ray diffraction study of crystalline amino-acid-transfer ribonucleic acid. *Nature* **194,** 1014–1020.

Strzelecka, T. E., Davidson, M. W., and Rill, R. L. (1988). Multiple liquid crystal phases of DNA at high concentrations. *Nature* **331,** 457–460.

Weiner, S., and Traub, W. (1992). Bone structure: From angströms to microns. *FASEB J.* **6,** 879–885.

Werbowyj, R. S., and Gray, D. G. (1976). Liquid crystalline structure in aqueous hydroxy-propyl cellulose solutions. *Mol. Cryst. Liq. Cryst.* **34,** 97–103.

Wharton, D. A., and Jenkins, T. (1978). Structure and chemistry of the eggshell of a nematode (*Tricchuris suis*). *Tissue & Cell* **10,** 427–440.

Wilkins, A. S. (1985). The limits of molecular biology. *Bioessays* **3,** 3.

Willison, J. H. M., and Abeysekera, R. M. (1988). A liquid crystal containing cellulose in living plant tissue. *J. Polymer Sci.* **26,** 71–75.

Zylberberg, L., Bereiter-Hahn, J., and Sire, J. Y. 1988. Cytoskeletal organization and collagen orientation in the fish scales. *Cell Tissue Res.* **253,** 597–607.

Vitellogenin Receptors: Oocyte-Specific Members of the Low-Density Lipoprotein Receptor Supergene Family

Wolfgang Johann Schneider

Department of Molecular Genetics, Biocenter and University of Vienna, A-1030 Vienna, Austria

Receptors that transport vitellogenin (VTG) into oocytes are of vital importance to egg-laying species, because they mediate a key step of oocyte maturation, a prerequisite to reproduction. Vitellogenins are lipophosphoglycoproteins that are produced under female hormonal control in large central organs (fat body in insects; liver in higher animals) and are transported in the circulation to the female gonads. VTG receptors localized in coated pits on the surface of growth-competent oocytes are able to accumulate in the yolk high concentrations of VTG and other ligands they recognize. The study of VTG receptors and their ligands has identified genes that specify related ligands, and a family of receptors. To date, all molecularly characterized VTG receptors belong to the low-density lipoprotein receptor supergene family, which ranges from a 600-kDa receptor in *Caenorhabditis elegans* to the 100-kDa so-called very-low-density lipoprotein receptors in mammals. These receptors, by and large, recognize ligands with similarities in structural elements first defined in the human apolipoproteins B-100 and E. Recent studies on the receptor family have added VTG and lipoprotein lipase to the list of co-evolved ligands and have revealed that VTG receptors are able to interact with ligands other than VTG and also with some unrelated to lipoprotein metabolism. For example, the chicken VTG receptor also imports very-low-density lipoprotein, riboflavin-binding protein, and alpha-2-macroglobulin into growing oocytes. Such multifunctionality of receptors is likely the result of evolutionary pressure to provide the female germ cell with a highly economical machinery for vitellogenesis.

KEY WORDS: Apolipoprotein, Oocyte, Receptors, Reproduction, Supergene family, Vitellogenin.

I. Introduction

Embryonic development of oviparous animals occurs entirely within the confines of the laid egg. As a consequence, the macro- and micronutrients provided by the egg contents must guarantee the production of viable offspring. The major source of the components required for embryo growth is the yolk (i.e., the complex storage compartment of the oocyte within the egg). This yolk reserve is formed during the growth of the oocyte by deposition of massive amounts of plasma- or hemolymph-derived molecules—the process of *vitellogenesis*. It is the current notion that practically all yolk components are synthesized outside the oocyte, namely, by the liver in vertebrates and various other organs in invertebrates. The dramatically induced synthesis of yolk precursors is under the control of female hormones such as estrogens or, in insects, ecdysone. Following their secretion into the circulatory system, yolk precursors are transported to the oocyte proper, where they are recognized by oocyte surface receptors that mediate their endocytosis. Receptor-mediated endocytosis assures selective, efficient, and accumulative uptake of yolk components and mediates oocyte growth. Thus it is the key regulatory mechanism underlying vitellogenesis.

The evolution of vitellogenin and vitellogenin receptors is a prime model for the emergence of ligand/receptor systems designed to sustain the reproductive effort of many species. This evolution has occurred in concert with that of similar ligand/receptor systems important for normal physiology of animals that do not rely on oviposition. As will become apparent, vitellogenin receptor genes belong to the supergene family of low-density-lipoprotein receptor (LDLR) related proteins (LRPs) that appear to have co-evolved in egg-laying and viviparous animals. Although it is less well established, vitellogenins share certain structural features with proteins that are ligands of LRPs.

In this review I provide a synopsis of aspects of oocyte development, a brief description of vitellogenins and their genes, and, in greater detail, our knowledge of the receptors for vitellogenins in the context of the LDLR supergene family. The review of information on vitellogenin receptors will emphasize molecular biological studies but will also deal with biochemical results.

II. Oocyte Growth

A. Oogenesis

Besides being essential to the reproductive effort of all oviparous animals, the generation of mature oocytes is one of the most striking examples of cell growth regulation. The sequence of events leading to a mature oocyte is very similar in a wide variety of species (Wasserman and Smith, 1978; Wallace, 1985), allowing their synoptic description. Where appropriate, I shall concentrate on birds as the highest oviparous phylum. Oocytes are derived from oogonia, mitotic cells that develop from primordial germ cells migrating into the ovary early in embryogenesis. In teleosts the transformation from oogonia to oocytes appears to occur within the germinal regions of the luminal ovarian epithelium (Selman and Wallace, 1989). After various periods of mitotic proliferation within the ovary, oogonia enter meiosis. The cells arrest in prophase of division I of meiosis, and the chromosomes take on a "lampbrush" configuration (diplotene) with high levels of RNA synthesis. These meiotically arrested female germ cells, termed *primary oocytes,* then enter a period of growth varying in length depending on the species. During this time primary oocytes accumulate all components of the machinery that will later be required for embryo development, such as ribosomes, tRNAs, cytoplasmic organelles, mRNAs, and "early" yolk. In birds this period consists of an initial slow phase (months to years) in which no yolk is deposited, followed by slow yolk deposition (approximately 2 months), and preceding ovulation, a final 1-week endspurt during which the oocyte's volume increases a further 60- to 100-fold. This latter phase, vitellogenesis, requires the expression of yolk precursor receptors on the surface of the oocytes.

B. Vitellogenesis

During vitellogenesis the cells accumulate large amounts of macro- and micronutrients required for the embryo. Chicken oocytes, for example, incorporate up to 1.5 g of protein per day during the last few days before oviposition (Gilbert, 1981; Johnson, 1986), resulting in a total weight gain of 10 to 13 g during vitellogenesis. Synthesis of the major yolk precursors is under hormonal control; in vertebrates estrogenic hormones play the key role (Clemens, 1974), and in insects, the juvenile hormone ecdysone (Bell and Barth, 1971). In vertebrates the ovary produces estrogens in response to pituitary-derived gonadotropins; in turn, the prime target of estrogens is the liver, where they dramatically induce the synthesis of yolk precursor macromolecules (Bergink *et al.,* 1974). In almost all oviparous animals the most prominent yolk component is vitellogenin (VTG), a lipophosphoglycoprotein described in more detail here and in Section III. In chickens, VTG protein levels in the circulation reach concentrations of 25 g/liter. Vertebrate VTGs circulate as dimers of individual 180- to 240-kDa subunits (Bergink *et al.,* 1974; Deeley *et al.,* 1075). The available sequence information (Spieth and Blumenthal, 1985; Gerber-Huber *et al.,* 1987; van het Schip *et al.,* 1987; Sharrock *et al.,* 1992; Trewitt *et al.,* 1992; Chen *et al.,* 1994; Yano *et al.,* 1994) clearly demonstrates that VTGs have been conserved during evolution. As is also discussed, the structural similarities in

VTGs are related to their fate, namely, proteolytic fragmentation, following uptake into the oocyte.

In amphibia, VTG provides 80–85% of the total yolk proteins (Wallace, 1985); in birds, triglyceride-rich very-low-density lipoproteins (VLDL) and VTG provide the bulk of yolk (Perry and Gilbert, 1979). Many additional components isolatable from yolk share structural and immunological properties with corresponding plasma proteins and are likely taken up into oocytes via specific receptors. Most of these have been identified in the chicken and are discussed in Section VI,A.

In some invertebrates (e.g., *Drosophila melanogaster),* VTG can be synthesized within the ovary by somatic cells in the vicinity of the oocyte. These cells, generally termed *follicle cells,* are an integral part of a complex oocyte-support structure, the follicle. Differentiation and growth of follicles are tightly linked to the growth of oocytes, from early oogenesis to just prior to ovulation (i.e., release of the oocyte from the follicle).

C. Folliculogenesis and Oocyte Maturation

At different stages of oogenesis, and in different ways, depending on the species (Richards, 1979; Selman and Wallace, 1989; Bahr, 1990; Etches and Petitte, 1990) oocytes and accessory somatic cells or cell layers become juxtaposed. These somatic cells support the growth, contribute to the mechanical stabilization, and participate in the regulation of maturation of the oocyte, the final stage of oogenesis, and ovulation. In *D. melanogaster* four mitotic divisions of the oogonium generate the 15 so-called nurse cells and one oocyte. In the pipefish the luminal epithelium is precursor to both primary oocytes and prefollicle cells (Selman and Wallace, 1989).

In additon to follicle cells and the oocyte, vertebrate ovarian follicles contain several other cell types. In the chicken, for instance, even before the vitellogenic phase the layer of follicle cells, here called *granulosa cells,* is joined by connective tissue cells (thecal cells), smooth muscle cells, and neuronal cells (Bahr, 1990; Etches and Petitte, 1990). The theca is highly vascularized. Two acellular structures are important as well: (i) the basement membrane, to which the granulosa cells are attached and that separates them from the theca, and (ii) the perivitelline layer, which forms a mechanical support network of fibers between the oocyte and the granulosa cells. At ovulation the oocyte, surrounded by the perivitelline layer, is released from the follicle and starts its descent into the oviduct.

Before this event, the follicle cells perform their most important job: they regulate the release of oocytes from their arrest in prophase of meiotic division I (Section II,A). Resumption of meiosis triggers the final stage of oocyte development, termed *oocyte maturation,* leading to arrest as a secondary oocyte in metaphase of division II of meiosis until fertilization. This process is a prime experimental system for studying cell cycle control, one of the fastest moving and expanding fields in biology. The reader is referred to excellent articles about this complex topic (e.g., Murray and Hunt, 1993; Murray 1994), since a discussion of these processes would by far exceed the limits of this review.

III. Vitellogenin in Oocyte Growth

A. From Serum to Yolk: Precursor–Product Relationships

The discovery of the relationships between circulating vitellogenins and certain proteins found in yolk is an example of tough biochemical work on an even tougher problem. Key hurdles have been the inherent lability and complex chemical properties of circulatory VTGs (Wiley *et al.*, 1979) and the fact that only certain proteolytic fragments of the VTGs can be recovered from yolk. In the end, molecular biological approaches have confirmed, but also extended, the chemists' results. Only the findings pertinent to the present review of receptors for VTGs are discussed here. As it turns out, the basic and characteristic features of VTGs and related circulatory protein complexes are shared by many species (Fig. 1).

It now appears that in almost all oviparous species, VTGs are synthesized as large (about 200 kDa) primary (pre–pro) translation products of a single contiguous mRNA. This has also been shown directly for animals other than frogs and chickens (Bose and Raikhel, 1988; Chen *et al.*, 1994; Yano *et al.*, 1994). The polypeptide backbone is extensively modified posttranslationally at the site of synthesis; these modifications include glycosylation, phosphorylation, sulfation, proteolytic cleavage into subunits (e.g., in the mosquito (Chen *et al.*, 1994) and in the chicken (Nimpf *et al.*, 1989b; Retzek *et al.*, 1992) and references therein), and association with lipids.

In vertebrates intact serum-derived VTGs are cleaved only inside the oocyte into several fragments, many of which can be isolated from yolk. The high-density fraction of various egg yolks contains components termed *lipovitellins, vitellins, phosvitins,* and *phosvettes* (Gerber-Huber *et al.*, 1987), usually themselves made up of a group of polypeptides derived from precursor VTGs. The reason for this complex situation is that these yolk proteins are derived from more than one VTG, the number depending on the species. In chicken, for instance, there are three VTGs (VTG I; II, the most abundant form; and III (Wang and Williams, 1982)); in *Xenopus laevis,* there are possibly four (Wahli *et al.*, 1979; Wahli *et al.*, 1981).

B. Genes, Structure, and Receptor Binding

The cloning of an increasing, but still relatively small, number of VTGs has allowed the definition of common features of their protein structures. The identification of conserved structural elements, in turn, may provide us with insights into their functionally important regions. VTGs likely arose from a common ancestral gene (Wahli, 1988), parts of which appear to have evolved into genes also found in mammals, such as those for lipoprotein lipase (Baker, 1988; Persson *et al.*, 1989), apolipoprotein B (Baker, 1988), apolipoprotein E, and von Willebrand factor (Byrne *et al.*, 1989; Perez *et al.*, 1991), and possibly, microsomal triglyceride transfer protein (Shoulders *et al.*, 1993), and are otherwise conserved in limited fashion.

The primary sequences derived from cloning of the cDNAs or genes for at least one of the VTGs found in various animals such as the nematode *Caenorhabditis elegans* (Spieth and Blumenthal, 1985), chicken *Gallus domesticus* (van het Schip *et al.*, 1987), frog *Xenopus laevis* (Gerber-Huber *et al.*, 1987), lamprey *Ichthyomyzon unicuspis* (Sharrock *et al.*, 1992), boll weevil *Anthonomus grandis* (Trewitt *et al.*, 1992), silkworm *Bombyx mori* (Yano *et al.*, 1994), and mosquito *Aedes aegypti* (Chen *et al.*, 1994) have in fact revealed frequently found regions. These structural elements, present in various sizes and numbers, are: (i) highly phosphorylated polyserine runs, often located in the central portion but also found close to the carboxy-terminus of the holo-VTG molecule; released from the rest of the molecule by intraoocytic proteolysis, these domains give rise to the aforementioned phosvitins and phosvettes in yolk; (ii) lipid-binding region(s), most often in the aminoterminal half of the molecule; in chicken and frog VTGs the intraoocytic cleavage products are termed lipovitellins I (LV-I) (Gerber-Huber *et al.*, 1987; van het Schip *et al.*, 1987); (iii) small but highly charged domains that can serve as cleavage signals either for posttranslational proteolysis (e.g., in the mosquito (Chen *et al.*, 1994) or in the silkworm (Yano *et al.*, 1994)) or, in vertebrates, for intraoocytic fragmentation (Nimpf *et al.*, 1989b; Retzek *et al.*, 1992); and (iv) clusters of positively charged and hydrophobic residues, typically substructures of the LV-I regions that have been identified as receptor-binding domains in vertebrate systems (Steyrer *et al.*, 1990; Stifani *et al.*, 1990b; Barber *et al.*, 1991). The key features of chicken VTG are schematically represented in Fig. 1.

As will become evident in the main part of this chapter, localization of the domains that mediate binding of the lipophosphoglycoprotein VTG to its receptor has been facilitated by the homology of ligand and receptor to a growing number of molecularly characterized vertebrate receptors for lipoproteins. The process underlying lipoprotein transport into cells (i.e., receptor-mediated endocytosis) has been delineated in studies of the best-known (but youngest!) member of the LDL receptor supergene family, the

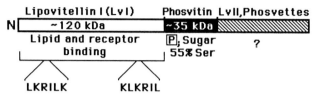

FIG. 1   Key features of vitellogenins, represented by the chicken proteins. As described in Section III, there are several more or less similar VTG genes and proteins in each species. In the chicken, three very similar genes and products exist; their "average" structure is presented. Four key features relevant to the physiology of VTGs are listed above the schematic structure. In the aminoterminal (N) LvI portion, two regions with presumed receptor-binding capacity, based on direct evidence (Steyrer *et al.,* 1990; Stifani *et al.,* 1990a,b) and sequence similarity to ligands of the mammalian LDLR family members (see text), are indicated.

human LDL receptor. Thus, a description of the process and its key player will introduce general operating principles and structural elements shared by lipoprotein receptors of mammalian and oviparous animals.

IV. Mechanism and Components of Lipoprotein Transport

A. Receptor-Mediated Endocytosis

When compared with other plasma proteins, such as albumin, VTG is taken up by growing oocytes 20–25 times more rapidly (i.e., is specific and is saturable) (Opresko and Wiley, 1987a,1987b). These are exactly the macroscopic properties of receptor-mediated endocytosis, the process first delineated and later extensively investigated in the human LDLR system (Goldstein *et al.,* 1985). The sequence of events taking place when LDL is taken up by a human fibroblast is likely identical to that in VTG transport across the plasma membrane, at least up to a certain point.

Shortly after binding of a ligand to its specific cell surface receptor, the ligand/receptor complex is internalized. This is achieved via the invagination of the plasma membrane region where the complex had formed, followed by pinching-off of the invagination as a vesicle into the interior of the cell. However, to provide control over the traffic from the surface into the cell, this internalization event can occur only when the receptor is located in specialized regions of the plasma membrane. These regions make up about 2% of the cell surface of a human fibroblast and contain an estimated 80% of the total number of receptors of the cell. Due to their pit-shaped appearance and an electron-dense cytoplasmic coat, these regions are termed *coated pits*. The nature of the major coat protein, clathrin, as far as required by the scope of this review, was described previously (Goldstein *et al.,* 1985). The vesicles generated from the coated pits, termed *coated vesicles,* carry ligand/receptor complexes into the cell, become uncoated, fuse with other uncoated vesicles to form primary or early "endosomes," and are acidified. Within the endosomes, LDL and LDLR dissociate due to the acidity; LDL is delivered to and degraded within lysosomes, and the receptor returns to the cell surface ("recycles"). An important point is that receptors act, in the context of the coated pit–coated vesicle system, similiar to an escalator: internalization and recycling take place regardless of occupancy.

In oocytes, which in fact are the cells where coated pits and vesicles were first observed and morphologically described (Roth and Porter, 1964), we encounter a slight variation in the final step described for the LDLR in fibroblasts: there are no true lysosomes in oocytes; thus the final target for the ligands is the endosomal compartment. These endosomes, membrane-enclosed organelles with a proteolytic capacity that is very low in comparison with that of true lysosomes (Wall and Meleka, 1985; Wall and Patel, 1987a,1987b; Retzek *et al.,* 1992), form the bulk of yolk. Nevertheless, given the enormous receptor-mediated endocytotic activity of a rapidly growing oocyte, receptors must be reutilized via recycling, in this case out of endosomes. It has been suggested that in frog oocytes the recyling rate could be influenced by ligand concentration and/or hormones (Opresko and Wiley, 1987a,1987b), in contrast with the LDLR system.

B. Vitellogenin within the Endosome

In oocytes of different species the yolk storage compartments, although all endosomal, are quite different in composition and appearance. This is a consequence of the different ligands extracted from the circulation in different animals but is also due to the oocytes' different sizes and environments. One common theme is the limited proteolysis within the oocyte, and there most likely in endosomes, of the endocytosed components. Often, but not always, the proteolysis results in the generation of insoluble fragments that form "yolk granules" or "yolk platelets" (Dumont, 1978; Opresko et al., 1980; Wall and Meleka, 1985). VTG-derived lipovitellins are considered the major components of these structures. In chicken oocytes the electron-dense granules form a separate phase within the fluid content of giant endosomes called "yolk spheres" (which reach a diameter of ~140 μm!) (Shen et al., 1993, and references therein). In frog oocytes, crystals consisting of VTG fragments are found within membrane-bound yolk platelets, de-rived from endocytic vesicles via multivesicular endosomes (Busson et al., 1989; Richter, 1989), also called *multivesicular bodies* (Wall and Patel, 1987b). It can be safely assumed that the specific intracellular processing of VTGs in the oocyte is tightly coupled to the event of receptor-mediated endocytosis. Such coupling has been clearly demonstrated for the lysosomal degradation of LDL in human fibroblasts, where the key player is the LDLR, the prototype of the LDL/VTGR family (Goldstein et al., 1985).

In the following section we describe the elements that characterize the members of this ever-expanding and important family of endocytotic recep-tors. The genes for these proteins have developed for more than 500 million years; the human LDLR therefore can be viewed as the most modern product of evolution of the family that is known.

V. The Mammalian LDL Receptor Supergene Family

As already mentioned, receptor-mediated endocytosis of LDL is one of the best characterized processes of macromolecular transport across the plasma membrane of eukaryotic cells. The following sections provide an overview of the physiological and biochemical properties of the LDL recep-tor and also describe the molecular basis for the monogenic disease familial hypercholesterolemia (FH). FH is the prime, and until recently only, exam-ple of severe phenotypic consequences of a defect in a receptor belonging to the LDLR supergene family. Now, as described in Section VI,A, studies in the chicken have revealed that a combination of female sterility and FH are the deleterious effects of a mutation that disrupts the function of an oocytic VTG receptor.

A. The LDL Receptor Pathway

The LDL receptor pathway encompasses the regulatory events associated with receptor-mediated endocytosis of LDL that allow cells to control their cholesterol content (Goldstein *et al.*, 1985). The cholesterol liberated by the lysosomal hydrolysis of LDL-cholesterylesters, or possibly oxidized sterols derived therefrom, mediate a complex series of feedback control mechanisms that protect the cell from overaccumulation of cholesterol. First, LDL-derived sterols suppress the activities of 3-hydroxy-3-methylglutaryl-CoA (HMG-CoA) synthase and HMG-CoA reductase, two key enzymes in cellular cholesterol biosynthesis. Second, the cholesterol activates a cholesterol-esterifying enzyme called acyl-CoA:cholesterol acyltransferase (ACAT; E.C.2.3.1.26), which allows the cells to store excess cholesterol in the form of cholesterylester droplets. Third, the synthesis of new LDL receptors is suppressed, preventing further cellular entry of LDL and thus cholesterol overloading.

The overall benefits from, and consequences of, this LDL receptor-mediated regulatory system are the coordination of the utilization of intra- and extracellular sources of cholesterol. Human fibroblasts and other mammalian cells are able to subsist in the absence of lipoproteins because they can synthesize cholesterol from acetyl-CoA. When LDL is available, however, the cells primarily use the LDL receptor to import LDL and keep their own synthetic activity suppressed. Thus, a constant level of cholesterol is maintained within the cell while the external supply in the form of lipoproteins can undergo large fluctuation. The main task of LDL receptors *in vivo* is to supply cells with cholesterol, thereby mediating the removal of cholesterol-rich lipoprotein particles from the bloodstream.

Oocytes express specific receptors to accumulate vast amounts of cholesterol-carrying circulatory components (e.g., in chicken, VLDL) for the purpose of storage and not out of cellular regulatory requirements. Thus, oocyte receptor levels do of course not succumb to feedback suppression by the endocytosed lipoproteins but are likely regulated through hormonal signals. Nevertheless, receptor defects can have severe consequences for lipid homeostasis at the cellular and systemic levels (see Sections VI,A and C), processes that we would neither appreciate nor understand without knowledge of the LDLR pathway and LDLR defects in patients with FH. Thus, a summary of findings on the LDLR in FH is provided in the following section; more detailed descriptions can be found in Hobbs *et al.* (1990), Westhuyzen *et al.* (1990), and Schneider (1991).

B. Familial Hypercholesterolemia: What Mutations Can Teach Us about the LDL Receptor Gene Family

Much of our knowledge about the normal receptor emerged from studies of patients with abnormal phenotypes specified by the genetic disease familial hypercholesterolemia (FH), for which mutations affecting the function of the LDL receptor are responsible. Intensive studies at the cell biological, biochemical, and molecular biological levels have identified many different mutant alleles at the LDL receptor locus, outlined later in the section. Clinically, FH is characterized by three cardinal features: (i) selective elevation in the plasma levels of LDL; (ii) cholesterol depositions in abnormal sites, in particular in tendons (formation of xanthomata) and in arteries (atheromata); and (iii) inheritance as an autosomal dominant trait with a gene dosage effect.

The presence of one mutant allele is found in 1 in about 500 persons, whereas one individual among about 1 million carries two mutant alleles at the LDL receptor locus. Severely affected offspring of a marriage between two heterozygotes can be either true homozygotes (observed most often in consanguineous marriages) or heteroallelic genetic compounds (the more frequent occurrence). Nevertheless, the term *FH homozygote* is used generally to describe all patients with two mutant alleles, as it is a clinically convenient and relevant classification.

Experimentally, the biochemical defect in FH was delineated by studies utilizing cultured diploid fibroblasts from patients with the phenotype of homozygous FH. Comparison of LDL metabolism, regulation of HMG-CoA reductase, and ACAT activity in FH versus normal cells revealed the following abnormalities in patients' cells: (i) they failed to bind and internalize LDL with high affinity; (ii) as a consequence, they failed to hydrolyze both the protein and lipid portion of LDL; and (iii) incubation of FH cells with LDL did not suppress HMG-CoA reductase or affect ACAT activity. In subsequent studies on more than 100 cell lines derived from FH homozygotes, it became apparent that there are several groups of mutant alleles at the LDL receptor locus in this disease; these are discussed in some detail in Section V,B,2. The identification of these groups of classes of mutations resulted from a combined effort at the cell biological and biochemical levels and was facilitated by the successful purification of the normal LDL receptor and production of a monoclonal anti–receptor antibody. Immunoprecipitation of newly synthesized radiolabeled receptor molecules from normal and mutant cell lines gave the first clue to the diversity of mutations underlying the FH phenotype. These results are described in the following section.

1. Biosynthesis and Structure of the LDL Receptor

The biosynthetic pathway of the normal LDL receptor encompasses the following steps: (i) synthesis of the polypeptide backbone containing N-linked and O-linked core sugars; this precursor molecule is localized to the endoplasmic reticulum and migrates on SDS–polyacrylamide gels to the position of a protein with an apparent M_r of 120,0000; (ii) transport to the Golgi apparatus, where the precursor carbohydrate chains are processed to their mature form; this maturation results in a dramatic reduction in the migration rate of the receptor in SDS gels—its apparent M_r becomes 160,000; (iii) incorporation of the finished product into the plasma membrane and localization to coated pits. For a better understanding of the normal receptor structure and the various abnormalities resulting from mutations in FH, details of the normal LDL receptor molecule, and differences identified in VTGRs, as far as known, are discussed next.

Studies at the protein chemistry, molecular biology, and cell biology levels have led to the elucidation of the LDL receptor structure from several species. The receptor is a highly conserved integral membrane glycoprotein consisting of five domains after cleavage of the signal sequence (21 residues in the human receptor) (Fig. 2). In order of appearance from the amino terminus these domains are (i) the ligand binding domain, (ii) a domain

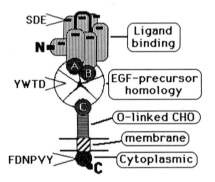

FIG. 2   Domain model of the human LDL receptor. The five domains of the mature protein (839 residues), from the N-terminus (N) to the carboxyterminus (C), are: (1) the ligand binding domain, composed of seven cystein-rich repeats, folded such that each cluster of negatively charged residues $(-)$ is displayed on the surface. The core of the charged sequence contains the consensus tripeptide Ser-Asp-Glu (SDE); the participation of the aminoterminal repeat in ligand binding is not established; repeat 5 (outlined in bold) is required for binding of apo E, and repeats 2–7 cooperatively bind apo B; (2) the EGF precursor homology region, consisting of ~400 amino acid residues; adjacent to the ligand binding domain and at the carboxyterminus of this region are located three cysteine-rich repeats homologous to repeats found in the precursor to epidermal growth factor (A, B, and C, respectively). The remaining portion of this domain consists of five internally homologous stretches of approximately 50 residues each that contain tetrapeptides with the consensus sequence Tyr-Trp-Thr-Asp (YWTD); (3) the O-linked carbohydrate (CHO) region, consisting of 58 amino acids with 18 Ser and Thr residues containing O-linked CHO chains; (4) a single membrane-spanning domain; and (5) the cytoplasmic tail with 50 amino acid residues containing the internalization signal, in LDLRs Phe-Asp-Asn-Pro-Val-Tyr (FDNPVY). This model is deduced from structure/function studies on the LDL receptor as discussed in the text. (Adapted from Schneider (1991)).

that has a high degree of homology with the precursor to the epidermal growth factor (EGF), (iii) a domain that contains a cluster of O-linked carbohydrate chains, (iv) a transmembrane domain, and (v) a short cytoplasmic region. Until direct information on the three-dimensional structure of the 839-residue receptor becomes available, an arrangement of these domains as presented in Fig. 2 may serve as a useful model.

a. The Ligand Binding Domain   The ligand binding domain mediates the interaction between the receptor and lipoproteins containing apo B and/or apo E (Esser *et al.*, 1988). The function is localized to a stretch of 292 amino acid residues at the amino terminus of the receptor, comprising seven repeats of approximately 40 residues each. These seven repeats are arranged in head-to-tail fashion; their high content of cysteines presumably mediates folding of the domain into a rigid structure with clusters of negatively charged residues on its surface ($-$, SDE in Fig. 2). The negative charges are found at the carboxyterminus of each of the seven repeat units, with repeats 2–7 thought to participate in the binding of lipoprotein(s) via positively charged residues on apo B or apo E: the fifth repeat is required for apo E binding, whereas repeats 2–7 cooperatively recognize apo B. The sequence Ser-Asp-Glu (SDE) forms the core of the negative charge cluster in all seven repeats. As is seen in Section VI, VTGRs appear to require cluster(s) of at least eight such tandemly arranged ligand binding repeats in order to bind their ligands.

b. The EGF Precursor Homology Domain   This domain of the LDL receptor lies adjacent to the LDL binding site and comprises approximately 400 amino acids; the outstanding feature is the sequence similarity of this region to parts of the EGF precursor, itself a membrane-bound protein. Optimal alignment of the relevant primary sequences results in 35% of the residues being identical in the two proteins. Of particular interest is the high degree of homology in three regions termed *growth factor repeats,* which are coded for by three of the five exons shared by the genes for the LDL receptor and the EGF precursor, respectively. Two of these repeats (A and B in Fig. 2) are located in tandem at the amino terminus, and the other (C) is at the carboxyterminus of the precursor homology region of the LDL receptor. Repeat A may have a supportive role in binding of apo B to the 40-residue repeats in the NH_2-terminal domain (Esser *et al.*, 1988). The remainder of this domain is made up of five ~50-residue stretches that contain tetrapeptide sequences with a consensus of Tyr-Trp-Thr-Asp (YWTD in Fig. 2). By deletion of the entire EGF precursor domain via site-specific mutagenesis, evidence for a possible involvement of the region in the receptor's acid-dependent dissociation from LDL and its subsequent recycling was obtained. All members of the LDLR family elucidated contain this domain in one or several copies (Fig. 3).

c. The O-Linked Sugar Domain   The O-linked sugar domain of the human LDL receptor is a 58–amino acid stretch highly enriched in serine and threonine residues, located just outside the plasma membrane. Most, if not all, of the 18 hydroxylated amino acid side chains are glycosylated. The O-linked oligosaccharides undergo posttranslational elongation in the course of receptor maturation: when the receptor leaves the endoplasmic reticulum, N-acetylgalactosamine is the sole O-linked sugar present, and upon its processing in the Golgi, galactosyl and sialyl residues are added. The acquisition of galactose and charged carbohydrate dramatically alters the electrophoretic migration of the LDL receptor in SDS–polyacrylamide gels: whereas the precursor molecule has an apparent M_r of 120,000, the mature receptor's apparent M_r is 160,000. As described next, this characteristic change in mobility has served as a useful diagnostic tool in the analysis of mutant receptors. Despite the detailed structural knowledge of this region, its function remains elusive. It may simply serve to provide a "stalk" for the protrusion of the binding domain out into the extracellular milieu to facilitate steric access of large lipoprotein particles. Splice variant forms of the chicken VTGR, a close relative of the LDLR, affecting this domain have recently been identified (see Section VI,A).

d. The Membrane Anchoring Domain   The membrane anchoring domain of the LDL receptor lies carboxyterminal to the O-linked carbohydrate cluster. It consists of 22–25 hydrophobic amino acids, the sequence of which is the least conserved of all receptor domains in seven mammalian species. This speaks against a specific function other than anchoring, such as signal transduction across the plasma membrane elicited by ligand binding, or formation of an ion channel. Further support for this conclusion can be derived from the fact that recycling of the receptor occurs constitutively in the absence of LDL. As would be expected, the deletion of this domain in certain naturally occurring mutations leads to secretion of truncated receptors from mutant cell lines.

e. The Cytoplasmic Tail   The cytoplasmic tail of the LDL receptor constitutes a short stretch of 50 amino acid residues (790–839), clearly involved in the targeting of LDL receptors to coated pits. The internalization-defective phenotypes of FH, characterized by lack of receptor localization to coated pits, are due to mutations affecting this region of the receptor. Furthermore, site-specific mutagenesis (Davis *et al.*, 1987) has shown the amino acids between residues 791 and 812 to be crucial, and residues 812–839 not to be required, for internalization of the LDL receptor. At positions 802–807 an "internalization signal" has been identified as Phe-Asp-Asn-Pro-Val-Tyr (FDNPVY in Fig. 2) in the human LDL receptor, and the tetrapeptide NPXY (where X denotes any amino acid) has now been found in all (more than 20) structurally elucidated members of the LDLR family.

A summary of the structural and functional features of the human LDL receptor is schematically represented in Fig. 2. As is apparent in the following section, the delineation of the biosynthetic pathway and of the pentapartite receptor structure have been greatly enhanced by the analysis of molecular defects in LDL receptors of FH patients.

2. Molecular Defects in LDL Receptors of Patients with Familial Hypercholesterolemia

Twenty years after the first report, the search for and analyses of mutant LDL receptors are still the topic of investigations in a large number of laboratories. The number of currently identified alleles is likely on the order of 200. Despite the elucidation of an ever-growing number of mutations, the originally reported classification of the mutants into four general classes (Schneider, 1991) is still valid. The LDL receptor locus (for details on the gene's structure see Schneider (1991)) spans about 50 kb of DNA, is made up of 18 exons, and is localized on the distal short arm of chromosome 19. There is a very strong correlation between the presumed structural organization of the protein and the exon organization in the gene; in essence, the LDL receptor gene is a compound of coding sequences that are shared by the growing LDLR superfamily. Because VTGRs are members of this gene family, it is important to realize that many of the features delineated in studies of FH are highly relevant to the receptors for VTG. Several characteristics of the gene structure also appear to predispose the LDLR to particular mutational events; examples for this hypothesis are found in each of the four classes of mutations. Typical examples are very briefly summarized next; more details on these can be found in Schneider (1991). We might expect that VTGRs also could be prone to certain mutations resulting in diminished or abolished function. One already identified example in support of this notion is the absence of a VTGR in a nonlaying mutant chicken strain suffering from female sterility and hyperlipidemia (see Section VI,A).

a. Class 1: No Detectable Precursor These so-called null alleles (R-0) form the most common class of mutations: half of all defective alleles belong to this group. These alleles fail to express receptor proteins as determined by functional assays (binding of ^{125}I-LDL) or attempts to identify proteins with a variety of monoclonal and polyclonal antibodies directed against the LDL receptor. The spectrum of mutations resulting in the absence of receptor synthesis, or at least in the failure to produce receptor protein that can be immunoprecipitated with the available antibodies, include point mutations that introduce premature termination codons early in the protein coding region, mutations in the promoter region that block transcription, mutations that lead to abnormal splicing of the mRNA, and large deletions.

b. Class 2: Slow or Absent Processing of Precursor   Class 2 alleles specify transport-deficient receptor precursors; they fail to move from the endoplasmic reticulum to and through the Golgi compartment(s) and on to the cell surface with normal rates. As a consequence, the typical sudden increase in apparent M_r from 120,000 to 160,000 observed during biosynthesis of the normal receptor is lacking. Most of these mutations are complete: there is a *total* absence of transport from the endoplasmic reticulum, and receptors never reach the cell surface. In a variation of class 2 alleles, precursors of normal size are processed but *much more slowly* than normal. In WHHL rabbits, an animal model for FH, cloning has revealed an in-frame 12-nucleotide deletion in the mutant allele that eliminates four amino acids (Asp-Gly-Ser-Asp) from the third cysteine-rich repeat in the binding domain of the rabbit receptor. The phenotype thus is not due to a mutation directly affecting an *O*- or *N*-linked glycosylation site, as might have been expected (Yamamoto *et al.*, 1986). Other class 2 mutations encompass a variety of defects; the molecular abnormalities do not appear to result from major deletions but rather from subtle changes in gene structure.

c. Class 3: Abnormal Ligand Binding   LDL receptors synthesized from class 3 alleles reach the cell surface at normal rates but are unable to bind LDL efficiently. Most of the class 3 receptors have normal M_r in SDS–PAGE. However, receptors with altered electrophoretic mobility have also been identified in this group. For example, allele R-210 b⁻ specifies a precursor of M_r 170,000 that is posttranslationally modified to a surface protein of M_r 210,000. The fact that the receptors produced by this class of alleles all undergo the normal maturation process suggests that abnormal M_r values and lack of binding capacity are due to alterations in amino acid sequence and not to carbohydrate content.

   In one of the mutant alleles a deletion arose from homologous recombination between *Alu* sequences in intron 4 and intron 5, resulting in a transcript containing exon 4 directly joined to exon 6. Inasmuch as the ends of exons 4, 5, and 6 all occur in the same reading frame, the translation product is exactly 41 amino acids (the sixth ligand binding repeat, encoded by exon 5) shorter than normal. This mutant receptor has lost its ability to bind LDL but has apparently retained activity toward apo E, as demonstrated by its capacity to bind β-VLDL, an apo E–and apo B–containing class of lipoprotein particles. It is believed that β-VLDL particles bind to the LDL receptor via apoE, and it is therefore conceivable that the deletion of the sixth cysteine-rich repeat in the binding domain has abolished its recognition of apo B but not of apo E. As is discussed in Section VI,A, the chicken VTGR is capable of interacting with apo E, despite the absence of apo E in this species. This has led to the proposal that VTG and apo E have evolved in parallel to the diversification of LDLRs versus VTGRs (Steyrer *et al.*, 1990; Schneider and Nimpf, 1993; Schneider, 1995).

Overall, the elucidation of class 3 mutations has resulted in a better understanding of functional aspects of the ligand binding domain.

d. Class 4: Internalization Defective   In class 4 mutations one of the prerequisites for effective ligand internalization—the clustering of LDL receptors in coated pits—is not fulfilled. The failure of these internalization-defective receptors to localize to coated structures results from mutations that directly or indirectly disrupt the carboxyterminal domain of the receptor, in particular, the sequence FDNPXY (see Section V,B,1,e). All the mutant receptors in this group undergo posttranslational processing and reach the cell surface; in some instances carboxyterminally truncated receptors are secreted from the cells.

One of the mutations elucidated is that in the paternal allele of patient J. D. (FH 380), the first patient identified with this phenotype (Brown and Goldstein, 1976). A single base change results in the substitution of a cysteine for a tyrosine at position 807, located 18 residues into the cytoplasm. The direct effect on the cytoplasmic domain suggests that the structural disruption elicits the receptor's failure to migrate or be collected into coated pits, a notion further strengthened by expression studies of mutant alleles constructed *in vitro* (Davis *et al.*, 1987). The normal mechanisms that cause LDL receptors (or any coated pit receptors, for that matter) to be clustered in coated pits still are largely unknown. The single amino acid substitution in J. D.'s receptor would be expected to have a profound effect on the three-dimensional structure of the cytoplasmic tail. However, before we learn the exact tertiary structure of at least this domain of the normal receptors, such considerations must remain speculative.

In summary, the molecular details of LDLR-mediated removal of VLDL-derived lipoproteins from plasma are extremely well understood. The occurrence and delineation of natural mutations in the LDLR gene, and their reproduction and expansion via directed mutagenesis, have provided almost complete insight into structural as well as regulatory features of the LDLR.

The definition of conserved structure–function relationships in the LDLR, in turn, has been most important for the characterization of VTGRs. We describe a few of the VTGRs known in molecular detail via cloning and then provide a briefer overview of receptors characterized biochemically. The review deals exclusively with systems in which investigations have generated at least some insights into structural aspects of the VTGRs. Pathways putatively involving VTG receptors, but lacking any biochemical description of them, are generally not described. Some relevant general aspects of the biochemistry and cell biology of VTGs are dealt with in Sections II,B–IV,B.

VI. Vitellogenin Receptors: Molecular Biological Studies

A. Chicken Oocyte Vitellogenin Receptors: Multifunctional Members of the LDL Receptor Family

The discovery and cloning of the chicken VTGR was the result of efforts to understand the key steps involved in regulation of oocyte growth via receptor-mediated endocytosis. The molecules involved in the transport of the lipoproteins VLDL and VTG into oocytes could be expected to be distinct from those regulating systemic transport and homeostasis. Thus, these studies were to yield novel information on receptors, ligands, and physiological aspects of oviparous species, with possible implications for our views on processes in viviparous species.

Because of the vast knowledge about the human LDLR, the search was begun with attempts to identify its chicken homologue; the result was an unexpected sudden growth of the LDLR gene family. Before 1986, molecular information on any of the proteins involved in oocyte growth and systemic lipoprotein transport in the laying hen was lacking. However, the presence of bona fide lipoprotein particles in yolk was highly suggestive of a lipoprotein receptor–mediated process for yolk deposition into the female germ cell (Perry *et al.,* 1978, 1984). Also, there was a priori no reason to believe that somatic cell lipid homeostasis in birds should have a different mode of regulation from that in mammals (i.e., via the LDLR pathway). Indeed, since then studies on yolk precursor transport in the laying hen have provided proof for these notions but also have revealed some surprising new aspects of lipoprotein receptor biology.

A single 95-kDa protein in the plasma membrane of the oocyte binds both major yolk lipoproteins, VLDL and VTG (Stifani *et al.,* 1990; Schneider and Nimpf, 1993). An oocyte membrane protein with the same properties has been identified in quail. Although it has not yet been characterized as thoroughly as the chicken receptor, it can safely be assumed that the degree of identity with the chicken protein is very high (Elkin *et al.,* 1995). The chicken receptor reacts with antibodies to mammalian LDL receptors

(LDLRs) (Hayashi *et al.*, 1989) and, surprisingly, recognizes apo E (Steyrer *et al.*, 1990), an apolipoprotein not known to be produced in birds. These hallmark properties predicted that the *o*ocyte VLDL and *V*TG *r*eceptor (OVR) would be a homologue of mammalian LDL receptors.

As already described, LDLRs are characterized by the presence of seven complement-type repeats that make up the ligand binding domain. In contrast the so-called VLDLRs, concisely reviewed by Jingami and Yamamoto (1995), are characterized by a cluster of *eight* such repeats (Gåfvels *et al.*, 1993, 1994; Oka *et al.*, 1994a, 1994b; Sakai *et al.*, 1994; Webb *et al.*, 1994; Jingami and Yamamoto, 1995). Molecular cloning has revealed that OVR has such an eight-repeat ligand binding domain (Bujo *et al.*, 1994). Besides the amazingly high degree of sequence similarity between the eight-repeat receptors (in general, around 85%, which in fact is greater than that among LDLRs), the absence from chicken OVR of a serine- and threonine-rich domain thought to normally carry *O*-linked carbohydrate chains (the *O*-linked sugar domain; see Section V,B,1,c) appears significant, as discussed below.

One would have to argue that the physiological function(s) of mammalian VLDLRs are not clearly established (Jokinen *et al.*, 1994). Fortunately, however, the role of OVR is documented both by biochemical and genetic evidence (Nimpf *et al.*, 1989a): it mediates a key step in the reproductive effort of the hen (i.e., normal oocyte growth). This conclusion is based on the fact that its functional absence leads to failure of oocytes to enter the rapid growth phase and, consequently, the absence of egg laying and failure to produce offspring. A chicken strain carrying a single mutation at the *ovr* locus (the "*R*estricted *O*vulator" (R/O) strain) cannot lay eggs (Ho *et al.*, 1974; Nimpf *et al.*, 1989a). As a consequence of the failure to deposit into their oocytes VLDL and VTG, which are produced at normal levels, the mutant females develop severe hyperlipidemia and features of atherosclerosis (Bujo *et al.* (1994) and references therein). The gene for OVR is located on the sex chromosome, Z, in concordance with the results of breeding studies (male birds are specified by ZZ, and hens by WZ) (Ho *et al.*, 1974; Bujo *et al.*, 1994). ovr^-/ovr^+ (carrier) roosters have normal lipid metabolism, in agreement with our finding that *ovr* is expressed exclusively in oocytes. $-/ovr^-$ females, which in fact represent a model for oocyte-specific familial hypercholesterolemia, are sterile due to nonlaying. The ovr^--mutation has now been identified (Bujo *et al.*, 1995). Due to a point mutation the first of the six cysteine residues in the EGF precursor homology domain (corresponding to the one labeled C in Fig. 2) is replaced by a serine. This causes protein misfolding and subsequent severely diminished surface expression of the recptor, typical of class 2 mutations originally defined in the LDLR in FH (Section V,B,2,b). As we speculated previously, such mutation may well be favored by structural features of the OVR gene, in analogy with the situation found in the human LDLR gene.

OVR expressed in COS-7 cells binds VLDL and VTG with high affinity (Bujo *et al.*, 1994). Mammalian VLDLRs are thought to have preference for apo E–containing VLDL and/or chylomicron remnants (Jingami and Yamamoto, 1995). The properties of OVR support this notion but also strengthen the hypothesis (Steyrer *et al.*, 1990) that the avian receptor is the product of an ancient gene that has retained the ability to interact with many, if not all, ligands of younger relatives of the LDLR superfamily. In this context VTG, which is absent from mammals, and apo E, not found in birds, have certain common biochemical properties and regions of sequence similarities and have been suggested to be functional analogues (Steyrer *et al.*, 1990). Thus, if triglyceride-rich particles would indeed turn out to be the physiological substrate for mammalian VLDLRs (Takahashi *et al.*, 1992), they could transport triglycerides into metabolically active tissues (such as muscle, where receptors are abundant), whereas in avian oocytes VTG and VLDL are taken up to provide nutrients and energy for the developing embryo. Furthermore, there is extensive overlap in ligand recognition patterns with other lipoprotein receptors, in particular with LDL receptor–related proteins/*alpha-2*-macroglobulin (a2M) receptors (LRPs) (Schneider and Nimpf, 1993).

In the course of studies on yolk precursor uptake in the chicken oocyte, it was discovered that chicken OVR transports not only major yolk components but also nonlipoproteins. Riboflavin binding protein (ribBP) is a 29-kDa phosphoglycoprotein that is synthesized (induced by estrogen just like VTG or VLDL (Mac Lachlan *et al.*, 1993)) in liver and oviduct. It is the major carrier for the vitamin in serum (where it derives from the liver), in egg white (when produced by the oviduct), and in yolk (taken up from the serum). Interestingly, serum ribBP associates with VTG in the circulation, and this complex is recognized, via VTG, by OVR (Mac Lachlan *et al.*, 1994). Thus, riboflavin is incorporated into yolk via a piggyback mechanism whose key component is OVR; similar to the oocytic accumulation of VTG, this transport system achieves a roughly ninefold higher concentration of ribBP in yolk than in serum. Whether a similar piggyback mechanism is maintained for vitamin transport in other oviparous animals, or operates via analogous ligands in mammals, is not known.

The most recently discovered trick mastered by the multifunctional oocyte receptor is its involvement in the transport of a2M. There is a special twist again: whereas LRPs are known to recognize a2M only following so-called activation (i.e., production of a2M* through cleavage by proteases *in vivo*), OVR interacts with the native ligand as well (Jacobsen *et al.*, 1995). This finding emphasizes one of the principles of metabolic dichotomy in the laying hen: that is, one form of the ligand (a2M*) is destined for systemic (hepatic) clearance, whereas the other form (native a2M) is targeted for uptake into oocytes, where it can function in embryogenesis. As outlined next, a second operating principle is receptor dichotomy (i.e., ligand targeting via cell-type-specific expression of receptor genes).

B. Homologous Receptors in Somatic Cells of the Chicken

In the laying hen, LDLR homologues come at least in pairs, and sometimes in groups. They are products of genes expressed in mutually exclusive fashion in somatic cells and germ cells, respectively. One biological explanation for this is that oocyte growth must not be feedback-inhibited by intracellular accumulation of ligands, as is the case for cholesterol in somatic cells (Hayashi et al., 1989). Thus, the regulation of the two receptor gene groups must be different. The functional counterpart of OVR in somatic cells is a 130-kDa sterol-regulated LDLR, structural details of which are yet unknown. The chicken LDL (apo B) R cannot bind apo E or VTG, further indicating the relationship between these two mammalian and avian LDLR family ligands, respectively. The chicken LDLR is thus the only known LDLR that interacts with apo B but not apo E. It will be interesting to determine the structural element(s) responsible for this unique property.

OVR itself seems to share one additional feature with its mammalian VLDLR relatives: the expression of splice-variant forms (i.e., the production of mRNAs with or without the exon coding for a serine- and threonine-rich domain). As stated previously, the oocyte receptor lacks this region, but analysis by PCR of mRNAs from several somatic tissues is highly suggestive of a longer mature transcript, consistent with the presence of an O-linked sugar domain. This VLDLR form appears to be expressed in addition to the LDLR, but as a member of the VLDLR group, it would not be expected to be sterol-regulated (Gåfvels et al., 1993, 1994; Oka et al., 1994a,b; Sakai et al., 1994; Webb et al., 1994; Jingami and Yamamoto, 1995).

The aforementioned LDL receptor–related proteins/a2M receptors (LRPs) are large (~600-kDa) relatives of the LDL/VLDLRs (for a schematic representation of all known receptor structures, see Fig. 3).

There is at least one pair of LRPs in chickens—one germ-cell-specific and one found only in somatic cells, mainly in the liver (Schneider and Nimpf, 1993; Schneider, 1995). Compared with human LRP, 83% of the amino acids in the somatic-cell-specific chicken protein (4522 residues) are identical (Nimpf et al., 1994a). Such tremendous conservation points to an important biological function common to mammals and birds; the hepatic catabolism of a2M* possibly is one of them. Also, based on a proposed function of mammalian LRP, the avian hepatic LRP could be responsible for the clearance of "portomicrons". Namely, in birds, intestinally synthesized triglyceride-rich particles are thought to be secreted directly into the portal vein (i.e., not as chylomicrons into the lymphatic system, as in mammals) and to be immediately and efficiently taken up by the liver. However, such a pathway has not been demonstrated unequivocally. Nevertheless, somatic LRPs from different animal kingdoms have common structures: as first revealed by studies on avian LDLR family members, they also may share physiological roles.

Less is known about the oocyte-specific LRP (Stifani *et al.*, 1991). It is smaller and has not yet been characterized at the molecular level. Because the oocyte-specific LRP is a major component of the oocyte's plasma membrane and its ligand binding properties seem to overlap those of other LDLR family members (clearly, it binds VTG), it might serve a backup role for OVR in oocyte growth (Stifani *et al.*, 1991). However, the fact that this receptor cannot compensate for the lack of OVR in the OVR-deficient R/O hens argues against this hypothesis, unless there is a conditional order of expression such that the absence of OVR-mediated early oocyte growth precludes subsequent expression of the germ-cell-specific LRP. Current studies in our laboratory address this interesting possibility.

It is hoped that these results in the chicken system will lead us to an understanding of the biochemical, cell biological, and molecular genetic aspects of oocyte growth via yolk deposition. Clearly, VTG receptors play a major role in this process, but what mechanisms orchestrate the correct tissue distribution of those ligands that are recognized by receptors in oocytes *and* somatic cells? For lipoproteins the answer is cell-type-specific expression of the receptors. For instance, the lack of recognition of VTGs

---

FIG. 3   Schematic representation of the LDLR superfamily proteins. The domains (for comparison, see Fig. 2) indicated are: negatively charged ligand binding repeats with six cysteines each; EGF precursor homology repeats (there are two subtypes, termed B1 and B2 (Herz *et al.*, 1988); they also contain six cysteines each); the YWTD repeats of the EGF precursor homology domain; the O-linked sugar domains, just outside the plasma membrane, missing in some of the proteins; and the consensus internalization signals, (FD)NPXY. LDLR represents the mammalian LDLRs with seven ligand binding repeats; VLDLR represents the VLDLRs of mammals, and OVR, the chicken oocyte receptor for VTG and VLDL (Bujo *et al.*, 1994), all of which contain eight ligand binding repeats; Yl, the putative VTGR of *Drosophila melanogaster* (Schonbaum *et al.*, 1995); LRP/a2MR, the LDLR-related proteins from chicken (Nimpf *et al.*, 1994b) and humans (Herz *et al.*, 1988), which are identical to the long-studied hepatic receptor for a2M (Strickland *et al.*, 1990); the putative protein product of a gene for an LRP-like protein in *Caenorhabditis elegans* (Yochem and Greenwald, 1993); and gp330, more recently, after its complete cloning, called megalin (Saito *et al.*, 1994), also a member of the LRP group with very similar properties to LRP (Willnow, 1992). A posttranslational cleavage site in LRP/a2MR is indicated (see text). The total sizes of the proteins are approximately 100, 200, and 600 kDa, respectively.

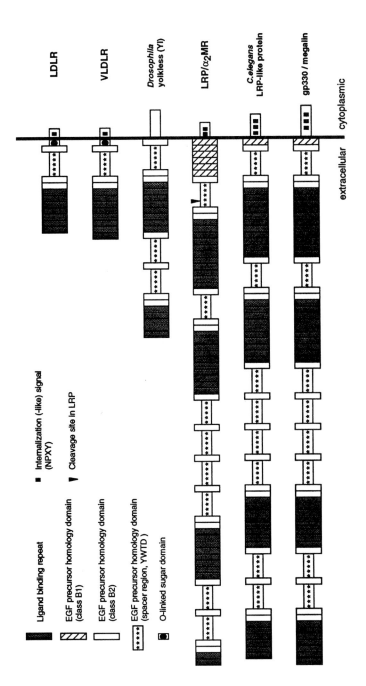

LDLR

VLDLR

Drosophila yolkless (Yl)

LRP/α_2MR

C. elegans LRP-like protein

gp330 / megalin

cytoplasmic

extracellular

- Ligand binding repeat
- EGF precursor homology domain (class B1)
- EGF precursor homology domain (class B2)
- EGF precursor homology domain (spacer region, YWTD)
- O-linked sugar domain

- Internalization (-like) signal (NPXY)
- Cleavage site in LRP

by somatic cell receptors is essential for their targeting to oocytes. We cannot exclude dual expression and similar roles of receptors for other components, because lipoprotein receptors are, at least under certain physiological conditions, capable of transporting ligands quite different from lipoproteins (Herz, 1993; Schneider and Nimpf, 1993). Even certain cold viruses can gain cellular entry via LDLR family members (Hofer *et al.*, 1994). As an alternative to receptor distribution, subtle structural differences in ligands could direct them to selected target cells, as is the case for a2M versus a2M*.

In the system described next, a putative VTGR was identified and molecularly characterized because its absence leads to disruption of yolk deposition and sterility. In fact, despite the dissimilarities of the animal suffering from this problem, the situation is rather analogous to that in the R/O chicken. However, the genetic approach to cloning a VTGR described next is restricted to genetically easily manipulable species.

C. The "Yolkless" Gene of *Drosophila melanogaster*

Sterile mutant female fruit flies carrying the *yolkless (yl)* gene fail to accumulate yolk protein(s) in their oocytes (Schonbaum *et al.*, 1995, and references therein). The germ-line-dependent phenotype offered the chance to test whether the mutation indeed disrupts the expression of a VTGR. By a combination of complementation mapping, chromosome walking, and Northern blotting, a female- and ovary-specific 6.5-kb transcript was identified. When a genomic DNA fragment encompassing this mRNA was introduced into *yl* flies, yolk uptake and fertility were restored (Schonbaum *et al.*, 1995).

The translation product of the 5844-nt open reading frame encoded for by the *yl* cDNA is a 1984-residue protein that contains several repeats of the structural elements that typify the LDLR supergene family members in mammals and the chicken (Fig. 3). There are 13 complement-type binding repeats in two clusters of 5 and 8 repeats each, respectively; 7 type-B repeats that flank three groups of 5 YWXD repeats each; and a single membrane-spanning domain. A sequence comparison of the 8-repeat clusters in the chicken OVR (Bujo *et al.*, 1994) and *Drosophila yl* reveals, however, that only 38% of the residues are identical. Between rabbit and chicken VLDLR homologues, the identity in this region is 84%. Furthermore, in the presumptive intracellular carboxyterminal domain, the Yl protein differs extensively from its relatives. Instead of the typical NPXY motif, the sequences NMHF and NDTF, as well as FAAQQF, are proposed as serving as potential internalization signals. Direct proof for this hypothesis must be awaited.

The size and structure of the putative *Drosophila* VTGR puts it in between the LDLR/VLDLR groups with 7 or 8 binding repeats (such as chicken OVR), and the LRPs with 31 to 36 repeats, respectively. As pointed out by Schonbaum *et al.,* (1995), Yl does not appear to be more closely related to OVR than to other LDLR family members. Nevertheless, there is a clear relationship between all these receptors and putative receptors, because there is similarity between their ligands as well. As pointed out in Section III, the VTGs of higher diptera, such as *D. melanogaster,* show similarity to lipoprotein lipases of higher animals, which are known to bind to LRPs (Beisiegel *et al.,* 1991; Nykjaer *et al.,* 1993). Before Yl can be shown directly and unambiguously to act as receptor for VTG in the fly, it remains possible that Yl mediates uptake of (an)other yolk component(s) with structural similarity to VTG; such ligand(s) could nevertheless be crucial for normal oocyte development. In all oviparous animals, as in chicken (which expresses the oocyte-specific OVR as well as LRP), more than one receptor of the LDLR superfamily may be responsible for the coordinated uptake of individual or structurally similar groups of yolk precursors. It now appears that there are three groups of LDLR family genes: small (represented by LDLRs and VLDLRs), middle (Yl and possibly others), and large (LRPs). Delineation of the expression patterns of these genes in species spanning the whole spectrum of evolution can be expected to shed light on their being redundant and/or cooperative.

In this context, an even more primitive animal than *Drosophila* has been found to harbor a gene clearly belonging to the large LDLR family members.

D. The *Caenorhabditis elegans* LRP Gene

A 23-kb gene with structural elements and organization clearly defining it as an LDLR supergene family member was isolated from a *C. elegans* genomic library by screening with cDNA fragments coding for portions of the nematode genes *lin*-12 and *glp*-1 (Yochem and Greenwald, 1989, 1993). These probes contained elements corresponding to EGF precursor homologous repeats and detected three cross-hybridizing genes in *C. elegans* (Yochem and Greenwald, 1993), one of which was the LRP-like gene. The predicted protein, whose expression, however, has not been shown, would be 4753 residues long. Its domain structure (Fig. 3) reveals 35 complement-type repeats, arranged in four clusters of various sizes; 16 copies of EGF-type repeats; 8 groups of five YWTD consensus elements exactly arranged as in human LRP; one membrane-spanning domain; and a 157-residue putative cytoplasmic domain with two potential internalization signals, NPVY and NPLY. Interestingly, as do human LRP (Herz *et al.,* 1988,

1990) and the somatic-cell-specific chicken LRP (Nimpf *et al.,* 1994b), the nematode LRP sequence contains a tetrabasic sequence, here RKKR, close to the membrane-spanning domain. Human LRP is posttranslationally cleaved at this site, and the two chains thus generated in the endoplasmic reticulum remain noncovalently associated until they are inserted into the plasma membrane (Herz *et al.,* 1990). The reason for this apparently futile proteolysis is unknown, and thus it would be interesting to study its significance. To this end, an easily manipulated system such as *C. elegans* appears ideal (Wood, 1988); unfortunately, following the report of an LRP gene in this animal no further studies on it have been published. Thus, we do not know whether the LRP protein is actually produced and if it is, where, so hypothetical considerations about its physiological role, such as in VTG transport, must remain just that.

In any case, the presence of the gene in the nematode, and its amazing similarity to mammalian and avian genes, further emphasizes the important role(s) that these genes must play *in vivo.* Reproduction certainly could be one of the processes requiring the members of the LDLR gene family.

In addition to these studies that have provided structural information on bona fide and putative VTGRs and/or members of the LDLR family through molecular biological experimentation, several biochemical studies have been directed toward identification of *oocyte* VTGRs or related proteins in a variety of species. These are summarized next, albeit more briefly than the studies at the molecular level, because other excellent and more comprehensive reviews have dealt with VTGRs (e.g., insects) (Raikhel and Dhadialla (1992), Sappington and Raikhel (1995), and references therein), in which vitellogenesis has been studied most intensively by biochemical means. Studies that report binding assays without visualization of the putatively involved receptor(s) or binding protein(s) are not presented.

VII. Receptors for Vitellogenin:
Biochemical Studies

Most of the biochemical experiments on the characterization of VTG receptors have been, and still are, based on two assumptions; fortunately, these have become increasingly validated through the cited studies. First, these receptors are presumed to share structural and thus immunological properties with identified VTGRs, and second, they can be visualized by the technique of ligand blotting. This is a technique originally developed for the visualization of the human LDL receptor (Daniel *et al.,* 1983). It relies on the ability of receptors with a cysteine-rich ligand binding domain to interact with their ligands following SDS–PAGE under nonreducing conditions. The ligands are labeled with radioiodine (George *et al.*, 1987; Barber *et al.*, 1991) or by biotinylation (Elkin and Schneider, 1994) and used in incubations of nitrocellulose sheets that contain SDS–PAGE–separated oocyte membrane proteins following electrophoretic transfer. Unbound ligand is removed by careful washing, and proteins specifically interacting with the ligands are visualized based on their label. In the following sections, putative VTGRs identified by immunological and/or ligand blotting approaches are described.

A. Putative Vitellogenin Receptors in Frog and Fish

Radiolabeled frog (*Xenopus laevis*) VTG was bound to a single 95-kDa membrane protein in chicken oocytes (Stifani *et al.*, 1990a); this protein corresponds to chicken OVR. Similarly, only one membrane protein from *Xenopus* oocytes, having an M_r of approximately 115 kDa, interacted with ^{125}I-labeled *Xenopus* VTG. Importantly, the binding of radiolabeled VTG to both chicken and *Xenopus* oocyte membrane proteins was totally abolished when membrane proteins were exposed to disulfide-bond reducing agents during gel electrophoresis, in agreement with the fact that intrachain disulfide bonds within the receptor molecule are necessary for retention of its biological activity (Daniel *et al.*, 1983; George *et al.*, 1987; Barber *et al.*, 1991). Based on these results, the VTG receptor in *Xenopus* oocytes has an apparent M_r of 115,000 and recognizes avian and reptilian VTGs.

In the immunological part of these studies an antibody raised against the pure bovine LDLR and shown to cross-react with chicken OVR, as well as an antibody raised against OVR, recognized the same bands as those identified by ligand blotting. These results were the first to suggest that oocyte VTGRs not only are related among oviparous species but also share structural features with lipoprotein receptors in mammalian species, a contention now strongly confirmed by the cloning efforts described in the previous sections.

These studies also directly addressed for the first time which portion of VTG mediates receptor binding. A novel approach, termed *reverse ligand blotting,* was developed. The isolated lipovitellin (LV) fraction from the yolk of chicken eggs, displaying the 120- to 125-kDa LV-I (see Fig. 1), proteins of 100/105, 30 (tentatively identified as a light chain derived from the carboxyterminal portion of VTG), 38, 41, and 78 kDa, was subjected to SDS–PAGE and transferred to nitrocellulose. The replicas were then incubated with or without detergent-solubilized chicken oocyte membrane proteins in buffer previously shown to allow ligand–receptor interactions in direct ligand blots. The soluble VTGR molecules that had bound to LV subunit(s) were visualized by incubation with anti-VTGR antibody followed by incubation with ^{125}I-Protein A. The receptor binding fragments in yolk were compatible with LV-I as mediator of receptor binding and demonstrated that the phosvitin portion is not involved in the ligand/receptor interaction (for a detailed discussion, see Stifani *et al.*, 1990b).

The receptor of a fish, the coho salmon (*Oncorhynchus kisutch*), has been studied by similar approaches (Stifani *et al.*, 1990a). It was identified as a 100-kDa protein in oocyte membrane extracts by ligand blotting. Underscoring the close relationship of all VTGRs studied so far, the piscine receptor showed cross-reactivity with the chicken and frog systems both in terms of ligand recognition and immunoreactivity. Binding of isolated chicken lipovitellin to chicken OVR and the fish VTGR confirmed the results obtained with reverse ligand blotting performed with the *Xenopus* receptor (Stifani *et al.*, 1990a, 1990b); namely, the LV-I portion of VTG harbors the receptor binding domain of VTG.

In these experiments, suramin, a polysulfated polycyclic hydrocarbon (also known as Germanin or Bayer 205), originally described as an antitrypanosomal agent, was shown to abolish VTG binding to the fish receptor, in agreement with previous results on human LDLR (Schneider *et al.*, 1982) and chicken OVR (Stifani *et al.*, 1988), as well as on the VTGR from mosquito (Dhadialla *et al.*, 1992) (see next paragraph) and the putative locust VTGR (Röhrkasten and Ferenz, 1987). Inhibition by suramin is thought to be caused by the high negative charge density of the compound, likely similar to that found on the surface of the binding domains of receptors of the LDLR superfamily (see Fig. 2). Suramin can be added to the list of useful tools in ligand blot and solid-phase binding analysis (not considered here) of VTGRs.

B. Vitellogenin Receptors from Mosquito, Locust, and Cockroach

Ligand blotting and/or immunoblotting with antibodies raised against heterologous VTGRs or VTG-binding proteins have been employed in a few other systems. Excellent reviews of biochemical studies in insect vitellogenesis, covering the receptors as well as additional aspects, are available (Raikhel and Dhadialla, 1992; Sappington and Raikhel, 1995).

The mosquito VTGR is of particular interest, because the mosquito oocyte was the cell in which coated pits and coated vesicle structures were first described by Roth and Porter (1964). The receptor from the yellow fever mosquito, *Aedes aegypti,* was thus identified as a 205-kDa glycoprotein, was isolated, and was shown to bind both VTG and its yolk storage form, vitellin (Dhadialla *et al.*, 1992). Interestingly, the receptor displayed no detectable immunoreactivity with antibodies to chicken OVR or the putative locust VTGR.

In the cockroach *Nauphoeta cinerea,* a 200-kDa protein was characterized as a potential VTGR (Indrasith *et al.,* 1990). Thus, the VTGRs from fly (see Section VI,C), cockroach, and mosquito fall into a ~200-kDa group of LDLR-related proteins. The apparent M_r of the locust receptor apparently is somewhat smaller, in the range of 156–190 kDa (Roehrkasten *et al.,* 1989; Hafer and Ferenz, 1991, 1994). In all reported studies, the ligand binding was destroyed when the receptor was reduced, in agreement with the lability toward reduction of the cysteine-rich binding domains of the LDLR family. In any case, vertebrate VTG receptors (chicken, fish, frog) collectively are probably about half the size of the insect VTG receptors, which in turn are half the size of the ubiquitous LRPs, certain of which, however, also can bind VTGs.

VIII. Future Aspects of Oocyte Vitellogenin Receptor Biology

Above all, we must make progress in obtaining detailed structural information on VTGRs from a much wider range of species through molecular biological approaches. The clarification of the relationships among the 100-, 200-, and 600-kDa members, and possibly the discovery of yet unknown relatives of the LDLR supergene family, will unravel the history and principles of evolution of these biologically very important genes. May the ancestor step forward!

As we have seen, many animals express several of these related genes (i.e., produce several similar proteins), even in the same cell type. Redundancy in gene products (provision of backup systems) is one of the biological means of protecting from genetic disease and/or of equipping for specialized tasks through similar proteins. The studies described in this review reveal the roles of multiple and possibly redundant genes in oocyte growth, the key to reproduction in oviparous species. We have gained some insights into, but must learn much more about, the regulation of the individual genes to be able to understand exactly why several similar gene products are expressed in a single organism. Specifically, do common regulatory elements exist in genes of receptors that trigger growth of the germ cell? Which of these receptors are expressed during embryonic development? Do posttranslational modifications of ligands determine their cellular fate (that is, receptor recognition)? What, if anything, can the similarity of VTGs to certain mammalian proteins tell us? What is the biological purpose of limited intraoocytic proteolysis instead of complete degradation of VTG?

It is hoped that we will soon be able to answer some of these questions by studying oocyte-specific events in heterologous cell systems amenable to manipulation or in transgenic oviparous animals.

References

Bahr, J. (1990). The avian ovary: Model for endocrine studies. *J. Exp. Zool.* **4,** (suppl.), 192–194.

Baker, M. E. (1988). Is vitellogenin an ancestor of apolipoprotein B-100 of human low-density lipoprotein and human lipoprotein lipase? *Biochem. J.* **225,** 1057–1060.

Barber, D. L., Sanders, E. J., Aebersold, R., and Schneider, W. J. (1991). The receptor for yolk lipoprotein deposition in the chicken oocyte. *J. Biol. Chem.* **266,** 18,761–18,770.

Beisiegel, U., Weber, W., and Bengtsson-Olivecrona, G. B. (1991). Lipoprotein lipase enhances the binding of chylomicrons to low density lipoprotein receptor–related protein. *Proc. Natl. Acad. Sci. U S A* **88,** 8342–8346.

Bell, D. J., and Barth, R. H. (1971). Initiation of yolk deposition by juvenile hormone. *Nature* **230,** 220–222.

Bergink, E. W., Wallace, R. A., van de Berg, J. A., Bos, E. S., Gruber, M., and AB, G. (1974). Estrogen-induced synthesis of yolk proteins in roosters. *Amer. Zool.* **14,** 1177–1193.

Bose, S. G., and Raikhel, A. S. (1988). Mosquito vitellogenin subunits originate from a common precursor. *Biochem. Biophys. Res. Comm.* **155,** 436–442.

Brown, M. S., and Goldstein, J. L. (1976). Analysis of a mutant strain of human fibroblasts with a defect in the internalization of receptor-bound LDL. *Cell* **9,** 663–674.

Bujo, H., Hermann, M., Kaderli, M. O., Jacobsen, L., Sugawara, S., Nimpf, J., Yamamoto, T., and Schneider, W. J. (1994). Chicken oocyte growth is mediated by an eight ligand binding repeat member of the LDL receptor family. *EMBO J.* **13,** 5165–5175.

Bujo, H., Yamamoto, T., Hayashi, T., Hermann, M., Nimpf, J., and Schneider, W. J. (1995). Mutant oocytic LDL receptor gene family member causes atherosclerosis and female sterility. *Proc. Natl. Acad. Sci. U S A* **92,** 9905–9909.

Busson, S., Ovtracht, L., and Gounon, P. (1989). Pathway and kinetics of vitellogenin–gold internalization in the *Xenopus* oocyte. *Biol. Cell* **67,** 37–49.

Byrne, B. M., Gruber, M., and AB, G. (1989). The evolution of egg yolk proteins. *Progr. Biophys. Mol. Biol.* **53,** 33–69.

Chen, J.-S., Cho, W.-L., and Raikhel, A. S. (1994). Analysis of mosquito vitellogenin cDNA. Similarity with vertebrate phosvitins and arthropod serum proteins. *J. Mol. Biol.* **237,** 641–647.

Clemens, M. J. (1974). The regulation of egg yolk protein synthesis by steroid hormones. *Prog. Biophys. Mol. Biol.* **28,** 71–107.

Daniel, T. O., Schneider, W. J., Goldstein, J. L., and Brown, M. S. (1983). Visualization of lipoprotein receptors by ligand blotting. *J. Biol. Chem.* **258,** 4606–4611.

Davis, C. G., van Driel, I. R., Russell, D. W., Brown, M. S., and Goldstein, J. L. (1987). The low density lipoprotein receptor. Identification of amino acids in cytoplasmic domain required for rapid endocytosis. *J. Biol. Chem.* **263,** 21–24.

Deeley, R. G., Mullinix, K. P., Wetekam, W., Kronenberg, H. M., Meyers, M., Eldridge, J. D., and Goldberger, R. F. (1975). Vitellogenin synthesis in the avian liver. Vitellogenin is the precursor of the egg yolk phosphoproteins. *J. Biol. Chem.* **250,** 9060–9066.

Dhadialla, T. S., Hays, A. R., and Raikhel, A. S. (1992). Characterization of the solubilized mosquito vitellogenin receptor. *Insect Biochem. Molec. Biol.* **22,** 803–816.

Dumont, J. N. (1978). Oogenesis in *Xenopus laevis* (Daudin). VI. The route of injected tracer transport in the follicle and developing oocyte. *J. Exp. Zool.* **204,** 193–218.

Elkin, R. G., Mac Lachlan, I., Hermann, M., and Schneider, W. J. (1995). Characterization of the Japanese quail ooocyte receptor for very low density lipoprotein and vitellogenin. *J. Nutr.* **125**, 1258–1266.

Elkin, R. G., and Schneider, W. J. (1994). Visualization of the chicken oocyte lipoprotein receptor by ligand blotting with biotinylated plasma and yolk very low density lipoproteins. *Poultry Sci.* **73**, 1127–1136.

Esser, V., Limbird, L. E., Brown, M. S., Goldstein, J. L., and Russell, D. W. (1988). Mutational analysis of the ligand binding domain of the low density lipoprotein receptor. *J. Biol. Chem.* **263**, 13,282–13,290.

Etches, R. J., and Petitte, J. N. (1990). Reptilian and avian follicular hierarchies: Models for the study of ovarian development. *J. Exp. Zool.* (suppl. 4), 112–122.

Gåfvels, M. E., Caird, M., Britt, D., Jackson, C. L., Patterson, D., and Strauss, J. F., III (1993). Cloning of a cDNA encoding a putative VLDL/apolipoprotein E receptor and assignment of the gene to chromosome 9pter–p23. *Somat. Cell Mol. Genet.* **6**, 557–569.

Gåfvels, M. E., Paavola, L. G., Boyd, C. O., Nolan, P. M., Wittmaack, F., Chawla, A., Lazar, M. A., Bucan, M., Angelin, B., and Strauss, J. F., III (1994). Cloning of a cDNA encoding the murine homolog of the VLDL/apolipoprotein-E receptor: Expression pattern and assignment of the gene to mouse chromosome 19. *Endocrinology* **135**, 387–394.

George, R., Barber, D. L., and Schneider, W. J. (1987). Characterization of the chicken oocyte receptor for low and very low density lipoproteins. *J. Biol. Chem.* **262**, 16838–16847.

Gerber-Huber, S., Nardelli, D., Haefliger, J.-A., Cooper, D. N., Givel, F., Germond, J.-E., Engel, J., Green, N. M., and Wahli, W. (1987). Precursor–product relationship between vitellogenin and the yolk proteins as derived from the complete sequence of a *Xenopus* vitellogenin gene. *Nucl. Acids Res.* **15**, 4737–4760.

Gilbert, A. B. (1981). The ovary. *In* "Physiology and Biochemistry of the Domestic Fowl" (D. J. Bell and G. M. Freeman, eds.), pp. 1163–1208. Academic Press, London.

Goldstein, J. L., Brown, M. S., Anderson, R. G. W., Russell, D. W., and Schneider, W. J. (1985). Receptor-mediated endocytosis: Concepts emeging from the LDL receptor system. *Ann. Rev. Cell Biol.* **1**, 1–39.

Hafer, J., and Ferenz, H.-J. (1991). Locust vitellogenin receptor: An acidic glycoprotein with *N*- and *O*-linked oligosaccharides. *Comp. Biochem. Physiol.* **100B**, 579–586.

Hafer, J., and Ferenz, H.-J. (1994). Yolk formation in *Locusta migratoria* and *Schistocerca gregaria*: Related ligands and oocyte receptors. *Arch. Insect Biochem. Physiol.* **25**, 107–120.

Hayashi, K., Nimpf, J., and Schneider, W. J. (1989). Chicken oocytes and fibroblasts express different apolipoprotein-B-specific receptors. *J. Biol. Chem.* **264**, 3131–3139.

Herz, J. (1993). The LDL-receptor–related protein-portrait of a multifunctional receptor. *Curr. Opin. in Lipidol.* **4**, 107–113.

Herz, J., Hamann, U., Rogne, S., Myklebost, O., Gausepohl, H., and Stanley, K. K. (1988). Surface location with high affinity for calcium of a 500-kd liver membrane protein closely related to the LDL receptor suggest a physiological role as a lipoprotein receptor. *EMBO J.* **7**, 4119–4127.

Herz, J., Kowal, R. C., Goldstein, J. L., and Brown, M. S. (1990). Proteolytic processing of the 600-kd low density lipoprotein receptor related protein (LRP) occurs in a trans-Golgi compartment. *EMBO J.* **9**, 1769–1776.

Ho, K.-J., Lawrence, W. D., Lewis, L. A., Liu, L. B., and Taylor, C. B. (1974). Hereditary hyperlipidemia in nonlaying chickens. *Arch. Pathol.* **98**, 161–172.

Hobbs, H. H. Russel, D. W., Browm, M. S., and Goldstein, J. L. (1990). LDL receptor. *Annu. Rev. Genet.* **24**, 133–170.

Hofer, F., Gruenberger, M., Kowalski, H., Machat, H., Huettinger, M., Kuechler, E., and Blaas, D. (1994). Members of the LDLR family mediate cell entry of a minor-group common cold virus. *Proc. Natl. Acad. Sci. U S A* **91**, 1839–1842.

Indrasith, L. S., Kindle, H., and Lanzrein, B. (1990). Solubilization, identification and localization of vitellogenin-binding sites in follicles of the cockroach *Nauphoeta cinerea*. *Arch Insect Biochem. Physiol.* **15,** 213–228.

Jacobsen, L., Vieira, P. M., Schneider, W. J., and Nimpf, J. (1995). The chicken oocyte receptor for lipoprotein deposition recognizes alpha2-macroglobulin. *J. Biol. Chem.* **270,** 6468–6475.

Jingami, H., and Yamamoto, T. (1995). The VLDL receptor: Wayward brother of the LDL receptor. *Curr. Opin. Lipidol.* **6,** 104–108.

Johnson, A. L. (1986). Reproduction in the female. *In* "Avian Physiology" (P. D. Sturkie, ed.), pp. 403–431. Springer-Verlag, New York.

Jokinen, E. V., Landschulz, K. T., Wyne, K. L., Ho, Y. K., Frykman, P. K., and Hobbs, H. H. (1994). Regulation of the very low density lipoprotein receptor by thyroid hormone in rat skeletal muscle. *J. Biol. Chem.* **269,** 26411–26418.

Mac Lachlan, I., Nimpf, J., and Schneider, W. J. (1994). Avian riboflavin binding protein binds to lipoprotein receptors in association with vitellogenin. *J. Biol. Chem.* **269,** 24127–24132.

Mac Lachlan, I., Nimpf, J., White, H. B., III, and Schneider, W. J. (1993). Riboflavinuria in the rd Chicken: 5'-Splice site mutation in the gene for riboflavin-binding protein. *J. Biol. Chem.* **268,** 23222–23226.

Murray, A. (1994). Cell cycle checkpoints. *Curr. Opin. Cell Biol.* **6,** 872–876.

Murray, A., and Hunt, T. (1993). "The Cell Cycle: An Introduction." W. H. Freeman Co., San Francisco.

Nimpf, J., Radosavljevic, M., and Schneider, W. J. (1989a). Oocytes from the mutant "Restricted Ovulator" hen lack receptor for very low density lipoprotein. *J. Biol. Chem.* **264,** 1393–1398.

Nimpf, J., Radosavljevic, M. and Schneider, W. J. (1989b). Specific postendocytic proteolysis of apolipoprotein-B in oocytes does not abolish receptor recognition. *Proc. Natl. Acad. Sci. U S A* **86,** 906–910.

Nimpf, J., Stifani, S., Bilous, P. T., and Schneider, W. J. (1994). The somatic cell–specific LDL receptor–related protein of the chicken: Close kinship to mammalian LDL receptor gene family members. *J. Biol. Chem.* **269,** 212–219.

Nykjaer, A., Bengtsson-Olivecrona, G., Lookene, A., Moestrup, S., K., Petersen, C. M., Weber, W., Beisiegel, U., and Gliemann, J. (1993). LRP binds LPL and b-VLDL with LPL. *J. Biol. Chem.* **268,** 15048–15055.

Oka, K., Ishimura-Oka, K., Chu, M., Sullivan, M., Krushkal, J., Li, W.-H., and Chan, L. (1994a). Mouse very-low-density-lipoprotein receptor (VLDLR) cDNA cloning, tissue-specific expression and evolutionary relationship with the low-density-lipoprotein receptor. *Eur. J. Biochem.* **224,** 975–982.

Oka, K., Tzung, K.-W., Sullivan, M., Lindsay, E., Baldini, A., and Chan, L. (1994b). Human very-low-density lipoprotein receptor complementary DNA and deduced amino acid sequence and localization of its gene (VLDLR) to chromosome band 9p24 by fluorescence *in situ* hybridization. *Genomics* **20,** 298–300.

Opresko, L., Wiley, H. S., and Wallace, R. A. (1980). Differential postendocytotic compartmentation in *Xenopus* oocytes is mediated by a specifically bound ligand. *Cell* **22,** 47–57.

Opresko, L. K., and Wiley, H. S. (1987a). Receptor-mediated endocytosis in *Xenopus* oocytes. I. Characterization of the vitellogenin receptor system. *J. Biol. Chem.* **262,** 4109–4115.

Opresko, L. K., and Wiley, H. S. (1987b). Receptor-mediated endocytosis in *Xenopus* oocytes. II. Evidence for two novel mechanisms of hormonal regulation. *J. Biol. Chem.* **262,** 4116–4123.

Perez, L. E., Fenton, M. J., and Callard, I. P. (1991). Vitellogenin-homologs of mammalian apolipoproteins? *Comp. Biochem. Physiol.* **100B,** 821–826.

Perry, M. M., and Gilbert, A. B. (1979). Yolk transport in the ovarian follicle of the hen *(Gallus domesticus):* Lipoprotein-like particles at the periphery of the oocyte in the rapid growth phase. *J. Cell Sci.* **39**, 257–279.

Perry, M. M., Gilbert, A. B., and Evans, A. J. (1978). Electron microscope observations on the ovarian follicle of the domestic fowl during the rapid growth phase. *J. Anat.* **124**, 481–497.

Perry, M. M., Griffin, H. D., and Gilbert, A. B. (1984). The binding of very low density and low density lipoproteins to the plasma membrane of the hen's oocyte. *Exp. Cell Res.* **151**, 433–446.

Persson, B., Bengtsson-Olivecrona, G., Enerbäck, S., Olivecrona, T., and Jörnvall, H. (1989). Structural features of lipoprotein lipase. Lipase family relationships, binding interactions, non-equivalence of lipase cofactors, vitellogenin similarities and functional subdivision of lipoprotein lipase. *Eur. J. Biochem.* **219**, 39–45.

Raikhel, A. S., and Dhadialla, T. S. (1992). Accumulation of yolk proteins in insect oocytes. *Annu. Rev. Entomol.* **37**, 217–251.

Retzek, H., Steyrer, E., Sanders, E. J., Nimpf, J., and Schneider, W. J. (1992). Cathepsin D: A key enzyme for yolk formation in the chicken oocyte. *DNA Cell Biol.* **11**, 661–672.

Richards, J. S. (1979). Hormonal control of ovarian follicular development. *Recent Prog. Horm. Res.* **35**, 343–373.

Richter, H.-P. (1989). Yolk organelles and their membranes during vitellogenesis of *Xenopus* oocytes. *Roux's Arch. Dev. Biol.* **198**, 92–102.

Roehrkasten, A., Ferenz, H.-J., Buschmann-Gebhardt, B., and Hafer, J. (1989). Isolation of the vitellogenin-binding protein from locust ovaries. *Arch. Insect Biochem. Physiol.* **10**, 141–149.

Röhrkasten, A., and Ferenz, H. (1987). Inhibition of yolk formation in locust oocytes by trypan blue and suramin. *Roux's Arch. Dev. Biol.* **196**, 165–168.

Roth, T. F., and Porter, K. R. (1964). Yolk protein uptake in the oocyte of the mosquito *Aedes aegypti* L. *J. Cell Biol.* **20**, 313–332.

Saito, A., Pietromonaco, S., Kwor-Chieh Loo, A., and Farquhar, M. G. (1994). Complete cloning and sequencing of rat gp 330/"megalin," a distinctive member of the low density lipoprotein receptor gene family. *Proc. Natl. Acad. Sci. U S A* **91**, 9725–9729.

Sakai, J., Hoshino, A., Takahashi, S., Miura, Y., Ishii, H., Suzuki, H., Kawarabayasi, Y., and Yamamoto, T. (1994). Structure, chromosome location, and expression of the human very low density lipoprotein receptor gene. *J. Biol. Chem.* **269**, 2173–2182.

Sappington, T. W., and Raikhel, A. S. (1995). Receptor-mediated endocytosis of yolk proteins by insect oocytes. *In* "Recent Advances in Insect Biochemistry and Molecular Biology" (S. Y. Takahashi, ed.). Nagoya University Press, Japan.

Schneider, W. J. (1991). Removal of lipoproteins from plasma. *In* "Biochemistry of Lipids, Lipoproteins and Membranes" (D. E. Vance, and J. Vance, eds.), pp. 461–487. Elsevier, Amsterdam.

Schneider, W. J. (1995). Yolk precursor transport in the laying hen. *Curr. Opin. Lipidol.* **6**, 92–96.

Schneider, W. J., Beisiegel, U., Goldstein, J. L., and Brown, M. S. (1982). Purification of the low density lipoprotein receptor, an acidic glycoprotein of 164,000 molecular weight. *J. Biol. Chem.* **257**, 2664–2673.

Schneider, W. J., and Nimpf, J. (1993). Lipoprotein receptors: Old relatives and new arrivals. *Curr. Opin. Lipidol.* **4**, 205–209.

Schonbaum, C. P., Lee, S., and Mahowald, A. P. (1995). The *Drosophila* yolkless gene encodes a vitellogenin receptor belonging to the low density lipoprotein receptor superfamily. *Proc. Natl. Acad. Sci. U S A* **92**, 1485–1489.

Selman, K., and Wallace, R. A. (1989). Cellular aspects of oocyte growth in teleosts. *Zool. Sci.* **6**, 211–231.

Sharrock, W. J., Rosenwasser, T. A., Gould, J., Knott, J., Hussey, D., Gordon, J. I., and Banaszak, L. (1992). Sequence of lamprey vitellogenin. *J. Mol. Biol.* **226**, 903–907.

Shen, X., Steyrer, E., Retzek, H., Sanders, E. J., and Schneider, W. J. (1993). Chicken oocyte growth: Receptor-mediated yolk deposition. *Cell Tiss. Res.* **272**, 459–471.

Shoulders, C. C., Brett, D. J., Bayliss, J. D., Narcisi, T. M. E., Jarmuz, A., Grantham, T. T., Leoni, P. R. D., Bhattacharya, S., Pease, R. J., Cullen, P. M., Levi, S., Byfield, P. G. H., Purkiss, P., and Scott, J. (1993). Abetalipoproteinemia is caused by defects of the gene encoding the 97 kDa subunit of a microsomal triglyceride transfer protein. *Hum. Mol. Gen.* **2**, 2109–2116.

Spieth, J., and Blumenthal, T. (1985). The *Caenorhabditis elegans* vitellogenin gene family includes a gene encoding a distantly related protein. *Mol. Cell. Biol.* **5**, 2495–2501.

Steyrer, E., Barber, D. L., and Schneider, W. J. (1990). Evolution of lipoprotein receptors: The chicken oocyte receptor for very low density lipoprotein and vitellogenin binds the mammalian ligand, apolipoprotein E. *J. Biol. chem.* **265**, 19575–19581.

Stifani, S., Barber, D. L., Aebersold, R., Steyrer, E. Shen, X., Nimpf, J., and Schneider, W. J. (1991). The laying hen expresses two different low density lipoprotein receptor–related proteins. *J. Biol. Chem.* **266**, 19079–19087.

Stifani, S., Barber, D. L., Nimpf, J., and Schneider, W. J. (1990). A single chicken oocyte plasma membrane protein mediates uptake of very low density lipoprotein and vitellogenin. *Proc. Natl. Acad. Sci. U S A* **87**, 1955–1959.

Stifani, S., George, R., and Schneider, W. J. (1988). Solubilization and characterization of the chicken oocyte vitellogenin receptor. *Biochem. J.* **250**, 467–475.

Stifani, S., Le Menn, F., Nunez Rodriguez, J., and Schneider, W. J. (1990a). Regulation of oogenesis: The piscine receptor for vitellogenin. *Biochim. Biophys. Acta* **1045**, 271–279.

Stifani, S., Nimpf, J., and Schneider, W. J. (1990b). Vitellogenesis in *Xenopus laevis* and chicken: Cognate ligands and receptors. *J. Biol. Chem.* **265**, 882–888.

Strickland, D. K., Ashcom, J. D., Williams, S., Burgess, W. H., Migliorini, M., and Argraves, W. S. (1990). Sequence identity between the alpha-2 macroglobulin receptor and low density lipoprotein receptor related protein suggests that this molecule is a multifunctional receptor. *J. Biol. Chem.* **265**, 17401–17404.

Takahashi, S., Kawarabayasi, Y., Nakai, T., Sakai, J., and Yamamoto, T. (1992). Rabbit very low density lipoprotein receptor: A low density lipoprotein receptor–like protein with distinct ligand specificity. *Proc. Natl. Acad. Sci. U S A* **89**, 9252–9256.

Trewitt, P. M., Heilmann, L. J., Degrugillier, S. S., and Kumaran, A. K. (1992). The boll weevil vitellogenin gene: Nucleotide squence, structure and evolutionary relationship to nematode and vertebrate vitellogenin genes. *J. Mol. Evol.* **34**, 478–492.

van het Schip, F. D., Samallo, J., Broos, J., Ophuis, J., Gruber, M. M. M., and AB, G. (1987). Nucleotide sequence of a chicken vitellogenin gene and derived amino acid sequence of the encoded yolk precursor protein. *J. Mol. Biol.* **196**, 245–260.

Wahli, W. (1988). Evolution and expression of vitellogenin genes. *Trends Genet.* **4**, 227–232.

Wahli, W., Dawid, G. B., Wyler, T., Jaggi, R. B., Weber, R., and Ryffel, G. U. (1979). Vitellogenin in *Xenopus laevis* is encoded in a small family of genes. *Cell* **16**, 535–549.

Wahli, W., Dawid, I. B., Ryffel, G. W., and Weber, R. (1981). Vitellogenesis and the vitellogenin gene family. *Science* **212**, 298–304.

Wall, D. A., and Meleka, I. (1985). An unusual lysosome compartment involved in vitellogenin endocytosis by *Xenopus* oocytes. *J. Cell Biol.* **101**, 1651–1664.

Wall, D. A., and Patel, S. (1987a). The intracellular fate of vitellogenin in *Xenopus* oocytes is determined by its extracellular concentration during endocytosis. *J. Biol. Chem.* **262**, 14779–14789.

Wall, D. A., and Patel, S. (1987b). Multivesicular bodies play a key role in vitellogenin endocytosis by *Xenopus* oocytes. *Dev. Biol.* **119**, 275–298.

Wallace, R. A. (1985). Vitellogenesis and oocyte growth in nonmammalian vertebrates. *In* "Developmental Biology" (L. W. Browder, ed.), pp. 127–177. Plenum Press, New York.

Wang, S., and Williams, D. L. (1982). Purification of avian vitellogenin III: Comparison with vitellogenins I and II. *Biochemistry* **22**, 6206–6212.

Wasserman, W. J., and Smith, L. D. (1978). Oocyte maturation in nonmammalian vertebrates. *In* "The Vertebrate Ovary" (R. E. Jones, ed.), pp. 443–468. Plenum Press, New York.

Webb, J. C., Patel, D. D., Jones, M. D., Knight, B. L., and Soutar, A. K. (1994). Characterization and tissue-specific expression of the human "VLDL receptor" mRNA. *Hum. Mol. Genet.* **3**, 531–537.

Westhuyzen, D. R. v. d., Fourie, A. M., Coetzee, G. A., and Gevers, W. (1990). The LDL receptor. *Curr. Opinion Lipidol.* **1**, 128–135.

Wiley, H. S., Opresko, L., and Wallace, R. A. (1979). New methods for purification of vertebrate vitellogenin. *Anal. Biochem.* **97**, 145–152.

Willnow, T. E., Goldstein, J. L., Orth, K., Brown, M. S., and Herz. J. (1992). LDL Receptor-related protein and gp 330 bind similar ligands, including plasminogen activator–inhibitor complexes and lactoferrin, an inhibitor of chylomicron remnant clearance. *J. Biol. Chem.* **267**, 26172–26180.

Wood, W. B. (1988). "The Nematode *Caenorhabditis elegans.*" Cold Spring Harbor Laboratories, Plainview, NY.

Yamamoto, T., Bishop, R. W., Brown, M. S., Goldstein, J. L., and Russell, D. W. (1986). Deletion in cysteine rich region of LDL receptor impedes transport to cell surface in WHHL rabbit. *Science* **232**, 1230–1237.

Yano, K.-i., Sakurai, M. T., Watabe, S., Izumi, S., and Tomino, S. (1994). Structure and expression of mRNA for vitellogenin in *Bombyx mori. Biochim. Biophys. Acta.* **1218**, 1–10.

Yochem, J., and Greenwald, I. (1989). glp-1 and lin-12, genes implicated in distinct cell–cell interactions in *C. elegans,* encode similar transmembrane proteins. *Cell* **58**, 553–563.

Yochem, J., and Greenwald, I. (1993). A gene for a low density lipoprotein receptor–related protein in the nematode *Caenorhabditis elegans. Proc. Natl. Acad. Sci. U S A* **90**, 4572–4576.

Follicular Dendritic Cells and Germinal Centers

Yong-Jun Liu, Géraldine Grouard, Odette de Bouteiller, and Jacques Banchereau
Schering-Plough, Laboratory for Immunological Research, 69571 Dardilly, France

Follicular dendritic cells (FDCs) are stromal cells unique to primary and secondary lymphoid follicles. Recirculating resting B cells migrate through the FDC networks, whereas antigen-activated B cells undergo clonal expansion within the FDC networks in a T cell–dependent fashion, thereby generating germinal centers. Here, B cells undergo somatic mutation, positive and negative selection, isotype switching and differentiation into high-affinity plasma cells and memory B cells. Since the discovery of FDCs by electron microscopy as long-term antigen-retaining cells 30 years ago isolation of FDCs and generation of FDC-like cells lines and of FDC-specific monoclonal antibodies have been achieved. FDCs express all three types of complement receptors as well as Ig–Fc receptors, through which antigen–antibody immune complexes are retained. However, the mechanism that prevents FDCs from internalizing the antigens and retaining them in native form for long periods of time remains obscure. Substantial evidence derived from cultures in vitro indicates that FDCs contribute directly to the survival and activation of peripheral B cells. The adhesion between FDCs and B cells is mediated by ICAM-1 (CD54)–LFA-1(CD11a) and VCAM–VLA-4. T cells may interact with FDCs in a CD40/CD40–ligand–dependent fashion. Whether FDCs originate from hematopoietic progenitors or from stromal elements is still a controversy. New evidence suggests the presence of two types of dendritic cells within human germinal centers: (i) the classic FDCs that express DRC-1, KiM4, and 7D6 antigens represent stromal cells; and (ii) the newly identified $CD3^-CD4^+CD11c^+$ germinal center dendritic cells (GCDC) represent hematopoietic cells that may be analogous to the antigen-transporting cells described in mice. Finally, FDCs appear to be involved in the growth of follicular lymphomas and in the pathogenesis of HIV infection.

KEY WORDS: Secondary lymphoid tissue, Germinal center, Follicular dendritic cell, Memory B cell, Somatic mutation, Lymphoma, HIV.

I. Introduction

The development of antibody diversity occurs mainly during the early B cell development in the bone marrow or fetal liver through combinatorial rearrangement of many different immunoglobulin gene segments (Rajewsky, 1992). About 10^9 potentially different complete IgV genes can be generated in the mouse (Jerne, 1984). Because one B cell normally makes only one type of antibody (Nossal and Lederberg, 1958), the limited number of B cells in a mouse (about 1.2×10^8) (Osmond, 1991) determines that the frequency of B cell specificity for a given antigen at any given time is very low. To make a prompt antibody response, the immune system relies on two strategies: (i) the generation of low-affinity polyspecific B cell clones that permit binding to the entire repertoire of antigens/pathogens (Kearney, 1989); and (ii) a powerful system within the peripheral lymphoid organs that allows rapid selection and expansion of the rare antigen-specific B cells following antigen intrusion (Burnet, 1959). All secondary lymphoid organs, including lymph nodes, spleen, and mucosal-associated lymphoid tissues, are composed of T cells and interdigitating cell-rich extrafollicular areas, and B cells and follicular dendritic cell-rich follicles. During humoral immune responses the rare antigen-specific B cells are activated and undergo a transient clonal expansion in the extrafollicular areas. This expansion is followed by a massive, rapid, and long-lasting clonal expansion of antigen-specific B cells within the follicles, forming the so called germinal centers.

A. Germinal Centers as Sites of Antigen-Driven Lymphocyte Production

Germinal centers (GCs) were identified by Flemming in 1884 as foci with high mitotic activities within the lymphoid follicles of peripheral lymph nodes. He suggested that GCs may represent major sites of production of naive lymphocytes (Flemming, 1885a, 1885b). Since then identifying the nature and function of GC has presented a challenge (Nieuwenhuis and Opstelten, 1984; Thorbecke, 1990; Thorbecke *et al.*, 1994).

The first experimental evidence suggesting that GCs may not be the sites for naive lymphocyte production came from Glimstedt (1936), who observed that the peripheral lymphoid tissues of germ-free guinea pigs did not have GCs but contained normal numbers of lymphocytes in the circulation (Glimstedt, 1936). Subsequent studies showed that GCs arise in primary follicles about 4 days after antigenic stimulation and after the onset of specific antibody production (Thorbecke, 1959; Langevoort *et al.*, 1963; Nieuwenhuis, 1969; White *et al.*, 1975; Kim *et al.*, 1994). The persistence of GC reactions in mucosal-associated lymphoid tissues—which are sites of constitutive antigenic stimulation—provides further evidence that GC reactions depend on antigens. Interestingly, antigens such as polysaccharide, which were later classified as T cell–independent antigens, were found unable to induce GC formation (Liu *et al.*, 1991b). In addition, congenitally athymic nude mice do not have GCs in their peripheral lymphoid tissues, and no GC could be induced in such nude animals by antigen administration (De Sousa *et al.*,1969; Takahashi *et al.*, 1971; Vos *et al.*, 1980). However, injection of syngenic thymus cells into nude mice restored their capacity to mount a GC reaction after antigenic stimulation (Jacobson *et al.*, 1974). The preceding experimental evidence firmly established that GCs are not the sites of primary lymphocyte production but the sites of antigen-driven thymus (T-cell)–dependent clonal expansion of lymphocytes.

B. Germinal Centers as Major Sites of Memory B Lymphocyte Production

Many studies have shown that GC reactions confer an animal with the ability to mount secondary antibody responses (Thorbecke, 1969). With the discovery that lymphoid follicles consist largely of B lymphocytes (Weissman, 1975), it became clear that GCs represent important sites of B cell proliferation that might be involved in memory B cell generation. In particular, inhibition of GC reactions during immune responses by depleting complement fragment C3 with cobra venom factor (CVF) resulted in the blocking of memory B cell generation (Klaus *et al.*, 1980). The observation that mouse GC B cells can bind peanut agglutinin with high affinity provided an important tool for separating germinal center B cells from other B cells (Rose *et al.*, 1980). Accordingly, B cell memory could be transferred into syngenic animals by transferring peanut agglutinin–binding germinal center B cells but not by transferring other B cells (Coico *et al.*, 1983). These two experiments established that one of the major function of GCs is the generation of B cell memory.

C. Structural Features of a Germinal Center and Kinetics of Synchronized Germinal Center Formation

Classically, GCs have been divided into a dark zone and a light zone (Nieuwenhuis and Opstelten, 1984; Szakal, 1989). Detailed immunohistological analysis of a wide range of molecules on human tonsillar sections further revealed four distinct compartments within a GC (Hardie *et al.*, 1993; MacLennan, 1994):

1. A dark zone that contains Ki67$^+$ proliferating centroblasts [Ki67 is a nuclear antigen expressed on proliferating cells; Fig. 1A (see color insert)], very few CD3$^+$CD4$^+$CD45RO$^+$CD40–ligand$^+$ GC T cells (Fig. 1C; see color insert), and fine follicular dendritic cell (FDC) processes expressing CD21, DRC-1, KiM4, and 7D6 (Fig. 1D; see color insert) but not CD23 (Fig. 1B; see color insert). (DRC-1, KiM4, and 7D6 are undefined pan-FDC-specific antigens.)

2. A basal light zone that contains large centrocytes recently generated from proliferating centroblasts. Many large centroblasts still express nuclear Ki67 antigen. The fine FDC processes do not express CD23, and there are very few GC T cells.

3. An apical light zone that contains small centrocytes and many CD3$^+$ CD4$^+$CD45RO$^+$CD40–ligand$^+$ GC T cells. It is characterized by dense FDC networks strongly expressing CD23 (Fig. 1B; see color insert).

4. An outer zone that is characterized by the accumulation of CD3$^+$CD4$^-$CD45RO$^+$CD40–ligand$^+$ T cells beneath the follicular mantle, forming a "T cell mantle" (Figs. 1C, 1E; see color insert). Fine CD23–FDC processes pass through the outer zone and penetrate into the follicular mantle.

In addition, GCs also contain tingible body macrophages, which have the capacity to phagocytose apoptotic cells. They are localized mainly in the dark zone and in the apical light zone.

The kinetics of the GC reaction has been studied in various animal models following different immunization methods (Kroese *et al.*, 1987; Jacob *et al.*, 1991a; Liu *et al.*, 1991b). The GC reaction is preceded by a brief and rapid antigen-specific B cell proliferation in the T cell and interdigitating cell-rich extrafollicular areas, forming the so-called foci. In addition to generating short-lived plasma cells, antigen-specific B cell proliferation within the foci also generates and recruits GC founder cells. A GC is initiated by the colonization of one to five founder cells into the FDC network of a primary follicle. A founder cell undergoes rapid proliferation, with a cell cycle time of 6–10 hours, and gives rise to about 5000 cells within 2–3 days (Zhang *et al.*, 1988; Liu *et al.*,1991b). At this stage the proliferating B blasts (centroblasts) form the dark zone of a GC. These cells differentiate into nonproliferating centrocytes that form the light zone of a GC. A typical GC reaction lasts for about 3 weeks after a single immunization. Thus, a GC founder cell may potentially undergo up to 50 divisions. A typical GC containing a dark zone and a light zone is finally replaced by a follicular focus consisting of a few antigen-specific B blasts.

D. Mechanisms and Steps of Memory B Cell Generation within Functional Compartments of Germinal Centers

The affinity of the antigen receptor represents a major difference between memory B cells and naive B cells, with those carried by memory B cells being of higher affinity than those carried by naive B cells (Weiss and Rajewsky, 1990). The mechanisms of affinity maturation of antibody responses have been analyzed by several groups (Manser *et al.*, 1985; Weigert, 1986; Berek and Milstein, 1988; Kocks and Rajewsky, 1988; Levy *et al.*, 1989). Basically, during the course of T cell–dependent antibody responses, a hypermutation mechanism directly targets the rearranged IgV gene segments. This results in an alteration of the affinity of antigen receptors on the responding B cells. The subsequent positive selection for high-affinity B cells leads to affinity maturation and memory B cell development.

In 1986 MacLennan and Gray proposed that (1) GCs are the anatomic structures where B cells undergo somatic mutation in their IgV genes during the course of clonal expansion within the GC dark zones, and (2) antigen retained by follicular dendritic cells within the GC light zones provides a selection machinery for high-affinity mutants (MacLennan and Gray, 1986). This fundamental hypothesis has caused many molecular immunologists to focus on the study of GC development. Subsequent sequence analysis of IgV genes from GC B cells isolated by flow cytometry according to peanut agglutinin–binding or directly picked from tissue sections indicated that somatic mutation is actually an ongoing process that occurs within GCs (Berek *et al.*,1991; Jacob *et al.*, 1991b; Leanderson *et al.*, 1992; Klein *et al.*, 1993; Küppers *et al.*, 1993; McHeyzer-Williams *et al.*, 1993; Ford *et al.*, 1994; Pascual *et al.*, 1994; Ziegner *et al.*, 1994).

After somatic mutation the generated mutants migrate into the basal light zone and undergo a positive selection process that leads to affinity maturation. In parallel with the positive selection processes occurring during the primary T cell development in the thymus and primary B cell development in the bone marrow, GC selection also relies on a mechanism that renders the candidate cells short-lived. GC B cells are characterized by (i) a down-regulation of survival gene Bcl-2 (Pezella *et al.*, 1990; Hockenberry *et al.*, 1991; Liu *et al.*, 1991a; Merino *et al.*, 1994), and (ii) an up-regulation of the death-related genes Fas, c-myc, P^{53}, and Bax (Liu *et al.*, 1995; Martinez-Valdez *et al.*, 1995). Therefore, GC B cells require stimulatory signals for their survival (Liu *et al.*, 1989). Various studies have demonstrated that antigen receptor triggering represents an important signal for the rescue of GC B cells from apoptosis (Liu *et al.*, 1989). This signal can be provided by antigen retained by FDCs (MacLennan and Gray, 1986). After high-affinity GC B cells pick the antigen, they process it, and then migrate into the apical light zone and outer zone, where they present specific antigen-derived peptides to GC T cells. Then, GC B cells receive another important survival proliferation signal via the CD40–ligand expressed on GC T cells (Liu *et al.*, 1989; Banchereau *et al.*, 1991; Madassery *et al.*, 1991; Holder *et al.*, 1993; Tsubata *et al.*, 1993; Clark and Ledbetter, 1994; Lagresle *et al.*, 1995; Merville *et al.*, 1995). This signal allows the selected high-affinity GC B cells to expand, switch isotype (Martinez-Valdez *et al.*, 1995), and ultimately differentiate into memory B cells (Arpin *et al.*, 1995).

Numerous questions are still pending regarding this mechanism. In particular, what are the molecular signals that (i) make GC B cells proliferate every 6 to 10 hours, (ii) turn on apoptosis, and (iii) switch on somatic mutation. To address these questions it is important to analyze further the cellular interactions within the GC microenvironment and to dissect each developmental stage of GC B cell development (Liu and Banchereau, 1995).

Because follicular dendritic cells represent a unique cell type that has been identified only in B lymphoid follicles (Szakal, 1989), this chapter provides an overview of the biology of this unique cell, in the context of its function in GC reactions, and in the pathogenesis of follicular lymphomas and HIV infection.

II. General Functions of Follicular Dendritic Cells

A. Identification

To understand how an antigen initiates an immune response in peripheral lymphoid tissues, it is important to know (i) where the invading antigens localize in the lymphoid tissues, (ii) what kinds of cells are involved in the antigen transport and handling, and (iii) what happens to the antigens. In 1950 Kaplan *et al.* observed that a fluorescent protein injected into animals localized for long periods of time within the lymphoid follicles (Kaplan *et al.*, 1950). To understand the cellular basis of follicular localization of antigens and its function in GC reactions, detailed electron microscopic analyses of lymphoid follicles have been carried out after injection ^{125}I-labeled proteins (Nossal *et al.*, 1968; Szakal and Hanna, 1968) or electron-dense tracers, such as colloidal carbon and horseradish peroxidase (Chen *et al.*, 1978; Mandel *et al.*, 1980). Under the electron microscope a new cell type was identified within both primary follicles and secondary follicles for its ability to trap antigens on the surface of complicated dendritic processes. The nucleus of these cells is characteristically large, irregularly shaped, and lacks heterochromatin. The cytoplasm extends in several directions, forming an intricate network of processes interdigitated with the fine processes of lymphocytes. Relatively few organelles can be identified in the cytoplasm, except for well-developed Golgi regions and some mitochondria. This unique cell type, which traps antigen for long periods of time within B lymphoid follicles, was named the *follicular dendritic cell* (FDC).

B. Mechanisms of Antigen Retention

In 1971 Nossal and Ada showed that the long-term antigen retention on FDCs requires antigen-specific antibodies that can either be produced during the immune response to the antigen or be passively injected (Nossal and Ada, 1971). The antigen retention depends on the Fc portion of Ig molecules. Also, in nude animals, protein antigens were shown to be localized within the primary follicles only if the animals received antibody passively (Tew et al., 1980). These experiments suggested that antigen is retained by FDCs via antibody in the form of immune complexes. This hypothesis is consistent with the identification of IgG and IgM within GCs of human lymph nodes (Gajl-Peczalska et al., 1969). The mechanisms of immune complex trapping thus became a question to address. In this context the complement fragment C3 became a candidate when it was detected within GCs (Gajl-Peczalska et al., 1969) and when it was found that mice deprived of C3 by treatment with CVF failed to trap ^{125}I-labeled aggregated human IgG in their splenic follicles (Dukor et al., 1970). Pryjma and Humphrey (1975) proposed that the first step of antigen trapping within follicles is the formation of immune complexes during the early stage of immune responses. Subsequently, the antigen–antibody complexes activate the complement system, and C3 fragments covalently bind to the immune complexes. Finally, antigen–antibody–C3 complexes are trapped by follicular dendritic cells through C3 receptors. As evidence for this mechanism it was recently shown that injection of monoclonal antibody against rat C3 results in a selective inhibition of immune complex trapping by follicular dendritic cells (Van den Berg et al., 1992). The later identification of Ig Fc receptors on the surface of FDCs also suggested their involvement in antigen trapping within follicles (Humphrey and Grennan, 1982; Schriever et al., 1989).

C. Function of Antigen–Antibody–C3 Complexes

1. Germinal Center B Cell Proliferation

GCs are sites of extensive B cell proliferation during T cell–dependent antibody responses. Direct evidence that GC reactions represent antigen-specific clonal expansion of B cells was provided by direct labeling of antigen receptors with antigen–enzyme conjugates on splenic sections (Jacob et al., 1991b; Liu et al., 1991b). In rats that have been immunized with DNP-KLH and OX-KLH, all GC B cells in the spleen were found to be either DNP-binding cells or OX-binding cells. The molecular mechanisms controlling B cell proliferation within germinal centers remain largely unidentified, yet antigen–antibody–C3 complexes (Ag–Ab–C3) have been considered to contribute to GC B cell proliferation. In particular, it has been shown that mice immunized with Ag–Ab complexes develop GCs faster than those given antigen alone (Laissue *et al.,* 1971; Klaus *et al.,* 1980). In addition, experiments *in vitro* have suggested that Ag–Ab–C3 complexes may induce lymphocyte proliferation (Bloch-Shtacher *et al.,* 1968; Soderberg and Coons, 1978). As evidence, human complement fragment C3 bound to particles induces activated mouse B cells to enter the S-phase of the cell cycle (Melchers *et al.,* 1985; Cahen-Kramer *et al.,* 1994). The hypothesis that Ag–Ab–C3 complexes are required for B cell proliferation within the follicles has been challenged by other observations *in vivo.* In particular, following primary immunization with antigens, follicular B cell proliferation was observed before follicular antigen trapping could be identified (Humphrey and Frank, 1967; van Rooijen, 1972). Furthermore, blocking of follicular antigen trapping through irradiation or C3 depletion using CVF did not prevent follicular B cell proliferation, although the GCs became significantly smaller (Kroese *et al.,* 1986). Thus, during the induction phase of GC reactions, the proliferation of antigen-specific B cells within the follicles may be independent of Ag–Ab–C3 complexes. The recruitment of antigen-specific B cells into the follicle may depend on their early activation in T cells and interdigitating cell-rich areas, resulting in the expression of receptors for homing into the follicles. Because selective survival of high-affinity mutants and their further clonal expansion depends on Ag–Ab–C3 complexes in the light zones of the GCs, depletion of Ag–Ab–C3 within the follicles prevents the positive selection of high-affinity B cells, thereby explaining the smaller GCs and the blocking of memory B cell development in mice treated by CVF.

2. Maintenance of Memory B Cell Clones and Antibody Response

In 1979 Tew *et al.* provided evidence that Ag–Ab–C3 complexes can persist for up to 18 months in lymphoid follicles. Because GC reactions last only about 3–4 weeks after a single immunization, the function of long-term immune complex retention was questioned (Tew *et al.,* 1980). In this context two interesting observations need to be reported. First, Gray *et al.* demonstrated that memory B cells transferred into irradiated congenic animals are lost within 4 weeks if no antigen is provided, whereas they can be transferred indefinitely into congenic animals that have been challenged with the relevant antigen (Gray and Skarvall, 1988). Second, hyperimmunized animals do not require additional antigen to maintain high circulating antibody levels (Richter *et al.,* 1964; Britton and Moller, 1968; Bystryn *et al.,* 1970; Weigle, 1975), and when large volumes of blood are removed, antibody levels promptly return to their initial levels. It has been proposed that the immunogenicity of the Ag–Ab–C3 complexes depends on the antigen–antibody ratio (Tew *et al.*, 1980). The alterations in serum antibody levels result in formation or dissociation of these complexes. When antibody levels in the circulation decline, antigenic determinants are exposed, memory B cells are stimulated, and a new generation of memory B cells and antibody-secreting plasma cells are produced. Three experimental sets of data support this model: (1) 6 weeks after immunization with DNP-KLH, rat DNP-binding proliferating B blasts were identified only within the follicular centers in association with FDC networks; (2) follicular B cell proliferation can be induced by removing serum antibodies after bleeding (Donaldson *et al.*, 1986); (3) only memory B cells but not naive B cells are able to respond to antigen localized on FDCs (Gray, 1988).

3. Selection and Maintenance of Memory T Cells

Like memory B cells, memory T cells were also shown to undergo rapid decay in the absence of antigen after transfer into congenic animals (Gray and Matzinger, 1991; Oehen *et al.*, 1992). Whether long-term antigen retention on FDCs within GCs provides a stimulation signal for the maintenance of memory T cell clones, either directly or indirectly, remains an enigma. In 1972, T cells were identified within mouse GCs (Gutman and Weissman, 1972). All these T cells, mainly found in the GC light zone express CD4, $CD8^+$ T cells being notoriously absent from normal GCs (Berman *et al.*, 1981). The presence of $CD4^+$ T cells in GC was further established in human lymphoid tissues (Rose *et al.*, 1980; Poppema *et al.*, 1981; Stein *et al.*, 1982; Roscoe and Whiteside, 1983; Janossy *et al.*, 1989; Bowen *et al.*, 1991) as well as in rat lymphoid tissues (Kroese *et al.*, 1985). In addition, GC CD4 T cells were all found to express exclusively CD45RO (Janossy *et al.*, 1989; Bowen *et al.*, 1991), whereas some also expressed CD57 (Porwit-Ksiazek *et al.*, 1983; Swerdlow and Murray, 1984; Bowen *et al.*, 1991). Adoptive transfers of antigen-specific T cells into congenic mouse with different Thy-1 allotypes have shown that the homing of T cells into GCs is directed by antigens (Michie and Rouse, 1988). In particular, GCs induced by cytochrome *c* in B10.A mice contained $V\alpha11$ T cells, whereas GCs induced by myelin basic protein in PL/J mice contained $V\beta8$ T cells (Fuller *et al.*, 1993). A proportion of antigen-specific GC T cells clearly had undergone proliferation during the 5–7 days that followed immunization. Interestingly, antigen-specific T cells directly picked from GCs appeared to express limited but significant somatic mutations in their TCR $V\alpha$ segments (Zheng *et al.*, 1994). In contrast, antigen-specific T cells isolated from mouse spleen at different times after immunization did not show mutations in their TCR (McHeyzer-Williams and Davis, 1995). These observations raise a question about the nature of the cell presenting antigen to GC T cells, because antigens retained by FDCs are essentially in an intact, unprocessed form. GC B cells are most likely to represent the major antigen-presenting cells for GC T cells, because they express MHC class II, CD80/B7.1, and CD86/B7.2 (Liu *et al.*, 1995) and have the capacity to capture antigen from FDCs (Kosco *et al.*, 1988). However, since FDCs were shown to have the capacity to trap MHC–class II peptide complexes shed by surrounding GC B cells (Gray *et al.*, 1991), FDCs may also contribute to T cell activation. In keeping with this hypothesis, highly purified FDCs isolated by FACS sorting with monoclonal antibody HJ2 were shown to induce the proliferation of allogeneic T cells or T cell lines (Butch *et al.*, 1994).

III. Isolation of Mouse FDCs and the Identification of an Alternative Antigen-Presenting Pathway Mediated by Iccosomes

Studies on isolated cells have been undertaken to further characterize the surface phenotype and function of FDCs. Because FDCs have long and delicate dendritic processes that intimately associate with surrounding cells, their isolation has represented a challenging and difficult task. Humphrey and Grennan (1982) reported the first isolation of FDC–B cell clusters from the spleens of mice that had intravenously received preformed FITC-labeled immune complexes. The purification required digestion of splenic tissue with collagenase and dispase and enrichment by repeated sedimentation at 1 g through fetal calf serum. The isolated living FDCs appeared as large multinucleated cells carrying fluorescein-labeled immune complexes. These FDCs were shown to express both Immunoglobulin (Ig) Fc receptors and C3 receptors and to lack phagocytic activity.

In 1985 Szakal *et al.* performed three-dimensional ultrastructural analysis of isolated mouse FDCs and demonstrated the presence of two FDC types (Szakal *et al.*, 1985). One type displays long filiform dendrites 15–20 μm in length and 0.1–0.3 μm in diameter; the other type has long dendrites made up of numerous interconnected 0.3- to 0.7-μm-diameter spherical or ellipsoidal segments. These globular structures have been called *iccosomes* for "immune complex coated bodies." Both dendrites and iccosomes have a highly ordered pattern of immune complex attachment on their surfaces. Kinetics studies (Szakal *et al.*, 1988) showed that iccosomes develop on FDCs 1 day after immunization. They appear as dispersed structures among surrounding lymphocytes and can be identified within the tingible body macrophages and GC lymphocytes in association with the Golgi apparatus between 3 to 5 days after immunization. By day 14, iccosomes cannot be identified anymore. Iccosomes are thought to facilitate antigen uptake by GC B cells through endocytosis, followed by antigen processing and presentation to antigen-specific GC T cells (Kosco *et al.*, 1988). As a result of this direct T–B cell interaction within GCs, high-affinity B cells differentiate into antibody-forming cells or memory B cells (Kosco *et al.*, 1988). By transmission electron microscopy we also observed "iccosome-like" structure on large tonsillar cells enriched for FDCs (Fig. 2).

IV. Isolation of Human FDCs and Their Detailed Phenotypic Analysis

A. Isolation of FDC–Lymphocyte Clusters by Albumin Gradients, Percoll Gradients, and MACS Sorting

Enriched human FDC–lymphocyte clusters have been isolated from human tonsils after tissue digestion with collagenase and DNase, followed by centrifugation of the cell suspension through either 35% Percoll gradient at 400 g or 1.5–7.5% BSA gradients at 10 g (Gerdes *et al.*, 1983; Heinen *et al.*, 1986; Ennas *et al.*, 1989; Schriever *et al.*, 1989; Sellheyer *et al.*, 1989; Petrasch *et al.*, 1990; Parmentier *et al.*, 1991; Grouard *et al.*, 1995), or using MACS sorting after staining with anti-FDC monoclonal antibody KiM4 (Schmitz *et al.*, 1993). A typical human FDC–lymphocyte cluster contains 1 to 3 FDCs and 5 to 20 lymphocytes (Fig. 3; see color insert). Paradoxically, isolated human FDCs have a morphology that appears to differ from that of isolated mouse FDCs. Isolated human FDCs do not have very long filiform dendrites or long dendrites with numerous interconnected iccosomes. Instead, human FDCs display a fibroblast-like morphology together with extensive cytoplasm extensions and foldings and contain from one to several large and round nuclei with decondensed chromatin and clear nucleoli.

B. Phenotype of Human FDCs

Because of the availability of numerous antibodies, the phenotype of human FDCs is better known than that of mouse or rat FDCs (Tew *et al.*, 1990; Dijkstra and Van den Berg, 1991; Schriever and Nadler, 1992). All FDCs express the monocyte marker CD14 (a receptor for LPS), the three types of complement receptors (CR1/CD35, CR2/CD21, CR3/CD11b), and IgFcγ receptor (CD32). A subset of FDCs in the GC light zone expresses the low-affinity receptor for IgE (Fcε RII, CD23), which also represents one

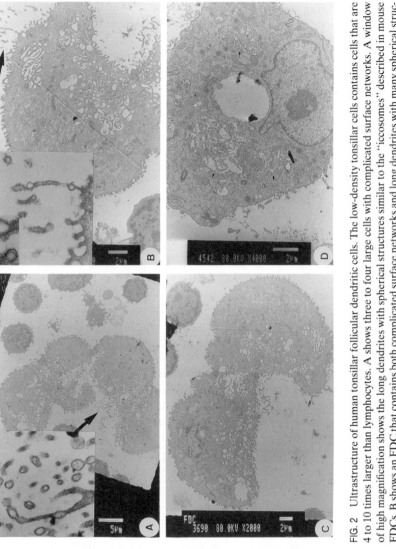

FIG. 2   Ultrastructure of human tonsillar follicular dendritic cells. The low-density tonsillar cells contains cells that are 4 to 10 times larger than lymphocytes. A shows three to four large cells with complicated surface networks. A window of high magnification shows the long dendrites with spherical structures similar to the "iccosomes" described in mouse FDCs. B shows an FDC that contains both complicated surface networks and long dendrites with many spherical structures. Many of the spherical structures seem to contain two layers of cytoplasmic membrane. C shows the close contact between two large cells. D shows a large cell with two clusters of dendritic networks (arrow). The elongated cytoplasm forms a circle, giving rise to a closed extracellular space. Many mitochondria are found within the cell cytoplasm.

of the ligands for CD21. Thus, these complement receptors and Fc receptors confer FDCs with an efficient mechanism for trapping Ag–Ab–C3 complexes (Fig. 4). FDCs express a wide range of adhesion molecules, including ICAM-1/CD54, VCAM-1, VLA-3, 4, 5, 6, and VLAβ chain. Experiments *in vitro* concluded that adhesion between B cells and FDCs is mediated by the interactions between ICAM-1/CD54 and LFA-1/CD11a as well as between VCAM-1 and VLA-4 (Freedman *et al.*, 1990; Koopman *et al.*, 1991). Interestingly, purified ICAM-1 molecules appear to deliver survival signals to human GC B cells through LFA-1 molecules, indicating that adhesion molecules may perform functions other than only holding the cells together (Koopman *et al.*, 1993). Immunohistological studies on tonsillar sections suggest that FDCs may express the low-affinity receptor for nerve growth factor (NGF). Detailed double staining on purified FDC clusters shows that the expression of the low-affinity receptor for NGF is restricted to a proportion of CD23$^+$ FDCs in the light zone of GCs (Liu, unpublished observation). FDCs express CD40 antigen, whose ligand on activated T cells was shown to be a key molecule in the activation of CD40-expressing cells including B cells, dendritic cells, monocytes, thymic epithelial cells, endothelial cells, and synovial fibroblasts from rheumatoid synovium (Bancher eau *et al.*, 1994). In particular, activated human T cells can induce a human FDC-like cell line to proliferate in a CD40–ligand–dependent fashion (Kim *et al.*, 1994), which suggests that CD40/CD40–ligand interactions may be involved in FDC–T cell interactions *in vivo* (Fig. 4).

The expression of Fc receptors on FDCs renders the phenotypic analysis difficult, and the expression, on human FDCs, of the B cell markers CD19, CD20, and CD24, the pan-leukocyte antigen CD45, and MHC class II antigens remains a subject of controversy. Yet, PCR analysis on single FDCs showed mRNA for CD21 but not for CD20, CD45, or TNF-α. In the mouse, adoptive transfer of B cells from MHC class II I^E transgenic mice into congenic mice has suggested that the host FDCs do not synthesize MHC class II antigens but rather capture the donor MHC class II I^E molecules shed by surrounding donor GC B cells (Gray *et al.*, 1994). Thus, a complete mRNA analysis of highly purified FDCs is presently required to definitively establish the full FDC phenotype.

C. Isolation of Highly Purified Single FDCs by FACS Sorting

Suspensions of highly purified single FDCs (98% pure) have been obtained by sorting CD14+ low-density large tonsillar cells (Schriever *et al.*, 1989) or by sorting low-density tonsillar cells using the FDC-specific monoclonal antibody HJ2 (Butch *et al.*, 1994). Highly purified (>90%) single FDC preparations have also been generated by initially sorting DRC-1$^+$CD4$^-$

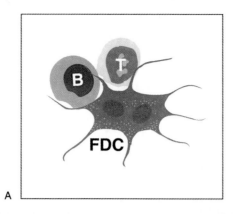

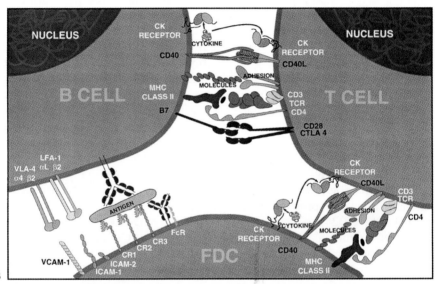

FIG. 4 Molecular interactions between B cells, T cells, and FDCs. A B cell–FDC–T cell cluster (A) may represent a functional unit of the germinal center light zone. The known molecules possibly involved in the cell–cell interaction are depicted in B. (i) FDC–B cell interaction: FDCs retain antigen–antibody immune complexes through complement receptors CR1, CR2, and CR3 and immunoglobulin Fc receptor. The immune complexes cross-link the B cell antigen receptors that trigger the activation, selection, and clonal expansion of antigen-specific B cells. Adhesion molecule pairs such as LFA-1–ICAM-1/ICAM-2 and VLA-4–VCAM-1 are involved in the B cell–FDC interaction. (ii) FDC–T cell interaction: FDCs may trap MHC class II–peptide complexes shed by other antigen-presenting cells. These complexes trigger the TCR complexes of T cells. The CD40–ligand and cytokines expressed and produced by germinal center T cells may have profound effects on the phenotype and function of FDCs. (iii) T–B cell interactions: Three pairs of surface molecules that play key roles in T–B cell interaction are MHC class II–peptide–TCR complexes, CD40/CD40–ligand, and B7-CD28/CTLA-4. Cytokine–cytokine receptors and adhesion molecules are also involved in T–B cell interactions.

low-density tonsillar cells, followed by sorting the DRC-1⁺CD4⁻ CD19⁻IgD⁻ cells (Stahmer *et al.*, 1991). According to the double expression of CD21 and CD14, we have been able to obtain 90% pure CD21⁺CD14⁺ single FDCs with a single sorting on low-density tonsillar cells (see Fig. 3). Unfortunately, these highly purified FDCs appear to be damaged inasmuch as they display cytoplasm losses (see Fig. 3) and are unable to support B cell growth *in vitro*. The loss of FDC viability may be the consequences of a physical problem due to the passing of large cells through a size-limited nozzle or of a physiological problem due to the breaking of tight cellular junctions between FDCs/B cells or FDCs/FDCs. However, such highly purified FDCs provided appropriate cellular materials for the generation of FDC-specific monoclonal antibodies (detailed later) and the analysis of cytokine gene expression in FDCs (Schriever *et al.*, 1991).

V. Generating FDC-Specific Monoclonal Antibodies and FDC-like Cell Lines

A. FDC-Specific Monoclonal Antibodies

Six monoclonal antibodies relatively specific for human FDCs have been reported. R4/23 (DRC-1), generated by immunizing mice with sheep red blood cells coated with complement fragments (Naiem *et al.*, 1983), also stains a subset of activated B cells. Ki-M4, generated against the moncytic cell line U937 (Parwaresch *et al.*, 1983), also stains a subset of peripheral blood monocytes. HJ2 was generated against human splenic cells (Butch *et al.*, 1994). Bu10/2BF11 was generated against leukemia cells (Johnson *et al.*, 1986). 12B1, generated against the erythroblast cell line K562, also stains adherent monocytes (Farace *et al.*, 1986). We have recently generated a series of monoclonal antibodies against human FDCs by immunizing mice with enriched FDC–B cell clusters or highly purified FDCs obtained by FACS-sorted CD21⁺CD14⁺ cells (Liu, unpublished observations). One of the antibodies, 7D6, was shown to stain FDCs only in spleens and tonsils without any other background stainings (see Fig. 3). It does not stain any cells in human thymus, fetal liver, bone marrow, or peripheral blood. When screened on more than 30 different human cell lines, 7D6 appears to be different from other antibodies such as DRC-1, Ki-M4, CD54, CD21, CD23, and VCAM. Finally, this antibody does not stain CD21 or CD23 transfected mouse L cells.

A rat anti-mouse monoclonal antibody, FDC-M1, developed by immunizing the mice with enriched FDC–B cell clusters stains both FDCs and tingible body macrophages (Kosco and Gray, 1992). In rats one monoclonal antibody, ED5, developed by immunizing mice with cells from rat spleen and Peyer's patches, stains a subset of FDCs in the light zone of GC (Jeurissen and Dijkstra, 1986). The molecular identification of FDC-specific antigens may ultimately permit us to understand the functions of FDCs.

B. FDC Cell Lines

Due to the technical difficulties encountered in isolating highly purified FDCs, the generation of functional FDC cell lines has become an alternative approach. Thus far, five reports have described the generation of FDC-like cell lines from purified human FDC–lymphocyte clusters:

1. Tsunoda *et al.* (1990) and Kim *et al.* (1994) have generated two comparable FDC-like cell lines by culturing enriched tonsillar FDCs in culture medium without exogenous cytokines. These cell lines are adherent fibroblastic cell lines that are stained with the anti-human fibroblast antibody 1B10. Such lines lose the FDC markers DRC-1, CD21, and CD23 very rapidly during establishment. Although the HK cell line shares some of the morphological and phenotypic characteristics of fibroblasts, it displays some unique phenotypic and functional features: (a) It expresses HJ2 (an FDC-specific marker), VCAM-1, and CD14 in contrast with either skin fibroblasts or fetal lung fibroblasts. (b) It produces 20-fold more IL-6 than do typical fibroblast cell lines in response to LPS. (c) Anti-CD3-activated T cells promote the growth of the HK cell line but not the growth of fibroblast cell lines. Such growth is blocked by soluble CD40–Ig chimeric molecules, indicating the involvement of CD40/CD40–ligand interaction. (d) The HK line, unlike other fibroblast cell lines, promotes human B cell growth together with anti-Ig, LPS, IL-4, and anti-CD40.

2. Another FDC-like cell line has been generated by Clark *et al.* (1992), who cultured enriched tonsillar FDCs with GM-CSF. This line has morphological and phenotypic features comparable to the preceding two FDC cell lines generated in the absence of cytokines. IFNγ induces this line to secrete large amounts of IL-6, whereas IL-4 up-regulates VCAM-1 expression, which suggests that T cells may play an important role in modulating FDC functions. B cells adhere to this FDC-like cell line via a VLA-4–dependent mechanism and are induced to proliferate and secrete Ig when CD40 triggering is provided.

3. An EBV-transformed FDC-like cell line was generated by Lindhout *et al.* (1994). This fibroblastic adherent cell line was shown to protect human GC B cells from apoptosis but to inhibit B cell proliferation induced by anti-IgM plus IL-2 or SAC+IL-2.

4. Another immortalized FDC-like cell line (FDC-H1) has been generated by fusion of enriched tonsillar FDCs with the mouse myeloma cell line SP2/0-Ag 14. The FDC-H1 cell line has unlimited growth potential and expresses the FDC-specific antigen KiM4. RT-PCR analysis indicates that the FDC-H1 cell line contains mRNAs for IL-1α, IL-6, and CD23 (Orscheschek *et al.*, 1994).

In summary, five FDC-like cell lines have been generated that represent important tools for studying the biology of FDCs. The similarities shared by fibroblasts, freshly isolated FDCs, and established FDC-like cell lines support the hypothesis that FDCs may be of fibroblastic reticular stromal origin. However, since all these FDCs cell lines have been derived from enriched cell populations containing less than 30% FDCs, the FDC origin of these cell lines remains to be firmly established.

VI. The Origin of FDCs

There are two schools of thought on the issue of the origin of FDCs. One school says that FDCs are of nonhematopoietic origin and are derived from mesenchymal reticular fibroblasts, for the following reasons: (i) Studies on the ontogenetic development of FDCs with the capacity to trap antigen have concluded that FDCs are dervied from follicular mesenchymal reticular cells 28 days after birth (Dijkstra *et al.*, 1984). (ii) Transferring mouse allogeneic bone marrow cells or splenic cells into congenic animals or SCID mice has resulted in the generation of FDCs of host origin but not donor origin (Humphrey *et al.*, 1984; Gray *et al.*, 1991; Yoshida *et al.*, 1995). In addition, transplantation of slices of spleen from one strain of mice into the host mice resulted in the population of host B cells into the donor FDC networks within the splenic transplant; host FDCs and donor FDCs did not migrate into the other site (Imazeki *et al.*, 1992). (iii) Freshly isolated human tonsillar FDCs and established FDC-like cell line (HK) are adherent fibroblastic cells, expressing fibroblast marker 1B10 (Kim *et al.*, 1994) and the cytoskeleton component vimentin (Tsunoda *et al.*, 1990). (iv) Human tonsillar FDCs have been shown by single-cell PCR not to express the leukocyte common antigen CD45 (Schriever *et al.*, 1991).

An opposite theory proposes that FDCs derive from hematopoietic progenitor, from the following observations: (i) The mouse anti-human FDC monoclonal antibody KiM4 was raised against the human monocytic cell line U937 (Parwaresch *et al.*, 1983). This antibody also recognizes a subset of peroxidase positive monocytes from peripheral blood. In addition, a monocytic cell line, THP-1, was induced to express KiM4 in response to soluble factors produced by activated T cells (Fliedner *et al.*, 1990). (ii) Administration of rat fetal liver or bone marrow, or mouse mismatched MHC class I bone marrow, to SCID mice resulted in the generation of FDCs with the donor phenotype (Kapasi *et al.*, 1993). (iii) Ultrastructural studies *in vivo* showed that antigen-transporting cells appear in the subcapsular sinus and then migrate into the follicular center and develop into FDCs. These antigen-transporting cells have a monocyte morphology and react with monoclonal antibody specific for mouse FDCs (FDC-M1) (Kapasi *et al.*, 1993).

The differences between these two theories may not be settled before the following issues are solved and considered: (i) Cells of hematopoietic origin and mesenchymal origin may share the expression of certain molecules. For example, the CD40 antigen and ICAM-1/CD54 have been shown to be expressed on a wide range of cell types, including lymphocytes, dendritic cells, fibroblasts, and carcinoma cells (Banchereau *et al.*, 1994). (ii) FDCs may passively capture many molecules from their surrounding cells, as described for CD45 (Schriever *et al.*, 1991) and MHC class II (Gray *et al.*, 1991). (iii) Two types of cells of different origin but with dendritic morphology may conexist within the follicles. In this context we describe in Section VIII the identification of a novel type of dendritic cells from human GCs. These cells, which express CD4, CD11C, and CD45, have the characteristics of monocytic dendritic cells and may correspond to the antigen-transporting cells described by Szakal (1989). These cells can be isolated independently from $CD14^+CD21^+$ FDCs that appear to display characteristics of fibroblasts.

VII. *In Vitro* Study of FDC–B Cell Interactions

A. FDCs Improve B Cell Survival and Promote B Cell Proliferation

GCs are characterized by a high rate of B cell proliferation within the FDC network. To determine whether FDCs directly contribute to B cell proliferation, isolated FDCs and B cells have been cultured *in vitro*. Schnizlein *et al.* (1985) demonstrated that addition of FDC-enriched preparations markedly enhanced proliferation of polyclonally stimulated mouse B cells. This finding was confirmed in the human system, where human tonsillar FDCs were shown to augment the proliferation of B cells stimulated by anti-IgD and PMA (Cormann *et al.*, 1986). Further experiments showed that depleting FDCs with a specific antibody FDC-M1, abolished the increase of B cell proliferation. Our own experiments using enriched human FDC clusters also confirm and further extend the stimulatory role of FDC on B cells earlier obtained in the mouse system. In particular, removal of FDCs from FDC clusters with the FDC-specific monoclonal antibody 7D6 also abolishes FDC-dependent B cell proliferation (Grouard *et al.*, 1995). The proliferating B cells are always found in close contact with FDCs (Petrasch *et al.*, 1991). The adhesion of B cells onto FDCs is mediated by adhesion molecules VLA-4 and VCAM-1 as well as LFA-1 and ICAM-1 (Freedman *et al.*, 1990; Koopman *et al.*, 1991; Kosco, 1991). Experiments on FDC–B cell coculture also show that FDCs can protect GC B cells from apoptotic cell death (Cormann *et al.*, 1986; Kosco and Gray, 1992; Lindhout *et al.*, 1993).

In contrast, FDCs have also been reported to inhibit SAC-dependent B cell proliferation (Freedman *et al.*, 1992), an observation further confirmed with EBV-transformed FDC cell lines (Lindhout *et al.*, 1994). In line with such results, we found that FDC clusters could weakly inhibit SAC-activated B cell proliferation in two experiments but weakly stimulated SAC-activated B cells in two other experiments (Grouard *et al.*, 1995). The significance of these discordant results remains to be elucidated, but it is unlikely to be of critical significance when antigen-dependent, T cell–dependent B cell responses are considered, because SAC may best mimic T cell–independent responses.

B. FDCs Enhance the Proliferation of B Cells Induced by Cytokines and CD40 Ligation

Because GCs contain cytokine-producing cells such as tingible body macrophages and T cells, a wide range of cytokines was analyzed for their effects on the proliferation of B cells cultured in the presence of FDCs (Grouard *et al.*, 1995). Among IL-1α, IL-2, 3, 4, 5, 6, 7, 9, 10, 13, GM-CSF, and IFNγ, only IL-2 was shown to induce a consistent and significant augmentation of B cell proliferation in B cell–FDCs cocultures.

The CD40/CD40–ligand interaction plays a key role in T–B cell interactions (Banchereau *et al.*, 1994; Clark and Ledbetter, 1994), and activated T cells that express CD40–ligand can be identified *in situ* within tonsil GCs (Lederman *et al.*, 1992; Casamayor-Palleja *et al.*, 1995). Mutations in the CD40–ligand gene locus result in the development of the hyper-IgM syndrome (Callard *et al.*, 1993), a disease characterized by the lack of memory B cells and the absence of GCs and FDCs in peripheral lymphoid tissues (Facchetti *et al.*, 1995). Further studies in mice have indicated that blocking CD40/CD40–ligand interaction *in vivo* by injecting soluble CD40 protein (Gray *et al.*, 1994) or anti–CD40–ligand antibody (Foy *et al.*, 1994) or by gene targeting (Kawabe *et al.*, 1994; Renshaw *et al.*, 1994; Xu *et al.*, 1994) abolished memory B cell development and GC reactions during T cell–dependent antibody responses. Taken together these observations indicate that CD40/CD40–ligand interactions play a key role in GC reactions. Accordingly, addition of anti-CD40 antibodies to the FDC–B cell cocultures results in greatly increased B cell proliferation. Maximal B cell proliferation was obtained when CD40-triggering +IL-2+IL-10+ FDCs or CD40-jtriggering +IL-4+IL-10+ FDCs were provided. However, only CD40-triggering +IL-2+IL-10+ FDCs induce high levels of Ig production (Grouard *et al.*, 1995). The preceding experiments suggest that both FDCs and GC T cells are required for B cell proliferation and differentiation within GCs (Fig. 4).

VIII. Identification of a New Dendritic Cell Type with Germinal Centers: Germinal Center Dendritic Cells

Many observations *in vivo* and *in vitro* suggest that both GC T cells and follicular dendritic cells are essential for strong B cell proliferation. It has therefore been puzzling why GC T cells and follicular dendritic cells are localized essentially in the light zone of GC, whereas the major proliferation activity of GC B cells is in the dark zone. Several possibilities may account for this fact, including (i) the intensive B cell proliferation in the GC dark zone, which pushes away the helper T cells and FDCs; (ii) the B cell proliferation, which is autonomous (independent of any other cell types); and (iii) an unidentified cell type, which may contribute to the B cell proliferation in the dark zone. As evidence for (iii), immunohistological analysis of a wide range of antigens recognized by monoclonal antibodies on human tonsil sections allowed us to identify a large CD3$^-$CD4$^+$ cell type within GCs (Grouard *et al.*, unpublished observation, 1995). Double immunoenzymatic stainings showed that these CD3$^-$CD4$^+$ cells are CD1a$^-$, CD40^{low}B7^{low}, and DRC-1$^-$, which indicates that these cells are not Langerhans cells (CD1a$^+$), or interdigitating cells (CD40^{high}, B7.1^{high}, B7.2^{high}), or FDCs (DRC-1$^+$, KIM4$^+$). These cells can be isolated from tonsillar mononuclear cells after magnetic bead depletion of CD20$^+$ and CD3$^+$ cells, followed by three-color FACS sorting of CD4$^+$CD11C$^+$CD3$^-$CD20$^-$CD1a$^-$ cells. The isolated cells are characterized by strong MHC class II expression and a morphology of monocyte–dendritic cells. Preliminary experiments showed that these cells are strong antigen-presenting cells to CD4$^+$ allogeneic T cells. The function of this cell type in GC B cell proliferation is currently under investigation.

IX. FDCs and Follicular Lymphomas

Follicular lymphomas represent the most common human lymphoma, characterized by nodular proliferation of neoplastic B cells that are of GC origin (Stein *et al.*, 1982). According to the morphologic criteria, follicular lymphomas are classified into cleaved centrocytic lymphomas and noncleaved centroblastic lymphomas or mixed centrocytic–centroblastic lymphomas (Lennert, 1978). The most striking genetic changes that occur in most follicular lymphomas is the chromosome translocations involving the immunoglobulin genes (on chromosomes 14, 18, or 2) and the bcl-2 gene (on chromosome 18) that generate a constitutively expressed bcl-2–immunoglobulin fusion gene (Tsujimoto and Croce, 1986). Thus, unlike normal GC B cells, follicular lymphoma cells express Bcl-2 protein and fail to undergo physiological cell death (apoptosis) (Pezella *et al.*, 1990; Hollowood and Macartney, 1991; Liu *et al.*, 1991a).

FDCs are also found in follicular lymphomas, forming a dense dendritic network in close association with malignant B cells, as in normal GCs (Stein *et al.*, 1982). Phenotypic analysis of FDCs *in situ* or in isolated form suggests that FDCs within follicular lymphomas have a phenotype similar to those from normal GCs (Stein *et al.*, 1982; Petrasch *et al.*, 1990). The adhesions between FDCs and B cells from either nonmalignant tissues or from malignant tissue are mediated by VLA-4 and VCAM-1 (Freedman *et al.*, 1992). *In vitro* culture of isolated FDC–lymphocyte clusters derived from follicular lymphoma suggests that FDCs form a microenvironment favorable for the growth of follicular lymphoma cells (Petrasch *et al.*, 1991).

Most follicular lymphomas are low-grade lymphomas, which grow within the dense FDC networks. However, 25% of low-grade follicular lymphomas may undergo transformation into large-cell diffuse lymphomas. This malignant progression requires a second genetic hit such as c-myc translocation. In the large-cell diffuse type of lymphoma, no FDCs or only scattered FDCs can be identified (Stein *et al.*, 1982), which suggests that the growth of these malignant B cells may become autonomous and independent of any FDC-dependent microenvironment. Thus, interrupting the FDC–lymphoma cell interaction at an early stage may represent a useful strategy for inhibiting malignant B cell growth before the malignancy progresses further.

FDCs were also recently identified in some T cell lymphomas. The expression of CD21, CD23, VCAM-1, and CD35 on FDCs from T cell lymphomas is similar to that on FDCs from B cell lymphomas (Suzuki *et al.*, 1995). Direct evidence for T lymphoma cell–FDC interactions is still lacking. However, experiments *in vivo* suggest that the development of FDCs in SCID mice or in irradiated animals requires the presence of both B cells and T cells (Kapasi et al., 1993; Yoshida et al., 1995). In addition, interactions between the FDC-like cell line and T cells have been demonstrated in vitro and have been shown to involve CD40/ CD40–ligand interactions (Kim et al., 1994). The detailed analysis of FDC–B lymphoma cell interactions as well as FDC–T lymphoma cell interactions may ultimately permit us to understand the microenvironmental control of malignant cell growth.

X. FDC and HIV Infection

A. FDCs Are Major Reservoirs of HIV

Pathological examination of sequential lymph node biopsies from HIV patients in the early 1980s revealed two remarkable stages of generalized lymphadenopathy: (a) follicular hyperplasia and (b) follicular atrophy and fibrosis (Racz et al., 1991). Soon after the isolation of the T lymphotropic retrovirus (HIV) from AIDS patients (Barre-Sinoussi et al., 1983; Levy et al., 1984; Popvic et al., 1984), retrovirus particles were observed by electron microscopy to be closely associated with FDCs in the lymph nodes of AIDS patients (Armstrong and Horne, 1984; Tenner-Racz et al., 1985). The association of HIV with FDCs within GCs was confirmed by immunohistology using antibodies against HIV–core protein or envelope protein and by in situ hybridization detection of mRNA encoding the HIV–envelope protein (Emilie et al., 1990; Fox et al., 1991; Spiegel et al., 1992). A series of systemic and quantitative studies on viral burden and viral replication in peripheral blood and lymph nodes of AIDS patients at different stages of disease (Pantaleo et al., 1994) established that FDCs, within the GCs of the secondary lymphoid organs, are major reservoirs of HIV. Such findings should contribute to our future understanding of (i) the pathological progression of HIV-associated disease, (ii) the mechanism of HIV interaction with the host immune system, and (iii) ways of designing a therapeutic intervention aimed at controlling HIV infections.

B. How Are HIV Particles Localized on FDCs within Germinal Centers?

A cardinal feature of FDCs is their capacity to trap antigen–antibody complexes through complement receptors (CR1, CR2, and CR3) and IgFc receptors. Thus, antibody-coated HIV-1 can bind to the FDC surface through these receptors and directly activate the human complement system via the classical pathway (Dierich *et al.*, 1993). HIV-infected cells can also activate the human complement system via the alternative pathway. Thus, HIV-infected cells and free HIV particles both have the ability to bind to FDCs via complement fragments–complement receptors in the absence of antibodies.

C. What Are the Pathological Consequences of HIV Trapping on FDCs within Germinal Centers?

1. Restriction of B Cell Repertoire

HIV on FDCs may lead to clonal overexpansion of HIV-specific B cells (Grimaldi *et al.*, 1988; Amadori *et al.*, 1990; D'Amelio *et al.*, 1992; Müller *et al.*, 1992), thereby restricting considerably the diversity of the B cell repertoire. Detailed immunohistological studies of lymph nodes from HIV-infected patients indicated that follicular hyperplasia is caused mainly by the expansion of the proliferating centroblasts and of the follicular dendritic cell networks (Janossy *et al.*, 1991). In addition, a decrease in the number of $CD4^+CD57^+CD45RO^+$ GC T cells is observed that is followed by an infiltration of many $CD8^+CD45RO^+$ T cells into the GCs (Janossy *et al.*, 1991). Thus, the primary signals for increased B cell expansion within GCs may be derived from the expanded FDC networks coated with HIV particles. A recent study suggests that the gp120 protein behaves like a B cell superantigen, which stimulates IgVH3-expressing B cells to undergo early clonal expansion (follicular hyperplasia) and then clonal deletion by apoptosis (follicular atrophy) (Berberian *et al.*, 1993). Thus, the initial VH3 clonal dominance in B cell repertoire may prevent the effective triggering of other clones that have the potential to respond to HIV mutants and other infectious microorganisms.

2. HIV on FDCs May Become More Infectious

As described in Section X,B, HIV may bind to FDCs through three mechanisms: (i) HIV–antibody–FcR; (ii) HIV–antibody–complement–complement receptors; and (iii) HIV–complement–complement receptors.

Normally, FDCs express high levels of CD59 (Butch *et al.*, 1994), a protein that protects cells from complement-mediated lysis. CD59 molecules have also been found on the surface of HIV particles, which may thus protect HIV from complement-mediated lysis (Davies *et al.*, 1994). Thus, following binding to the FDC surface, HIV particles may capture CD59 from FDCs and bind antibody and complement fragments that ultimately allows them a more efficient entry into their target cells.

Although antibodies and complement fragments have traditionally been considered as protective mechanisms in infectious diseases, recent data have set off an alarm indicating that neutralizing antibodies and complement may enhance HIV infection under certain circumstances (Takeda et al., 1988; Bakker et al., 1992; van de Wiel et al., 1994). In the absence of antibody, normal human serum enhances the capacity of free HIV-1 particles to infect cells that express complement receptors. Antibodies against CR3 inhibited the infection of these cells by HIV (Sölder et al., 1989; Larcher et al., 1990; Boyer et al., 1991; Gras and Dormont, 1991). Consistent with these experiments in vitro, a more rapid decrease in CR2-expressing $CD4^+$ T cells was observed in HIV-1 infected individuals (June et al., 1992). The generation of "enhancing antibodies" by vaccination that may facilitate HIV infection has been suggested by many experiments in vitro (Füst et al., 1994). Furthermore, the appearance of different anti-HIV antibodies has been correlated with a clinical progression or transmission of HIV infection (Füst et al.,1995). The possible mechanism by which antibodies facilitate HIV infection may be activation of the complement system by antibody–virus complexes. As evidence for this mechanism, macaques vaccinated with subunit SIV vaccine, or transfused with polyclonal anti-SIV antibodies, may show enhanced rates of infection and disease progression under certain conditions of virus challenge (Schwartz, 1994).

Finally, it has been demonstrated that FDCs directly isolated from the secondary lymphoid organs of SIV-infected animals or HIV-pulsed animals are highly infectious (Burton et al., 1995).

3. HIV on FDCs Finally Knocks Out Lymphoid Follicles

FDC, a stromal cell uniquely found in the B lymphocyte follicles, seems essential in maintaining the integrity of the lymphoid follicle architecture. In secondary lymphoid organs of HIV-infected patients and of animal models, the initial follicular hyperplasia was shown to be followed by follicular involution (Janossy et al., 1991; Pantaleo et al., 1994). The involution is characterized by the dissolution of the FDC network, which eventually leads to a destruction of the lymph node architecture and an inability to trap extracellular virions. The precise mechanisms through which the FDC network is ultimately destroyed remain to be determined. (i) FDCs may be killed by the infiltrating $CD8^+$ cytotoxic T cells, (ii) although experiments in vitro showed that FDCs may die as a consequence of HIV infection (Stahmer et al., 1991), direct evidence that FDCs are indeed infected in vivo is still lacking (Schmitz et al., 1994). (iii) Because the integrity of the

FDC network appears to be highly dependent on the presence of B cells and T cells (Kapasi *et al.*, 1993; Yoshida *et al.*, 1995), the depletion of $CD4^+$ T cells may result in the loss of important signals for the survival of FDCs. The destruction of the FDC network and the subsequent follicular involution correlate with the clinical progression of HIV disease (Pantaleo *et al.*, 1994).

In summary, HIV associates with the FDC network via complement receptors or Fc receptors, which are normally used by FDCs to trap immune complexes. HIV protects itself from complement-mediated lysis possibly by capturing CD59 molecules from FDCs. HIV facilitates its entry into target cells by activation of the complement system. Thus, HIV appears to use many strategies of the humoral immune system to facilitate its spread and finally destroy the structure of secondary lymphoid organs.

XI. Our Current Views of FDCs and Germinal Centers

1. Initiation of Antigen-Dependent Immune Responses; Extrafollicular Responses

All immune responses against foreign antigens start when antigens are captured by Langerhans cells or dendritic cells beneath the injured body linings (mucosa or skin). These Langerhans cells or dendritic cells process the antigens when they migrate into the secondary lymphoid tissues through the draining afferent lymph. These cells enter the paracortical areas of lymph nodes, where T cells accumulate while B cells are passing by, before entering primary follicles. The antigen-loaded Langerhans cells/dendritic cells, now called *interdigitating cells* (IDC), serve as a bridge that allows the encounter of the rare antigen-specific T cells and B cells. IDCs have a strong capacity for stimulating antigen-specific naive T cells to proliferate, secrete cytokines, and express CD40–ligand. These activated T cells subsequently stimulate antigen-specific naive B cells to proliferate and to form the so-called B cell proliferation foci that generate short-lived plasma cells and GC founder cells (Fig. 5). Isotype switch may occur within the foci in the absence of somatic mutation. Whether IDC directly interact with antigen-specific B cells remains to be determined.

2. Germinal Center Reaction

a. Generation of Centroblasts in the Dark Zone   Germinal center formation starts with the colonization of GC founder cells in the primary follicles and depends on both the transport of antigen onto FDCs and the presence of CD40–ligand–expressing T cells. Our hypothesis is that antigen-specific

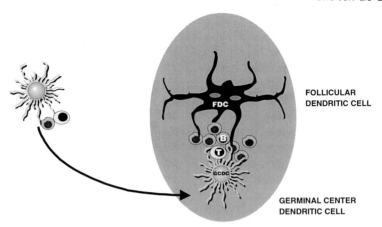

FOLLICULAR
DENDRITIC CELL

GERMINAL CENTER
DENDRITIC CELL

FIG. 5 Hypothetical model for the induction of a germinal center. An interdigitating cell primes an antigen-specific T cell and forms a bridge to bring an antigen-specific B cell into contact with an antigen-specific T cell. The activated B cells proliferate and generate the so-called foci. In addition to generating low-affinity plasma cells within the foci, some activated B cells migrate into a primary follicle together with interdigitating cells and T cells. The T cells become germinal center T cells, and interdigitating cells become germinal center dendritic cells (GCDC). The interplay among the four cell types (FDC, B cell, T cell, and GCDC) leads to the formation of a germinal center.

T cells and antigen-transporting dendritic cells (IDC) may migrate into the primary follicles together with GC founder B cells, forming a cell cluster consisting of GC founder B cells, antigen-specific T cells, and antigen-transporting dendritic cells (Fig. 5). The antigen-transporting cells described by Tew and Szakal (Tew *et al.,* 1980) in the mouse system may correspond to the $CD3^-CD4^+CD11C^+$ GC dendritic cells in human tonsils (Fig. 3).

The appearance of a classic GC is initiated by the unequal division of blasts, giving rise to a proliferating dark zone (which contains centroblasts) and a nonproliferating light zone (which contains centrocytes) (Fig. 6). The relatively low numbers of T cells in the dark zone compared with that in the light zone should not be considered as an indication that the proliferation and somatic mutation of centroblasts are T cell–independent. It may just reflect the dominant B cell proliferation that pushes away the helper T cells into the light zone. Indeed, a recent experiment *in vivo* suggests that the degree of somatic mutation is quantitatively dependent on the number of available antigen-specific T helper cells (Miller *et al.,* 1995). Lebecque and his colleagues in our laboratory have recently succeeded in inducing naive B cells to mutate their VH genes on *in vitro* culturing over anti-CD3 activated T cells (Lebecque, unpublished observation). These data indicate an essential role of T cells in the GC dark zone.

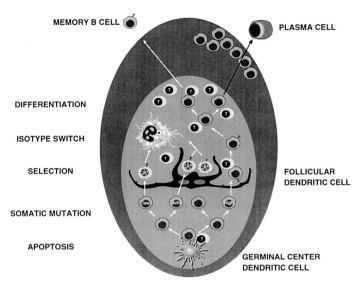

FIG. 6 Current model of affinity maturation within a germinal center. Apoptosis, somatic mutation, antigen-driven selection, isotype switching, and differentiation are sequentially triggered during a germinal center reaction. Apoptosis is switched on as soon as germinal center founder cells colonize the primary follicles. This allows for only those B cells that carry the high-affinity germ line IgV genes to undergo clonal expansion to form the dark zone. Somatic mutation is triggered in these selected proliferating centroblasts, which generate a large number of mutants (centrocytes) that may fall into three categories: (i) low affinity for the immunizing antigen; (ii) high affinity for self-antigen (auto-reactive clones); and (iii) high affinity for the immunizing antigen. Only the high-affinity mutants can be selected by antigens on FDCs. Both low-affinity clones and auto-reactive clones will die by apoptosis (detailed in the text). The high-affinity mutants retrieve the antigen from FDCs, process it, and present it to the GC T cells. During the cognate T–B interaction, T cells are induced to express CD40–ligand and secrete cytokines, which allow B cells to expand and to undergo isotype switching. The high-affinity isotype-switched B cells then differentiate into memory B cells in the presence of a CD40–ligand signal, or into plasma cells in the absence of CD40–ligand signaling.

The mechanism that limits the proliferation of activated GC T cells is unknown. It is likely that germinal center dendritic cells (GCDC) may turn on a T cell activation status that limits T cell growth but allows them to signal B cells to proliferate and mutate.

b. Selection of Centrocytes in the Light Zone Centrocytes expressing mutated antigen receptors are subsequently continuously generated from centroblasts to fill in the light zones. The fate of these mutated centrocytes depends on the specificity and affinity of the mutated antigen receptors (Fig. 6).

i. Centrocytes Display High-Affinity Antigen Receptors against Immunizing Antigens The centrocytes with high-affinity B cell antigen receptor (BCR) retrieve the antigen from the immune complexes bound to FDC. The FDC–B cell interaction provides a survival signal to the antigen-specific B cell, which is delivered through the triggering of the specific antigen receptor and the engagement of adhesion molecules such as VCAM-1 and LFA-1. B cells process the retrieved antigen and then present it to the antigen-specific T cells (Kosco *et al.*, 1989; Gray *et al.*, 1991).

This T cell–B cell dialogue permits the expansion of the uniquely produced centrocyte that expresses high-affinity BCR. This dialogue involves CD40/CD40-ligand interactions as shown by the presence of CD40-ligand interactions as shown by the presence of CD40-ligand positive T cells within germinal centers (Lederman *et al.*, 1992; Casamayor-Palleja *et al.*, 1995). The importance of this CD40/CD40-ligand interaction has also been demonstrated *in vivo* by the administration of soluble CD40 (a weak antagonist to CD40-ligand), which results in the blocking of the generation of B cell memory while leaving the formation of germinal centers unaffected (Gray *et al.*, 1994). The CD40/CD40-ligand interaction is most likely to be crucial at that level in turning on the isotype switch machinery which specificity is given by cytokines such as IL-4/IL-13, IL-13, or TGFβ. In fact, it is economical for the immune system to set up isotype switching after selection of high-affinity mutants rather than at the same time (e.g., in the dark zone) or even earlier (e.g., during the extrafollicular reaction). Furthermore, germinal center CD4$^+$ T cells isolated according to CD57 expression display IL-4 and IL-10, which represent switch factors toward IgG$_4$, IgE, IgG$_1$, IgG$_3$, and IgA (Butch *et al.*, 1993).

Interactions between T cells and FDCs are likely to occur at this stage. In particular, FDCs may be induced to proliferate, as indicated by the considerably denser FDC network in secondary follicles in contrast with those in primary follicles and by the existence of Ki67$^+$ FDCs in secondary follicles. FDC proliferation may indeed be triggered in response to T cell CD40-ligand signaling inasmuch as FDC-like cell lines proliferate in response to T cells in a CD40-dependent fashion and as fibroblasts are induced to proliferate in response to CD40 ligation (Fries *et al.*, 1995). The proliferating centrocytes mature into memory B cells in response to prolonged CD40 triggering and into plasma blasts when CD40 signaling is limiting and when IL-10 is available (Arpin *et al.*, 1995). In all cases the high-affinity B cell clone expansion is limited because prolonged contact between the antigen-specific T cell and the antigen-specific B cell results in cell death as a consequence of protracted Fas/Fas-L interactions (Garrone *et al.*, 1995). In fact, such a mechanism prevents the possibility that immune responses

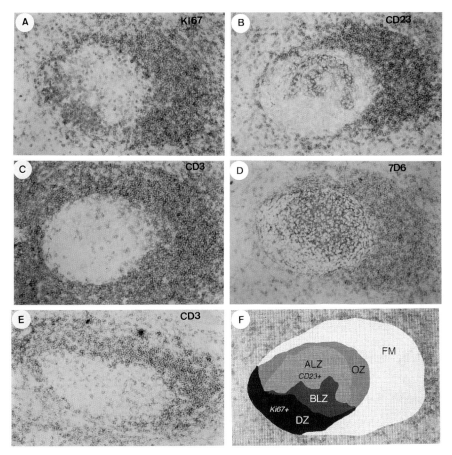

FIG. 1 Germinal center structure. The brown anti-IgD peroxidase staining shows the follicular mantles in A to E. The blue anti-Ki67 alkaline phosphatase staining shows the proliferating centroblasts in the dark zone (DZ) of a germinal center (A and F). The blue anti-CD23 staining shows the CD23$^+$ FDCs in the apical light zone (ALZ) of the same germinal center (B and F). The area between the dark zone and the apical light zone is the basal light zone (BLZ), containing CD23$^-$7D6$^+$ FDCs (F). The area between the apical light zone and the follicular mantle is the outer zone (OZ), containing many T cells, CD23$^-$7D6$^+$ FDCs, and many Ki67$^+$ B blasts (B and F). The blue anti-7D6 staining shows the entire FDC network within a germinal center (D), (anti-7D6 mAb is specific for human FDCs). The blue anti-CD3 staining shows that germinal center T cells are mainly accumulated in the outer zone and apical light zone in two germinal centers (C and E); a few T cells can be found within the proliferating dark zone. F is the schematic drawing of the different zones within a germinal center shown in A, B, C, and D. E represents another follicle that has been chosen for its typical expression of T cells in the outer zone.

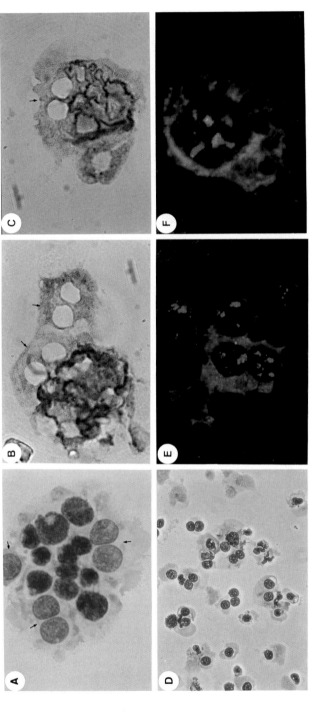

FIG. 3 Isolated human follicular dendritic cells. (A) Giemsa staining of a typical FDC–lymphocyte cluster, containing 3 FDCs (arrows) and 11 lymphocytes. (B) Staining of an FDC–lymphocyte cluster with 7D6, an anti-FDC specific monoclonal antibody (alkaline phosphatase staining). (C) Staining of an FDC–lymphocyte cluster with an anti-CD23 antibody. (D) Giemsa staining of a highly purified single FDC preparation isolated by FACS sorting of CD14⁺CD21⁺ low-density tonsillar cells. Note that many FDCs have multiple nuclei, and many have damaged cytoplasm. (E and F) Double immunofluorescence staining shows two BU10⁺ FDCs (green) that express the proliferation nuclear antigen Ki67 (red). (These two confocal images were generated in the Department of Immunology, The Medical School, University of Birmingham, UK, with kind help from Dr. G. D. Johnson and Professor Ian C. M. MacLennan.)

may be dominated by only a few B cell clones that would prevent the building of a repertoire vast enough to match the extreme diversity of potentially harmful antigens. Yet, it is interesting to note that BCR-triggered B cells are less sensitive to Fas/Fas-L–dependent B cell death than are B cells that have not engaged their BCR (Rothstein *et al.*, 1995; Garrone, unpublished observations). Thus, antigen-triggered B cells would have a survival advantage over non-antigen-triggered B cells that may interact with the activated T cells in a noncognate fashion.

ii. Centrocytes Display Low-Affinity Antigen Receptors against Immunizing Antigens Such cells receive no signals, either through their BCR or from their CD40 antigen, as no antigen-specific T cell is encountered. We propose that these B cells are eliminated by entering into apoptosis either spontaneously or following noncognate interaction with activated T cells. These latter cells may express Fas-L and thus may represent an ultimate selection gate to kill nonselected B cells in a Fas/Fas-L–dependent interaction.

iii. Centrocytes Display Antigen Receptors That Acquired an Affinity for Autoantigens A very large repertoire of autoantigens is likely to be present within the GC microenvironment. The centrocytes that have acquired BCR binding to such antigens undergo protracted BCR triggering in the absence of specific T cell help (autoreactive T cells are normally deleted in the thymus). As demonstrated *in vitro,* these cells enter apoptosis. Recent studies *in vivo* have indeed demonstrated that injection of high doses of soluble antigen induces apoptosis (and phagocytosis by tingible body macrophages) of antigen-specific B cells within GCs (Pulendran *et al.*, 1995; Shokat and Goodnow, 1995). It is not clear yet whether the autoantigen is presented specifically by FDC or by any other cell type. The B cells that escape this selection stage (e.g., by interacting in a noncognate fashion with activated T cells that express CD40-ligand and secrete IL-4) will be rescued and, thus, will represent the source of high-affinity autoantibodies. However, such an event is unlikely to result in autoimmunity as long as autoreactive T cells, necessary for launching a secondary humoral response, do not escape thymic selection.

XII. Future Prospects

The destruction of FDC networks and subsequent involution of GCs observed in HIV patients provides the best evidence that FDCs play an essential role to the integrity of the GC structure and functions. Retaining antigen–antibody immune complexes seems to be a major function of FDCs, which confer antigen specificity to GC reactions and provide the selection mechanism for affinity maturation and for maintaining the long-term memory B cell clones or T cell clones and serum antibody levels. However, detailed molecular mechanisms of how FDCs, either in addition to or in collaboration with other cell types within the GCs, contribute to the high rate of B cell proliferation, survival, somatic mutation, isotype switching and differentiation are still fragmentary. Improved techniques of obtaining highly purified healthy FDCs and the application of various molecular cloning techniques (PCR display, subtraction libraries, and expression cloning using monoclonal antibodies) are expected to allow the isolation of FDC-specific functional molecules in the near future.

In addition to the key role of FDCs in GC B cell proliferation and differentiation, data both *in vivo* and *in vitro* suggest the involvement of signals such CD40–ligand and cytokines provided by GC T cells. The observation that CD40 triggering together with IL-4 or IL-10 induces isotype switching but not somatic mutation in sIgD$^+$ naive B cells (Galibert *et al.,* 1995) suggests that another key molecule (somatic mutation trigger, whose identification remains a tantalizing objective) may play an important role in the GC reaction.

Reconstituting a GC *in vitro* has not been accomplished to date but this may now represent an achievable goal that will be required to elucidate the temporal interplay between at least four cell types: B cells, T cells, GCDCs, and FDCs.

Acknowledgments

The authors wish to thank Daniel Lepot for his invaluable help in drawing the figures, and Sandrine Bonnet-Arnaud, Nicole Courbière, and Muriel Vatan for their excellent editorial help.

References

Amadori, A., Gallo, P., Zamarchi, R., Veronese, M. L., De Rossi, A., Wolf, D., and Chieco-Bianchi, L. (1990). IgG oligoclonal bands in sera of HIV-1 infected patients are mainly directed against HIV-1 determinants. *AIDS Res. Hum. Retroviruses* **6,** 581–586.

Armstrong, J. A., and Horne, R. (1984). Follicular dendritic cells and virus-like particles in AIDS-related lymphadenopathy. *Lancet* **II,** 370–372.

Arpin, C., Dechanet, J., van Kooten, K., Merville, P., Grouard, G., Brière, F., Banchereau, J., and Liu, Y.-J. (1995). *In vitro* generation of memory B cells and plasma cells. *Science* **268,** 720–722.

Bakker, L. J., Nottet, H. S., de Vos, N. M., de Graaf, L., van Strijp, J. A., Visser, M. R., and Verhoef, J. (1992). Antibodies and complement enhance binding and uptake of HIV-1 by human monocytes. *AIDS* **6,** 35–41.

Banchereau, J., Bazan, F., Blanchard, D., Brière, F., Galizzi, J.-P. van Kooten, C., Liu, Y.-J., Rousset, F., and Saeland, S. (1994). The CD40 antigen and its ligand. *Ann. Rev. Immunol.* **12**, 881–922.

Banchereau, J., de Paoli, P., Vallé, A., Garcia, E., and Rousset, F. (1991). Long term human B cell lines dependent on interleukin 4 and antibody to CD40. *Science* **251**, 70–72.

Barre-Sinoussi, F., Chermann, J. C., Rey, F., Nugeyre, M. T., Chamaret, S., Gruest, J., Dauguet, C., Axler-Blin, C., Vezinet-Brun, F., Rouzioux, C., Rosenbaum, W., and Montagnier, L. (1983). Isolation of a T-lymphotropic retrovirus from a patient at risk for acquired immune deficiency syndrome (AIDS). *Science* **220**, 868–871.

Berberian, L., Goodglick, L., Kipps, T. J., and Braun, J. (1993). Immunoglobulin VH3 gene products: Natural ligands for HIV gp120. *Science* **261**, 1588–1591.

Berek, C., Berger, A., and Apel, M. (1991). Maturation of immune responses in germinal centers. *Cell* **67**, 1121–1129.

Berek, C., and Milstein, C. (1988). The dynamic nature of the antibody repertoire. *Immunol. Rev.* **105**, 1–26.

Berman, M. A., Rafiei, S., and Gutman, G. A. (1981). Association of T cells with proliferating cells in lymphoid follicles. *Transplantation* **32**, 426–430.

Bloch-Shtacher, N., Hirschhorn, K., and Uhr, J. W. (1968). The response of lymphocytes from non-immunized humans to antigen–antibody complexes. *Clin. Exp. Immunol.* **3**, 889–899.

Bowen, M. B., Butch, A. W., Parvin, C. A., Levine, A., and Nahm, M. H. (1991). Germinal center T cells are distinct helper–inducer T cells. *Human Immunol.* **31**, 67–75.

Boyer, V., Desgranges, C., Trabaud, M. A., Fischer, E., and Kazatchkine, M. D. (1991). Complement mediates human immunodeficiency virus type 1 infection of a human T cell line in a CD4- and antibody-independent fashion. *J. Exp. Med.* **173**, 1151–1158.

Britton, S., and Moller, G. (1968). Regulation of antibody synthesis against *Escherichia coli* endotoxin. I. Suppressive effect of endogenously produced and passively transferred antibodies. *J. Immunol.* **100**, 1326–1334.

Burnet, M. F. (1959). "The Clonal Selection Theory of Acquired Immunity." Cambridge University Press, Cambridge.

Burton, G. F., Heath, S. L., Tew, J. G., and Tew, J. G. (1995). HIV on follicular dendritic cells (FDC) is highly infectious. *J. Cell. Biochem. Suppl.* **21A** (*Abst.* **C1-401**), 33.

Butch, A. W., Chung, G.-H., Hoffmann, J. W., and Nahm, M. H. (1993). Cytokine expression by germinal center cells. *J. Immunol.* **150**, 39–47.

Butch, A. W., Hug, B. A., and Hahm, M. H. (1994). Properties of human follicular dendritic cells purified with HJ2, a new monoclonal antibody. *Cell. Immunol.* **155**, 27–41.

Bystryn, J. C., Graf, M. W., and Uhr, J. W. (1970). Regulation of antibody formation by serum antibody. II. Removal of specific antibody by means of exchange transfusion. *J. Exp. Med.* **132**, 1279–1287.

Cahen-Kramer, Y., Martensson, I. L., and Melchers, F. (1994). The structure of an alternate form of complement C3 that displays costimulatory growth factor activity for B lymphocytes. *J. Exp. Med.* **180**, 2079–2088.

Callard, R. E., Armitage, R. J., Fanslow, W. C., and Spriggs, M. K. (1993). CD40 ligand and its role in X-linked hyper-IgM syndrome. *Immunol. Today* **14**, 559–564.

Casamayor-Palleja, M., Khan, M., and MacLennan, I. C. M. (1995). A subset of CD4+ memory T cells contains preformed CD40 Ligand that is rapidly but transiently expressed on their surface after activation through the T cell receptor complex. *J. Exp. Med.* **181**, 1293–1301.

Chen, L. L., Adams, J. C., and Steinman, R. M. (1978). Anatomy of germinal centers in mouse spleen, with special reference to "follicular dendritic cells." *J. Cell Biol.* **77**, 148–164.

Clark, E. A., and Ledbetter, J. A. (1994). How B and T cells talk to each other. *Nature (London)* **367**, 425–428.

Clark, M. R., Campbell, K. S., Kazlauskas, A., Johnson, S. A., Hertz, M., Potter, T. A., Pleiman, C., and Cambier, J. C. (1992). The B cell antigen receptor complex: Association of Ig-α and Ig-β with distinct cytoplasmic effectors. *Science* **258**, 123–126.

Coico, R. F., Bhogal, B. S., and Thorbecke, G. J. (1983). Relationship of germinal centers in lymphoid tissue to immunologic memory. IV. Transfer of B cell memory with lymph node cells fractionated according to their receptors for peanut agglutinin. *J. Immunol.* **131,** 2254–2257.

Cormann, N., Lesage, F., Heinen, E., Schaaf-Lafontaine, N., Kinet-Denoel, C., and Simar, L. J. (1986). Isolation of follicular dendritic cells from human tonsils and adenoids. V. Effect on lymphocyte proliferation and differentiation. *Immunol. Letters* **14,** 29–35.

D'Amelio, R., Biselli, R., Nisini, R., Matricardi, P. M., Aiuti, A., Mezzaroma, I., Pinter, E., Pontesilli, O., and Aiuti, F. (1992). Spectrotype of anti-gp120 antibodies remains stable during the course of HIV disease. *J. Acquir. Immune Defic. Syndr.* **5,** 930–935.

Davies, A., Davis, D., and Lachmann, P. J. (1994). CD59 on the surface of human immunodeficiency virus protects against complement-mediated lysis. *Clin. Exp. Immunol.* **97,** (*Suppl. 2*), 21.

De Sousa, M. A., Parrott, D. M., and Pantelouris, E. M. (1969). The lymphoid tissues in mice with congenital aplasia of the thymus. *Clin. Exp. Immunol.* **4,** 637–644.

Dierich, M. P., Ebenbichler, C. F., Marschang, P., Füst, G., Thielens, N. M., and Arlaud, G. J. (1993). HIV and human complement: Mechanisms of interaction and biological implication. *Immunol. Today* **14,** 435–440.

Dijkstra, C. D., Kamperdijk, E. W., and Dopp, E. A. (1984). The ontogenetic development of the follicular dendritic cell: An ultrastructural study by means of intravenously injected horseradish peroxidase (HRP)–anti-HRP complexes as marker. *Cell Tissue Res.* **236,** 203–206.

Dijkstra, C. D., and Van den Berg, T. K. (1991). The follicular dendritic cell: Possible regulatory roles of associated molecules. *Res. Immunol.* **142,** 227–231.

Donaldson, S. L., Kosco, M. H., Szakal, A. K., and Tew, J. G. (1986). Localization of antibody-forming cells in draining lymphoid organs during long-term maintenance of the antibody response. *J. Leukoc. Biol.* **40,** 147–157.

Dukor, P., Bianco, C., and Nussenzweig, V. (1970). Tissue localization of lymphocytes bearing a membrane receptor for antigen–antibody complement complexes. *Proc. Natl. Acad. Sci. USA* **67,** 991–997.

Emilie, D., Peuchmaur, M., Maillot, M. C., Crevon, M. C., Brousse, N., Delfraissy, J. F., Dormont, J., and Galanaud, P. (1990). Production of interleukins in human immunodeficiency virus-1–replicating lymph nodes. *J. Clin. Invest.* **86,** 148–159.

Ennas, M. G., Chilosi, M., Scarpa, A., Lantini, M. S., Cadeddu, G., and Fiore-Donati, L. (1989). Isolation of multicellular complexes of follicular dendritic cells and lymphocytes: Immunophenotypical characterization, electron microscopy and culture studies. *Cell Tissue Res.* **257,** 9–15.

Facchetti, F., Appiani, Cl., Salvi, L., Levy, J., and Notarangelo, L. D. (1995). Immunohistologic analysis of ineffective CD40–CD40 ligand interaction in lymphoid tissues from patients with X-linked immunodeficiency with hyper-IgM. Abortive germinal center cell reaction and severe depletion of follicular dendritic cells. *J. Immunol.* **154,** 6624–6633.

Farace, F., Keiffer, N., Caillou, B., Vainchenker, W., Tursz, T., and Dokhelar, M. C. (1986). A 39-kDa glycoprotein expressed on human cultured monocytes and dendritic reticulum cells defined by an anti-K562 monoclonal antibody. *Eur. J. Immunol.* **16,** 1521–2526.

Flemming, W. (1885a). Schlussbemerkingen über die zellvermehrung in den lymphoiden drüsen. *Arch. F. Micr. Anat.* **24,** 355.

Flemming, W. (1885b). Studïen über regeneration der gewebe. *Arch. F. Micr. Anat.* **24,** 50.

Fliedner, A., Parwaresch, M. R., and Feller, A. C. (1990). Induction of antigen expression of follicular dendritic cells in a monoblastic cell line: A contribution to its cellular origin. *J. Pathol.* **161,** 71–77.

Ford, J. E., McHeyzer-Williams, M. G., and Lieber, M. R. (1994). Analysis of individual immunoglobulin λ light chain genes amplified from single cells is inconsistent with variable region gene conversion in germinal-center B cell somatic mutation. *Eur. J. Immunol.* **24,** 1816–1822.

Fox, C. H., Tenner-Racz, K., Racz, P., Firpo, P. A., and Fauci, A. S. (1991). Lymphoid germinal centers are reservoirs of human immunodeficiency virus type 1 RNA. *J. Infect. Dis.* **164,** 1051–1057.

Foy, T. M., Laman, J. D., Ledbetter, J. A., Aruffo, A., Claassen, E., and Noelle, R. J. (1994). gp39–CD40 interactions are essential for germinal center formation and the development of B cell memory. *J. Exp. Med.* **180,** 157–163.

Freedman, A. S., Munro, J. M., Rhynhart, K., Schow, P., Daley, J., Lee, N., Svahn, J., Eliseo, L., and Nadler, L. M. (1992). Follicular dendritic cells inhibit human B-lymphocyte proliferation. *Blood* **80,** 1284–1288.

Freedman, A. S., Munro, J. M., Rice, G. E., Bevilacqua, M. P., Morimoto, C., McIntyre, B. W., Rhynhart, K., Pober, J. S., and Nadler, L. M. (1990). Adhesion of human B cells to germinal centers *in vitro* involves VLA-4 and INCAM-110. *Science* **249,** 1030–1033.

Fries, K. M., Sempowski, G. D., Gaspari, A. A., Blieden, T., Looney, R. J., and Phipps, R. P. (1995). CD40 expression by human fibroblasts. *Clin. Immunol. and Immunopathol.* **77,** 42–51.

Fuller, K. A., Kanagawa, O., and Nahm, M. H. (1993). T cells within germinal centers are specific for the immunizing antigen. *J. Immunol.* **151,** 4505–4512.

Füst, G., Toth, F. D., Kiss, J., Ujhelyi, E., Nagy, I., and Banhegyi, D. (1994). Neutralizing and enhancing antibodies measured in complement-restored serum samples from HIV-1–infected individuals correlate with immunosuppression and disease. *AIDS* **8,** 603–609.

Gajl-Peczalska, K. J., Fish, A. J., Meuwissen, H. J., Frommel, D., and Good, R. A. (1969). Localization of immunological complexes fixing beta1C (C3) in germinal centers of lymph nodes. *J. Exp. Med.* **130,** 1367–1393.

Galibert, L., Van Dooren, J., Durand, I., Rousset, F., Jefferis, R., Banchereau, J., and Lebecque, S. (1995). Anti-CD40 plus interleukin-4–activated human naive B cell lines express unmutated immunoglobulin genes with intraclonal heavy chain isotype variability. *Eur. J. Immunol.* **25,** 733–737.

Garrone, P., Neidhardt, E. M., Garcia, E., Galibert, L., van Kooten, C., and Banchereau, J. (1995). Fas ligation induces apoptosis of CD40-activated human B lymphocytes. *J. Exp. Med.* **182,** 1265–1273.

Gerdes, J., Stein, J., Mason, D. Y., and Ziegler, A. (1983). Human dendritic reticulum cells of lymphoid follicles: Their antigenic profile and their identification as multinucleated giant cells. *Virchows Arch. Cell Path.* **42,** 161–172.

Glimstedt, G. (1936). Bacterienfreie meerschweinchen. *Acta Pathol. Microbiol. Scand. Suppl.* **30.**

Gras, G. S., and Dormont, D. (1991). Antibody-dependent and antibody-independent complement-mediated enhancement of human immunodeficiency virus type 1 infection in a human, Epstein-Barr virus–transformed B-lymphocytic cell line. *J. Virol.* **65,** 541–545.

Gray, D. (1988). Recruitment of virgin B cells into an immune response is restricted to activation outside lymphoid follicles. *Immunology* **65,** 73.

Gray, D., Dullforce, P., and Jainandunsing, S. (1994). Memory B cell development but not germinal center formation is impaired by *in vivo* blockade of CD40–CD40 ligand interaction. *J. Exp. Med.* **180,** 141–155.

Gray, D., Kosco, M., and Stockinger, B. (1991). Novel pathways of antigen presentation for the maintenance of memory. *Int. Immunol.* **3,** 141–148.

Gray, D., and Matzinger, P. (1991). T cell memory is short-lived in the absence of antigen. *J. Exp. Med.* **174,** 969–974.

Gray, D., and Skarvall, H. (1988). B-cell memory is short-lived in the absence of antigen. *Nature (London)* **336,** 70–73.

Grimaldi, L. M., Roos, R. P., Devare, S. G., Robey, W. G., Casey, J. M., Gurney, M. E., Apatoff, B. R., and Lazzarin, D. (1988). Restricted heterogeneity of antibody to gp120 and p24 in AIDS. *J. Immunol.* **141,** 114–117.

Grouard, G., de Bouteiller, O., Banchereau, J., and Liu, Y.-J. (1995). Human follicular dendritic cells enhance cytokine dependent growth and differentiation of CD40 activated B cells. *J. Immunol.,* in press.

Gutman, G. A., and Weissman, I. L. (1972). Lymphoid tissue architecture. Experimental analysis of the origin and distribution of T-cells and B-cells. *Immunology* **23,** 465–479.

Hardie, D. L., Johnson, G. D., Khan, M., and MacLennan, I. C. M. (1993). Quantitative analysis of molecules which distinguish functional compartments within germinal centers. *Eur. J. Immunol.* **23,** 997–1004.

Heinen, E., Cormann, N., Braun, M., Kinet-Denoël, C., Vanderschelden, J., and Simar, L. J. (1986). Isolation of follicular dendritic cells from human tonsils and adenoids. *Ann. Inst. Pasteur* **137D,** 369–382.

Hockenberry, D., Zutter, M., Kickey, W., Nahm, M., and Korsmeyer, S. J. (1991). Bcl-2 protein is topographically restricted in tissues characterized by apoptotic cell death. *Proc. Natl. Acad. Sci. U.S.A.* **88,** 6961–6965.

Holder, M. J., Wang, H., Milner, A. E., Casamayor, M., Armitage, R., Spriggs, M. K., Fanslow, W. C., MacLennan, I. C. M., Gregory, C. D., and Gordon, J. (1993). Suppression of apoptosis in normal and neoplastic human B lymphocytes by CD40 ligand is independent of Bcl-2 induction. *Eur. J. Immunol.* **23,** 2368–2371.

Hollowood, K., and Macartney, J. C. (1991). Reduced apoptotic cell death in follicular lymphoma. *J. Pathol.* **163,** 337–342.

Humphrey, J. H., and Frank, M. M. (1967). The localization of nonmicrobial antigens in the draining lymph nodes of tolerant, normal and primed rabbits. *Immunology* **13,** 87–100.

Humphrey, J. H., and Grennan, D. (1982). Isolation and properties of spleen follicular dendritic cells. *Adv. Exp. Med. Biol.* **149,** 823–827.

Humphrey, J. H., Grennan, D., and Sundaram, V. (1984). The origin of follicular dendritic cells in the mouse and the mechanism of trapping of immune complexes on them. *Eur. J. Immunol.* **14,** 859–864.

Imazeki, N., Senoo, A., and Fuse, Y. (1992). Is the follicular dendritic cell a primarily stationary cell? *Immunology* **76,** 508–510.

Jacob, J., Kassir, R., and Kelsoe, G. (1991a). *In situ* studies of the primary immune response to (4-hydroxy-3-nitrophenyl) acetyl. I. The architecture and dynamics of responding cell populations. *J. Exp. Med.* **173,** 1165–1175.

Jacob, J., Kelsoe, G., Rajewsky, K., and Weiss, U. (1991b). Intraclonal generation of antibody mutants in germinal centres. *Nature* **354,** 389–392.

Jacobson, E. B., Caporale, L. H., and Thorbecke, G. J. (1974). Effect of thymus cell injections on germinal center formation in lymphoid tissues of nude (thymusless) mice. *Cell. Immunol.* **13,** 416–430.

Janossy, G., Boffil, M., Johnson, M., and Racs, P. (1991). Changes of germinal center organization in HIV-1-positive lymph nodes. *In* "Accessory Cells in HIV and Other Retroviral Infections" (P. Racz, C. D. Dijkstra, and J.-C. Gluckman, eds.), pp. 111–118. Karger, Basel.

Janossy, G., Bofill, M., Rowe, D., Muir, J., and Beverley, P. C. (1989). The tissue distribution of T lymphocytes expressing different CD45 polypeptides. *Immunology* **66,** 517–525.

Jerne, N. K. (1984). Idiotypic networks and other preconceived ideas. *Immunol. Rev.* **79,** 5–24.

Jeurissen, S. H. M., and Dijkstra, C. D. (1986). Characteristics and functional aspects of non-lymphoid cells in rat germinal centers, recognized by two monoclonal antibodies ED5 and ED6. *Eur. J. Immunol.* **16,** 562–568.

Johnson, G. D., Hardie, D. L., Ling, N. R., and MacLennan, I. C. M. (1986). Human follicular dendritic cells (FDC): A study with monoclonal antibodies (MoAb). *Clin. Exp. Immunol.* **64,** 205–213.

June, R. A., Landay, A. L., Stefanik, K., Lint, T. F., and Spear, G. T. (1992). Phenotypic analysis of complement receptor 2+ T lymphocytes: Reduced expression on CD4+ cells in HIV-infected persons. *Immunology* **75**, 59–65.

Kapasi, Z. F., Kosco, M. H., Schultz, L. D., Tew, J. G., and Szakal, A. K. (1993). Cellular origin of follicular dendritic cells. *In* "*In vivo* Immunology" (E. Heineh, ed.), pp. 231–235. Plenum Press, New York.

Kaplan, M. H., Coons, A. H., and Derne, H. W. (1950). Localization of antigen in tissue cells. III. Cellular distribution of pneumococcal polysaccharides types II and III in the mouse. *J. Exp. Med.* **91**, 15.

Kawabe, T., Naka, T., Yoshida, K., Tanaka, T., Fujiwara, H., Suematsu, S., Yoshida, N., Kishimoto, T., and Kikutani, H. (1994). The immune response in CD40-deficient mice: Impaired immunoglobulin class switching and germinal center formation. *Immunity* **1**, 167–178.

Kearney, J. (1989). Development of the B cell repertoire and the ability to respond to bacteria associated antigens. *In* "Progress in Immunology" (F. Melchers, ed.), pp. 404–408. Springer-Verlag, Berlin.

Kim, H.-S., Zhang, X., and Choi, Y. S. (1994). Activation and proliferation of follicular dendritic cell-like cells by activated T lymphocytes. *J. Immunol.* **153**, 2951–2961.

Klaus, G. G. B. Humphrey, J. H., Kunkl, A., and Dongworth, D. W. (1980). The follicular dendritic cell: Its role in antigen presentation in the generation of immunological memory. *Immunol. Rev.* **53**, 3–38.

Klein, U., Küppers, R., and Rajewsky, K. (1993). Human IgM$^+$IgD$^+$ B cells, the major B cell subset in the peripheral blood, express $V\chi$ genes with no or little somatic mutation throughout life. *Eur. J. Immunol.* **23**, 3272–3277.

Kocks, C., and Rajewsky, K. (1988). Stepwise intraclonal maturation of antibody affinity through somatic hypermutation. *Proc. Natl. Acad. Sci. USA* **85**, 8206–8210.

Koopman, G., Keehnen, R. M., and Pals, S. T. (1993). Interaction through the LFA-1/ICAM-1 pathway prevents programmed cell death of germinal center B cells. *Adv. Exp. Med. Biol.* **329**, 387–392.

Koopman, G., Parmentier, H. K., Schuurman, H.-J., Newman, W., Meijer, C. J. L. M., and Pals, S. T. (1991). Adhesion of human B cells to follicular dendritic cells involves both the lymphocyte function–associated antigen 1/Intercellular adhesion molecule 1 and very late antigen 4/Vascular cell adhesion molecule 1 pathways. *J. Exp. Med.* **173**, 1297–1304.

Kosco, M. (1991). Germinal centers and the immune response. *Res. Immunol.* **142**, 219–282.

Kosco, M. H., Burton, G. F., Kapasi, Z. F., Szakal, A. K., and Tew, J. G. (1989). Antibody-forming cell induction during an early phase of germinal center development and its delay with ageing. *Immunology* **68**, 312–318.

Kosco, M. H., and Gray, D. (1992). Signals involved in germinal center reactions. *Immunol. Rev.* **126**, 63–76.

Kosco, M. H., Szakal, A. K., and Tew, J. G. (1988). *In vivo* obtained antigen presented by germinal center B cells to T cells *in vitro*. *J. Immunol.* **140**, 354–360.

Kroese, F. G., Wubbena, A. S., Joling, P., and Nieuwenhuis, P. (1985). T-lymphocytes in rat lymphoid follicles are a subset of T helper cells. *Adv. Exp. Med. Biol.* **186**, 443–449.

Kroese, F. G., Wubbena, A. S., and Nieuwenhuis, P. (1986). Germinal centre formation and follicular antigen trapping in the spleen of lethally X-irradiated and reconstituted rats. *Immunology* **57**, 99–104.

Kroese, F. G. M., Wubbena, A. S., Seijen, H. G., and Nieuwenhuis, P. (1987). Germinal centers develop oligoclonally. *Eur. J. Immunol.* **17**, 1069–1072.

Küppers, R., Zhao, M., Hansmann, M.-L., and Rajewsky, K. (1993). Tracing B cell development in human germinal centres by molecular analysis of single cells picked from histological sections. *EMBO J.* **12**, 4955–4967.

Lagresle, C., Bella, C., Daniel, T., Krammer, P. H., and Defrance, T. (1995). Regulation of germinal center B cell differentiation: Role of the human APO-1/Fas (CD95) molecule. *J. Immunol.* **154,** 5746–5756.

Laissue, J., Cottier, H., Hess, M. W., and Stoner, R. D. (1971). Early and enhanced germinal center formation and antibody responses in mice after primary stimulation with antigen-isologous antibody complexes as compared with antigen alone. *J. Immunol.* **107,** 822–831.

Langevoort, H. L., Asofsky, R. M., Jacobson, E. B., de Vries, I., and Thorbecke, G. J. (1963). Gamma globulin and antibody formation *in vitro.* II. Parallel observations on histologic changes and on antibody formation in the white and red pulp of the rabbit spleen during the primary response with special reference to the effect of endotoxin. *J. Immunol.* **90,** 60.

Larcher, C., Schulz, T. F., Hofbauer, J., Hengster, P., Romani, P., Wachter, H., and Dierich, M. P. (1990). Expression of the C3d/EBV receptor and of other cell membrane surface markers is altered upon HIV-1 infection of myeloid, T and B cells. *J. Acquir. Immune Defic. Syndr.* **3,** 103–108.

Leanderson, T., Källberg, E., and Gray, D. (1992). Expansion, selection and mutation of antigen-specific B cells in germinal centers. *Immunol. Rev.* **126,** 47—61.

Lederman, S., Yellin, M. J., Krichevsky, A., Belko, J., Lee, J. J., and Chess, L. (1992). Identification of a novel surface protein on activated CD4$^+$ T cells that induces contact-dependent B cell differentiation (Help). *J. Exp. Med.* **175,** 1091–1101.

Lennert, K. (1978). "Malignant, Lymphomas Other Than Hodgkin's Disease." Springer-Verlag, New York.

Levy, J. A., Hoffman, A. D., Kramer, S. M., Landis, J. A., Shimabukro, J. M., and Oshiro, L. S. (1984). Isolation of lymphocytopathic retroviruses from San Francisco patients with AIDS. *Science* **225,** 840–842.

Levy, N. S., Malipiero, U. V., Lebecque, S. G., and Gearhart, P. J. (1989). Early onset of somatic mutation to immunoglobulin Vh genes during the primary immune response. *J. Exp. Med.* **169,** 2007–2019.

Lindhout, E., Lakeman, A., Mevissen, M. L. C. M., and de Groot, C. (1994). Functionally active Esptein-Barr virus–transformed follicular dendritic cell-like cell lines. *J. Exp. Med.* **179,** 1173–1184.

Lindhout, E., Mevissen, M. L., Kwekkeboom, J., Tager, J. M., and de Groot, C. (1993). Direct evidence that human follicular dendritic cells (FDC) rescue germinal center cells from death by apoptosis. *Clin. Exp. Immunol.* **91,** 330–336.

Liu, Y.-J., and Banchereau, J. (1996). Human peripheral B cells subsets. *In* "Handbook of Experimental Immunology" (D. Weir, C. Blackwell, and L. Hersenberg, eds.), in press. Blackwell, Oxford.

Liu, Y.-J., Barthelemy, C., de Bouteiller, O., Arpin, C., Durand, I., and Banchereau. J. (1995). Memory B cells from human tonsils colonize mucosal epithelium and directly present antigen to T cells by rapid upregulation of B7.1 and B7.2 *Immunity* **2,** 238–248.

Liu, Y.-J., Joshua, D. E., Williams, G. T., Smith, C. A., Gordon, J., and MacLennan, I. C. M. (1989). Mechanisms of antigen-driven selection in germinal centers. *Nature (London)* **342,** 929–931.

Liu, Y.-J., Mason, D. Y., Johnson, G. D., Abbot, S., Gregory, C. D., Hardie, D. L., Gordon, J., and MacLennan, I. C. M. (1991a). Germinal center cells express *bcl*-2 protein after activation by signals which prevent their entry into apoptosis. *Eur. J. Immunol.* **21,** 1905–1910.

Liu, Y.-J., Zhang, J., Lane, P. J. L., Chan, E. Y.-T., and MacLennan, I. C. M. (1991b). Sites of specific B cell activation in primary and secondary responses to T cell–dependent and T cell–independent antigens. *Eur. J. Immunol.* **21,** 2951–2962.

MacLennan, I. C., and Gray, D. (1986). Antigen-driven selection of virgin and memory B cells. *Immunol. Rev.* **91,** 61–85.

MacLennan, I. C. M. (1994). Germinal centers. *Annu. Rev. Immunol.* **12,** 117–139.

Madassery, J. V., Gillard, B., Marcus, D. M., and Nahm, M. H. (1991). Subpopulations of B cells in germinal centers. III. HJ6, a monoclonal antibody, binds globoside and a subpopulation of germinal center B cells. *J. Immunol.* **147,** 823–829.

Mandel, T. E., Phipps, R. P., Abott, A., and Tew, J. G. (1980). The follicular dendritic cell: Long-term antigen retention during immunity. *Immunol. Rev.* **53,** 29–59.

Manser, T., Wysocki, L. J., Gridley, T., Near, R. I., and Gefter, M. L. (1985). The molecular evolution of the immune response. *Immunol. Today* **6,** 94–101.

Martinez-Valdez, H., Malisan, F., de Bouteiller, O., Guret, C., Banchereau, J., and Liu, Y. J. (1996). Molecular evidence that isotype switching occurs within the germinal centres. *In* "Annals of the New York Academy of Science" (P. Casali and L. Silberstein, eds.), in press. New York Academy of Science Press, New York.

McHeyzer-Williams, M. G., and Davis, M. M. (1995). Antigen-specific development of primary and memory T cells *in vivo. Science* **268,** 106–111.

McHeyzer-Williams, M. G., McLean, M. J., Lalor, P. A., and Nossal, G. J. V. (1993). Antigen-driven B cell differentiation *in vivo. J. Exp. Med.* **178,** 295–307.

Melchers, F., Erdei, A., Schulz, T., and Dierich, M. P. (1985). Growth control of activated, synchronized murine B cells by the C3d fragment of human complement. *Nature (London)* **317,** 264–267.

Merino, R., Ding, L., Veis, D. J., Korsmeyer, S. J., and Nunez, G. (1994). Developmental regulation of the Bcl-2 protein and susceptibility to cell death in B lymphocytes. *EMBO J.* **13,** 683–691.

Merville, P., Dechanet, J., Grouard, G., Durand, I., and Banchereau, J. (1995). T cell–induced B cell blasts differentiate into plasma cells when cultured on bone marrow stroma with IL3 and IL10. *Int. Immunol.* **7,** 635–643.

Michie, S. A., and Rouse, R. V. (1988). Study of murine T cell migration using the Thy-1 allotypic marker: Demonstration of antigen-specific homing to lymph node germinal centers. *Transplantation* **46,** 98–104.

Miller, C., Stedra, J., Kelsoe, G., and Cerny, J. (1995). Facultative role of germinal centers and T cells in the somatic diversification of IgVh genes. *J. Exp. Med.* **181,** 1319–1331.

Müller, S., Nara, P., D'Amelio, R., Biselli, R., Gold, D., Wang, H., Kohler, H., and Silverman, G. J. (1992). Clonal patterns in the human immune response to HIV-1 infection. *Int. Rev. Immunol.* **9,** 1–13.

Naiem, M., Gerdes, J., Abdulaziz, Z., Stein, H., and Mason, D. Y. (1983). Production of a monoclonal antibody reactive with human dendritic reticulum cells and its use in the immunohistological analysis of lymphoid tissue. *J. Clin. Pathol.* **36,** 167–175.

Nieuwenhuis, P. (1969). Histophysiology of germinal centers and their role in antibody response: An autoradiographic study in the rabbit. *Adv. Exp. Med. Biol.* **5,** 113.

Nieuwenhuis, P., and Opstelten, D. (1984). Functional anatomy of germinal centres. *Am. J. Anat.* **170,** 421.

Nossal, G. J. V., Abbot, A., Mitchell, J., and Lummus, Z. (1968). Antigens in immunity. XV. Ultrastructural features of antigen capture in primary and secondary lymphoid follicles. *J. Exp. Med.* **127,** 277–290.

Nossal, G. J. V., and Ada, G. L. (1971). "Antigens, Lymphoid Cells and the Immune Response." Academic Press, New York.

Nossal, G. J. V., and Lederberg, J. (1958). Antibody production by a single cell. *Nature (London)* **181,** 1419–1420.

Oehen, S., Waldner, H., Kündig, T. M., Hengartner, H., and Zinkernagel, R. (1992). Antivirally protective cytotoxic T cell memory to lymphocytic choriomeningitis virus is governed by persisting antigens. *J. Exp. Med.* **176,** 1273–1281.

Orscheschek, K., Merz, H., Schlegelberger, B., and Feller, A. C. (1994). An immortalized cell line with features of human follicular dendritic cells. Antigen and cytokine expression analysis. *Eur. J. Immunol.* **24,** 2682–2690.

Osmond, D. G. (1991). Proliferation kinetics and the lifespan of B cells in central and peripheral lymphoid organs. *Curr. Opin. Immunol.* **3,** 179–185.

Pantaleo, G., Graziosi, C., Demarest, J. F., Cohen, O. J., Vaccarezza, M., Gantt, K., Muro-Cacho, C., and Fauci, A. S. (1994). Role of lymphoid organs in the pathogenesis of human immunodeficiency virus (HIV) infection. *Immunol. Rev.* **140,** 105–130.

Parmentier, H. K., Van der Linden, J. A., Krijnen, J., Van Wichen, D. F., Rademakers, L. H. P. M., Bloem, A. C., and Schuurman, H.-J. (1991). Human follicular dendritic cells: Isolation and characteristics in situ and in suspension. *Scand. J. Immunol.* **33,** 441–452.

Parwaresch, M. R., Radzun, H. J., Feller, A. C., Peters, K. P., and Hansmann, M. L. (1983). Peroxidase-positive mononuclear leukocytes as possible precursors of human dendritic reticulum cells. *J. Immunol.* **131,** 2719–2725.

Pascual, V., Liu, Y.-J., Magalski, A., de Bouteiller, O., Banchereau, J., and Capra, J. D. (1994). Analysis of somatic mutation in five B cell subsets of human tonsil. *J. Exp. Med.* **180,** 329–339.

Petrasch, S., Perez-Alvarez, C., Schmitz, J., Kosco, M., and Brittinger, G. (1990). Antigenic phenotyping of human follicular dendritic cells isolated from nonmalignant and malignant lymphatic tissue. *Eur. J. Immunol.* **20,** 1013–1018.

Petrasch, S. G., Kosco, M. H., Perez-Alvarez, C. J., Schmitz, J., and Brittinger, G. (1991). Proliferation of germinal center B lymphocytes *in vitro* by direct membrane contact with follicular dendritic cells. *Immunobiology* **183,** 451–462.

Pezella, F., Tse, A., Cordell, J. L., Pulford, K., Gatter, K. C., and Mason, D. Y. (1990). Expression of the Bcl-2 oncogene is not specific for the 14;18 chromosomal translocation. *Am. J. Pathol.* **137,** 2115–2120.

Popovic, M., Sarngadhraran, M. G., Read, E., and Gallo, R. C. (1984). Detection, isolation and continuous production of cytopathic retroviruses (HTLV-III) from patients with AIDS and pre-AIDS. *Science* **224,** 497–500.

Poppema, S., Bhan, A. K., Reinherz, E. L., McCluskey, R. T., and Schlossman, S. F. (1981). Distribution of T cell subsets in human lymph nodes. *J. Exp. Med.* **153,** 30–41.

Porwit-Ksiazek, A., Aman, P., Ksiazek, T., and Riberfeld, P. (1983). Leu 7+ (HNK-1+) cells. I. Selective compartmentalization of Leu 7+ cells with different immunophenotypes in lymphatic tissues and blood. *Scand. J. Immunol.* **18,** 495–499.

Pryjma, J., and Humphrey, J. H. (1975). Prolonged C3 depletion by cobra venom factor in T-deprived mice and its implications for the role of C3 as an essential second signal for B cell triggering. *Immunology* **28,** 569–576.

Pulendran, B., Kannourakis, G., Nouri, S., Smith, K. G. C., and Nossal, G. J. V. (1995). Soluble antigen can cause enhanced apoptosis of germinal-centre B cells. *Nature (London)* **375,** 331–334.

Racz, P., Dijkstra, C. D., and Gluckman, J.-C. (1991). "Accessory Cells in HIV and Other Retroviral Infections." Karger, Basel.

Rajewsky, K. (1992). Early and late B-cell development in the mouse *Curr. Opin. Immunol.* **4,** 171–176.

Renshaw, B. R., Fanslow, W. C., III, Armitage, R. J., Campbell, K. A., Liggitt, D., Wright, B., Davison, B. L., and Maliszewski, C. R. (1994). Humoral immune responses in CD40 ligand–deficient mice. *J. Exp. Med.* **180,** 1889–1900.

Richter, M., Zimmerman, S., and Haurowitz, F. (1964). Relation of antibody titer to persistence of antigen. *J. Immunol.* **94,** 938.

Roscoe, G., and Whiteside, T. L. (1983). Selective distribution and quantitation of T-lymphocyte subsets in germinal centers of human tonsils: Definition by use of monoclonal antibodies. *Arch. Pathol. Lab. Med.* **107,** 228–231.

Rose, M. L., Birbeck, M. S. C., Wallis, V. J., Forrester, J. A., and Davies, A. J. S. (1980). Peanut lectin binding properties of germinal centres of mouse lymphoid tissue. *Nature (London)* **284,** 364–366.

Rothstein, T. L., Wang, J. K. M., Panka, D. J., Foote, L. C., Wang, Z., Stanger, B., Cui, H., Ju, S.-T., and Marshak-Rothstein, A. (1995). Protection against Fas-dependent Th1-mediated apoptosis by antigen receptor engagement in B cells. *Nature (London)* **374,** 163–165.

Schmitz, J., Petrasch, S., van Lunzen, J., Racz, P., Kleine, H.-D., Hufert, F., Kern, P., Schmitz, H., and Tenner-Racz, K. (1993). Optimizing follicular dendritic cell isolation by discontinuous gradient centrifugation and use of the magnetic cell sorter (MACS). *J. Immunol. Methods* **159,** 189–196.

Schmitz, J., van Lunzen, J., Tenner-Racz, K., Grobschupff, G., Racz, P., Schmitz, H., Dietrich, M., and Hufert, F. (1994). Follicular dendritic cells (FDC) are not productively infected with HIV-1 *in vivo. In* "*In Vivo* Immunology" (E. Heinen, ed.) pp. 165–168. Plenum Press, New York.

Schnizlein, C. T., Kosco, M. H., Szakal, A. K., and Tew, J. G. (1985). Follicular dendritic cells in suspension: Identification, enrichment, and initial characterization indicating immune complex trapping and lack of adherence and phagocytic activity. *J. Immunol.* **134,** 1360–1368.

Schriever, F., Freedman, A. S., Freeman, G., Messner, E., Lee, G., Daley, J., and Nadler, L. M. (1989). Isolated follicular dendritic cells display a unique antigenic phenotype. *J. Exp. Med.* **169,** 2043–2048.

Schriever, F., Freeman, G., and Nadler, L. M. (1991). Follicular dendritic cells contain a unique gene repertoire demonstrated by single-cell polymerase chain reaction. *Blood* **77,** 787–791.

Schriever, F., and Nadler, L. M. (1992). The central role of follicular dendritic cells in lymphoid tissues. *Adv. Immunol.* **51,** 243–284.

Schwartz, D. H. (1994). Potential pitfalls on the road to an effective HIV vaccine. *Immunol. Today* **15,** 54–57.

Sellheyer, K., Schwarting, R., and Stein, H. (1989). Isolation and antigenic profile of follicular dendritic cells. *Clin. Exp. Immunol.* **78,** 431–436.

Shokat, K. M., and Goodnow, C. C. (1995). Antigen-induced B-cell death and elimination during germinal-centre immune responses. *Nature* **375,** 334–338.

Soderberg, L. S., and Coons, A. H. (1978). Complement-dependent stimulation of normal lymphocytes by immune complexes. *J. Immunol.* **120,** 806–811.

Sölder, B. M., Schulz, T. F., Hengster, P., Lower, J., Larcher, C., Bitterlich, G., Kurth, R., Wachter, H., and Dierich, M. P. (1989). HIV and HIV-infected cells differentially activate the human complement system independent of antibody. *Immunol. Lett.* **22,** 135–145.

Spiegel, H., Herbst, H., Niedobitek, G., Foss, H. D., and Stein, H. (1992). Follicular dendritic cells are a major reservoir for human immunodeficiency virus type 1 in lymphoid tissues facilitating infection of $CD4^+$ T-helper cells. *Am. J. Pathol.* **140,** 15–22.

Stahmer, I., Zimmer, J. P., Ernst, M., Fenner, T., Finnern, R., Schmitz, H., Flad, H.-D., and Gerdes, J. (1991). Isolation of normal human follicular dendritic cells and CD4-independent *in vitro* infection by human immunodeficiency virus (HIV-1). *Eur. J. Immunol.* **21,** 1873–1878.

Stein, H., Gerdes J., and Mason, D. Y. (1982). The normal and malignant germinal centre. *Clin. Haematol.* **11,** 531–559.

Suzuki, S., Nakamura, N., Abe, M., and Wakasa, H. (1995). Distribution and characterization of follicular dendritic cells (FDCs) in non-Hodgkin's lymphomas. *In* "Dendritic Cells in Fundamental and Clinical Immunology" (J. Banchereau and D. Schmitt, eds.), pp. 297–300. Plenum, London.

Swerdlow, S. H., and Murray, L. J. (1984). Natural killer (Leu 7+) cells in reactive lymphoid tissues and malignant lymphomas. *Am. J. Clin. Pathol.* **81,** 459–463.

Szakal, A. K. (1989). Microanatomy of lymphoid tissue during humoral immune responses: Structure–function relationships. *Ann. Rev. Immunol.* **7,** 91–109.

Szakal, A. K., Gieringer, R. L., Kosco, M. H., and Tew, J. G. (1985). Isolated follicular dendritic cells: Cytochemical antigen localization. Nomarski, SEM, and TEM morphology. *J. Immunol.* **134,** 1349–1353.

Szakal, A. K., and Hanna, M. G. (1968). The ultrastructure of antigen localization and viruslike particles in mouse spleen germinal centers. *Exp. Mol. Pathol.* **8,** 75–89.

Szakal, A. K., Kosco, M. H., and Tew, J. G. (1988). A novel *in vivo* follicular dendritic cell–dependent iccosome-mediated mechanism for delivery of antigen to antigen-processing cells. *J. Immunol.* **140,** 341–353.

Takahashi, T., Mond, J. J., Carswell, E. A., and Thorbecke, G. J. (1971). The importance of theta and Ig hearing cells in the immune response to various antigens. *J. Immunol.* **107,** 1520–1526.

Takeda, A., Tuazon, C. U., and Ennis, F. A. (1988). Antibody-enhanced infection by HIV-1 via Fc receptor–mediated entry. *Science* **242,** 580–583.

Tenner-Racz, K., Racz, P., Dietrich, M., and Kern, P. (1985). Altered follicular dendritic cells and virus-like particles in AIDS and AIDS-related lymphadenopathy. *Lancet* **i,** 105–106.

Tew, J. G., Kosco, M. H., Burton, G. F., and Szakal, A. K. (1990). Follicular dendritic cells as accessory cells. *Immunol. Rev.* **117,** 185–211.

Tew, J. G., Phipps, R. P., and Mandel, T. E. (1980). The maintenance and regulation of the humoral immune response: Persisting antigen and the role of follicular antigen-binding dendritic cells as accessory cells. *Immunol. Rev.* **53,** 175–201.

Thorbecke, G. J. (1959). *Ann. NY Acad. Sci.* **78,** 69–73.

Thorbecke, G. J. (1969). Germinal centers and immunological memory. *In* "Lymphatic Tissue and Germinal Centers in Immune Response. Advances in Experimental Medicine and Biology" (L. Fiore-Donati and M. G. Hanna, eds.), pp. 83–92. Plenum Press, New York.

Thorbecke, G. J. (1990). Presidential address. Focusing: The dilemma of interpreting sharp images on a blurred background. *J. Immunol.* **145,** 2779–2790.

Thorbecke, G. J., Amin, A. R., and Tsiagbe, V. K. (1994). Biology of germinal centers in lymphoid tissue. *FASEB J.* **8,** 832–840.

Tsubata, T., Wu, J., and Honjo, T. (1993). B cell apoptosis induced by antigen receptor cross-linking is blocked by T cell signal through CD40. *Nature (London)* **364,** 645–648.

Tsujimoto, Y., and Croce, C. M. (1986). Analysis of the structure, transcripts, and protein products of bcl-2, the gene involved in human follicular lymphoma. *Proc. Natl. Acad. Sci. U.S.A.* **83,** 5224–5218.

Tsunoda, R., Nakayama, N., Onozaki, K., Heinen, E., Cormann, N., Kinet-Denoël, C., and Kojima, M. (1990). Isolation and long-term cultivation of human tonsil follicular dendritic cells. *Virchows Arch. B Cell Pathol.* **59,** 95–105.

Van de Wiel, B. A., Bakker, L. J., de Graaf, L., de Weger, R. A., Verhoef, J., van den Tweel, J. G., and Joling, P. (1994). Complement and antibody enhance binding and uptake of HIV-1 by bone marrow cells. *In* "*In Vivo* Immunology" (E. Heinen, ed.), pp. 159–164. Plenum Press, New York.

Van den Berg, T. K., Döpp, E. A., Daha, M. R., Kraal, G., and Dijkstra, C. D. (1992). Selective inhibition of immune complex trapping by follicular dendritic cells with monoclonal antibodies against rat C3. *Eur. J. Immunol.* **22,** 957–962.

Van Rooijen, N. (1972). Antigens in the spleen. The non-specificity of the follicles in the process of antigen trapping and the role of antibody. *Immunology.* **22,** 757–765.

Vos, J. G., Berkvens, J. M., and Kruijt, B. C. (1980). The athymic nude rat. I. Morphology of lymphoid and endocrine organs. *Clin. Immunol. Immunopathol.* **15,** 229–237.

Weigert, M. (1986). The influence of somatic mutation on the immune response. *In* "Progress in Immunology, 6th ed." (B. Cinader and R. G. Miller, eds.), pp. 138–144. Academic Press, New York.

Weigle, W. O. (1975). Cyclical production of antibody as a regulatory mechanism in the immune response. *Adv. Immunol.* **21,** 87–111.

Weiss, U., and Rajewsky, K. (1990). The repertoire of somatic antibody mutants accumulating in the memory compartment after primary immunization is restricted through affinity maturation and mirrors that expressed in the secondary response. *J. Exp. Med.* **172,** 1681–1689.

Weissman, I. L. (1975). Development and distribution of immunoglobulin-bearing cells in mice. *Transplant. Rev.* **24,** 159.

White, R. G., Henderson, D. C., Eslami, M. B., and Nielsen, K. (1975). Localization of a protein antigen in the chicken spleen. Effect of various manipulative procedures on the morphogenesis of the germinal centre. *Immunology* **28,** 1–21.

Xu, J., Foy, T. M., Laman, J. D., Elliott, E. A., Dunn, J. J., Waldschmidt, T. J. Elsemore, J., Noelle, R. J., and Flavell, R. A. (1994). Mice deficient for the CD40 ligand. *Immunity* **1,** 423–431.

Yoshida, K., Kaji, M., Takahashi, T., Van Den Berg, T. K., and Dijkstra, C. D. (1995). Host origin of follicular dendritic cells induced in the spleen of SCID mice after transfer of allogeneic lymphocytes. *Immunology* **84,** 117–126.

Zhang, J., MacLennan, I. C. M., Liu, Y.-J., and Lane, P. J. L. (1988). Is rapid proliferation in B centroblasts linked to somatic mutation in memory B cell clones? *Immunol. Lett.* **18,** 297–299.

Zheng, B., Xue, W., and Kelsoe, G. (1994). Locus-specific somatic hypermutation in germinal centre T cells. *Nature (London)* **372,** 556–559.

Ziegner, M., Steinhauser, G., and Berek, C. (1994). Development of antibody diversity in single germinal centers: Selective expansion of high-affinity variants. *Eur. J. Immunol.* **24,** 2393–2400.

Cell Biology of Kidney Glomerulus

Shinichi Ohno, Takeshi Baba, Nobuo Terada, Yasuhisa Fujii, and Hideho Ueda

Department of Anatomy, Yamanashi Medical University, Tamaho, Yamanashi 409-38, Japan

It has been accepted that some artifacts are inevitably produced by the conventional preparation steps for electron microscopy, including fixation, dehydration, embedding, ultrathin sectioning, and staining. Therefore, conventional ultrastructural findings on kidney glomeruli are hardly thought to be correlated with the physiological functions of kidneys *in vivo*. In this chapter, two preparation techniques, the quick-freezing and deep-etching (QF–DE) method or the quick-freezing and freeze-substitution (QF–FS) method, are presented and shown to be useful for clarifying the ultrastructures of kidney glomeruli more closely to structures *in vivo* with fewer artifacts. Moreover, the ultrastructures of glomerular capillary loops have been demonstrated by a new "*in vivo* cryotechnique," that shows that hemodynamic factors should be considered in the morphological study of glomerular functions.

KEY WORDS: Kidney glomerulus, Quick-freezing, Freeze-substitution, Deep-etching, Basement membrane, Slit diaphragm, Anionic site.

List of Abbreviations

DE: deep-etching
FS: freeze-substitution
GA: glutaraldehyde
GBM: glomerular basement membrane
GO: glutaraldehyde–osmium tetroxide
HSPG: heparan sulfate proteoglycan
LRE: lamina rara externa
LRI: lamina rara interna
MM: Mesangial matrix
PB: 0.1 M phosphate buffer
PEI: polyethyleneimine
PTA: phosphotungstic acid
QF: quick-freezing
STZ: streptozotocin
TGO: tannic acid–glutaraldehyde–osmium tetroxide

I. Introduction

The renal corpuscle contains abundant capillaries that branch off from the afferent artery and come together again to form the efferent artery. The kidney glomerulus consists mainly of three types of cells: endothelial cells, epithelial cells, and mesangial cells (Fig. 1). Simple endothelial cells with elongated nuclei bulge into the capillary lumen, covering the inner surface of the capillaries. Outer epithelial cells differentiate into podocytes, whose foot processes cover the urinary side of the glomerular basement membrane (GBM). The mesangial cells are localized between the GBM and the endothelial cells. Functionally, the kidney glomerulus filters many substances, such as small molecular proteins, lipid components, carbohydrates, salts, and vitamins through the GBM. It is generally accepted that the GBM serves as both a charge-selective and a size-selective molecular sieve capable of excluding most plasma proteins from the urine (Brenner *et al.,* 1976, 1977, 1978; Kanwar, 1984). Proteins larger than albumin are usually restricted from passing into the urinary space (Venkatachalam *et al.,* 1970b; Caulfield and Farquhar, 1974; Ryan and Karnovsky, 1976; Rennke and Venkatachalam, 1977; Bariety *et al.,* 1978; Olivetti *et al.,* 1981). To gain access to the urinary space, many small substances in capillary lumens must cross the glomerular capillary wall, passing sequentially through the fenestrated endothelium, the GBM, and the narrow epithelial slit.

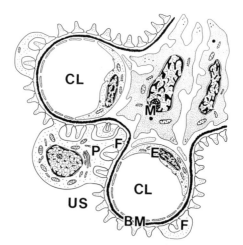

FIG. 1 Schematic representation of glomerular capillary loops. P, podocyte; M, mesangial cell; E, endothelial cell; BM, basement membrane; F, foot process; CL, capillary lumen; US, urinary space.

This chapter reviews the morphological and functional characteristics of kidney glomeruli under different conditions and summarizes what has recently been discovered about the kidney glomerular ultrastructure. In our research group the quick-freezing (QF) method has been used to examine various aspects of kidney glomeruli in normal and experimental animals (Hora *et al.*, 1990; Takami *et al.*, 1990; Furukawa *et al.*, 1991; Naramoto *et al.*, 1991a,b; Takami *et al.*, 1991; Yoshimura *et al.*, 1991; Nakazawa *et al.*, 1992; Naramoto *et al.*, 1992; Ohno *et al.*, 1992; Duan *et al.*, 1993; Moriya *et al.*, 1993; Otani *et al.*, 1993). The data reported here are largely from the most recent studies. However, the metabolic function of glomeruli has not yet been clarified in detail, so this review could contribute to elucidating the functional significance of the kidney glomeruli and could provide a guide for further experiments. Several other reviews on the kidney glomeruli cover virtually all aspects of glomerular structure and functions (Farquhar, 1975; Brenner *et al.*, 1976, 1977, 1978; Karnovsky, 1979; Michael *et al.*, 1980; Kanwar, 1984; Andrews, 1988). The reader is referred to any of the reviews for a discussion of the physiological functions of the kidney glomerulus.

II. Normal Structure and Function of Glomerulus

A. Conventional Methodological Aspects

The conventional procedure for preparing kidney samples for electron microscopic study includes the following steps: chemical fixation of kidney tissues, their dehydration and embedding in epoxy resin, ultrathin sectioning, and electron staining with heavy metals, including uranium and lead. The first step is commonly carried out by serial double fixations with glutaraldehyde (GA) and osmium tetroxide. These chemical fixatives are believed to preserve cellular ultrastructures *in situ*. It is generally accepted that the fixation mechanism of GA is a cross-linking of the extracellular and intracellular proteins via its bifunctional aldehyde residues and that osmium tetroxide binds with intracellular components, such as proteins and membrane lipids.

However, there are some disadvantages to using these chemical fixations to preserve the *in vivo* ultrastructures in cells and tissues. First, components such as ions and metals cannot be immobilized by the chemical cross-linking mechanism and are easily extracted during the subsequent buffer washing and dehydration steps. Second, fast cellular events cannot be captured by the chemical fixations, because the methods require at least several seconds to fix the cells and tissues. Finally, ultrastructures are unavoidably altered, and soluble proteins that are not associated with cytoskeletons and membranous organelles are translocated. Moreover the fixed kidney specimens are subjected to the subsequent dehydration and embedding steps. In the dehydration step, ethanol is generally used to substitute for any water in the specimens. However, considerable amounts of lipids and other ethanol-soluble components are easily extracted at this step. The dehydrated specimens are then embedded in viscous epoxy resin, which is polymerized by heating up to 60°C. The highly aqueous kidney tissues are thus embedded in the matrix of the resin network via these harsh manipulations. Next, the tissues are cut into sections 0.05 to 0.1μm thick with an ultramicrotome and mounted on copper grids. Thus, only a two-dimensional single plane is usually obtained from the three-dimensional cells and tissues and observed under an electron microscope. The ultrathin section staining with lead and uranium is necessary to enhance the contrast of ultrastructural images. In these steps we cannot avoid the artifacts produced by coating the ultrathin sections with heavy metals or the heat damage that occurs during the final electron microscopic observation under a high-vacuum condition.

Although conventional electron microscopy with ultrathin sections has the disadvantages described in the preceding paragraph, such studies have provided much important morphological information about kidneys that has greatly contributed to the progress in both basic and clinical nephrology. Meanwhile, some researchers have been trying to improve the sample preparation methods and have been looking for novel techniques to preserve the three-dimensional ultrastructures of kidneys *in vivo.*

1. Basement Membrane and Mesangial Matrix

The GBM is an extracellular structure that underlies both epithelium and endothelium. The scaffolding of the GBM is usually formed by a self-assembly mechanism of type IV collagen, laminin, and heparan sulfate proteoglycan (Yurchenco and Furthmayr, 1984; Yurchenco and Rubin, 1987; Grant *et al.,* 1989). The GBM usually provides the physical support for glomerular structures and epithelial or endothelial attachment. Moreover, the glomerular capillary wall restricts the transmural passage of large plasma proteins while offering little resistance to the filtration of water and small solutes (Chang *et al.,* 1975; Rennke *et al.,* 1975, 1978; Kaysen *et al.,* 1986). The conventional ultrathin section in the field of electron microscopy shows that the GBM is composed of three layers: an electron-dense lamina densa with a less dense zone on either side—the lamina rara externa (LRE) and the lamina rara interna (LRI) (Fig. 2). Fine fibrils with variable sizes have been described in all three layers of the GBM (Farquhar *et al.,* 1961; Rodewald and Karnovsky, 1974).

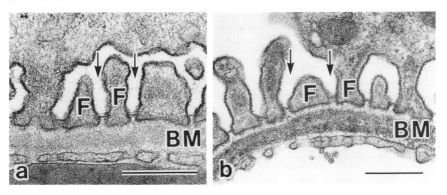

FIG. 2   Electron micrographs of glomerular capillary loops prepared by conventional methods, such as GO fixation (a) and TGO fixation (b). The spaces between foot processes (Fs) are mostly dilated (arrows). Three layers in the basement membrane (BM) are clearly identified. Bar = 0.5 μm.

When kidney tissues are prepared according to the conventional electron microscopic method, the GBM usually displays the lamina densa separated from epithelial or endothelial cells by the laminae rarae. However, the first fixation step with chemicals such as glutaraldehyde and osmium tetroxide has been known to occasionally disturb the native glomerular ultrastructure (Hora *et al.*, 1990; Yoshimura *et al.*, 1991; Reale and Luciano, 1993). Moreover, a rapid dehydration in organic solvents at room temperature has been found to induce tissue shrinkage (Furukawa *et al.*, 1991; Chan *et al.*, 1993). The occurrence of the laminae rarae in conventionally prepared specimens has been reported to be due to the fixation and rapid dehydration steps (Chan *et al.*, 1993; Reale and Luciano, 1993). Thus, it is probable that the laminae rarae are artifacts produced by the conventional preparation method, as described in Section II,C,1.

The mesangium is the central connective core of the individual lobules of glomerular tufts (Fig. 1). The mesangial cells are usually surrounded by matricial materials. This mesangial matrix (MM) is similar in appearance to but not identical with the peripheral GBM. The MM is composed of more coarsely fibrillar and slightly less electron dense materials than the GBM, as observed in conventional ultrathin sections by electron microscopy. The MM is of interest to many investigators because it is believed to play a role in the mediation of molecules passing through the glomerulus (Leiper *et al.*, 1977; Keane and Raij, 1980) and to be involved in a variety of kidney diseases (Kashgarian, 1985).

2. Anionic Sites

Specialized glomerular capillaries regulate protein filtration in normal and abnormal states (Barnes and Venkatachalam, 1985; Kaysen *et al.*, 1986).

The GBM has been identified as a size-restrictive structure (Caulfield and Farquhar, 1974; Bastford *et al.*, 1987), and the negative charge barrier of the GBM is also important as a regulator of proteinuria (Rennke *et al.*, 1975; Kanwar and Farquhar, 1979c). The principal constituent of the negative charge barrier, also called *anionic sites,* is believed to be heparan sulfate proteoglycan (HSPG), which is localized mainly in the LRE (Kanwar and Farquhar, 1979a, 1979b). It is also known that the destruction of glomerular anionic sites increases the permeability of GBM to macromolecules (Venkatachalam *et al.*, 1970a; Caulfield and Farquhar, 1978). The anionic sites in the GBM have been visualized by the use of various cationic probe molecules (Caulfield and Farquhar, 1976, 1978; Kanwar and Farquhar, 1979c). Polyethyleneimine (PEI) has been one of the most commonly used cationic tracer particles for electron microscopy (Schurer *et al.*, 1977, 1978). Renal tissue blocks are usually treated with PEI and immersed in a mixture of phosphotungstic acid (PTA) and GA to enhance contrast. After a routine buffer washing procedure, they are postfixed with osmium tetroxide, dehydrated in a graded series of ethanol solution, and embedded in epoxy resin. The blocks are processed for ultrathin sections, which are stained with uranyl acetate and lead citrate. In conventional ultrathin sections for visualization of anionic sites, the PEI particles are regularly arranged 45–65 nm apart (Fig. 3). The number of PEI particles ranges from 18 to 21 per 1000 nm of LRE, which is compatible with previously reported results (Schurer *et al.*, 1977, 1978). Although the PEI particles visualized in such prepared kidney tissues are organized mainly in dotted patterns with regular intervals along the LRE, there has been no clear explanation for this regular arrangement, which is discussed in Sections II,C,2 and II,D,2.

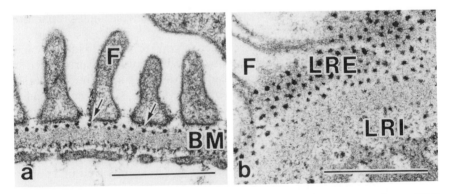

FIG. 3   Conventional ultrathin sections of anionic sites in the basement membrane (BM). Bar = 0.5 μm. (a) PEI particles (arrows) located in the LRE are regularly spaced. (b) Ultrathin sections obliquely cut to the BM. PEI particles are primarily distributed in the LRE and LRI. F, foot process.

3. Slit Diaphragm and Foot Process

To clarify the structure of slit diaphragms, renal cortical tissues are usually prepared by the conventional chemical fixation procedure. The tissues are perfused with tannic acid and GA in the buffer solution, cut into small pieces, and postfixed with osmium tetroxide. Such conventionally prepared specimens show the typical three-layered basement membrane (Fig. 2). Foot processes of the doubly fixed glomeruli also show shrunken cell surface contours. Widely opened epithelial clefts are formed between the glycocalyx layers of adjacent foot processes. Where the foot processes are cut vertically, some slit diaphragms can be distinguished as single lines between the foot processes. In addition, the slit diaphragms appear to be pulled apart, as reported by Furukawa *et al.* (1991).

In the past the slit diaphragm had been reported to have a filtering function for proteins in addition to that provided by the GBM (Graham and Karnovsky, 1966; Venkatachalam *et al.*, 1970b). However, its fully detailed description was not reported before 1974, because of the limited electron density of the structure after the routine fixation and staining techniques. In 1974 and 1975 the glomerular slit diaphragms in the normal rat, mouse, and human kidneys were reported to consist of regular patterns of zipperlike substructures with periodic cross-bridges extending from the foot processes to a central filament (Rodewald and Karnovsky, 1974; Karnovsky and Ryan, 1975; Ryan *et al.*, 1975; Schneeberger *et al.*, 1975). The specimens studied were fixed with tannic acid, glutaraldehyde, and osmium tetroxide (TGO). Fixation with tannic acid usually increases the electron density of extracellular structures in glomeruli and reveals the isoporous substructures of the slit diaphragm. However, their actual existence and structures have been questioned, because they change in appearance, depending on the techniques used for fixation and staining (Karnovsky and Ainsworth, 1972). It is suggested that the zipperlike substructures are artifacts resulting from shrinkage of adjacent foot processes during the conventional tissue preparations, as is described in Sections, II,C,3 and II,D,4. Some experiments with small molecular tracers demonstrated that there is a filtration barrier in the region of epithelial slit diaphragms as well as at the GBM (Graham and Karnovsky, 1966; Venkatachalam *et al.*, 1970b).

B. Quick-Freezing Method

A promising approach to studying less artificial glomerular morphology is the QF method, in which kidney tissues are usually brought into contact with copper or silver metal cooled in liquid helium or nitrogen (Harreveld and Crowell, 1964; Harreveld *et al.*, 1974; Dempsey and Bullivant, 1976;

Plattner and Backmann, 1982; Menco, 1986; Nicolas, 1991). The method has a high time resolution for dynamic ultrastructures, which is sufficient to capture rapid ultrastructural changes of cells and tissues (Harreveld *et al.*, 1974; Nicolas, 1991). It has been also accepted that the QF method is the best way to preserve antigenecity and chemical compositions of biological specimens (Zglinicki *et al.*, 1986; Hippe-Sanwald, 1993). Therefore, ultrastructural artifacts produced by the conventional chemical fixation and dehydration steps can be avoided by the QF method (Hora *et al.*, 1990; Furukawa *et al.*, 1991; Nicolas, 1991; Yoshimura *et al.*, 1991; Ohno *et al.*, 1992; Chan *et al.*, 1993; Chan and Inoue, 1994), whose goal is to preserve cellular ultrastructures and components in their natural states. Since the mid-1980s, the QF method has become a popular tool for the ultrastructural study of biological specimens, followed by deep-etching (DE) or freeze-substitution (FS). Much information has been published on the QF method, which includes plunging tissues into liquid cryogen, spraying liquid cryogen against the specimen, or slamming it against a precooled metal surface (Harreveld and Crowell, 1964; Dempsey and Bullivant, 1976; Jehl *et al.*, 1981; Plattner and Bachmann, 1982; Menco, 1986; Cole *et al.*, 1990; Takami *et al.*, 1990; Furukawa *et al.*, 1991; Takami *et al.*, 1991; Greene and Walsh, 1992).

C. Freeze-Substitution Method

Under sodium pentobarbital anesthesia, pieces of the renal cortex are usually excised from animal kidneys with a razor blade. The samples are immediately processed by the QF method after being diced into small pieces (Takami *et al.*, 1990, 1991). Such tissue fragments are quickly frozen by plunging them into a liquid isopentane–propane mixture (around $-193°C$) or slamming them against metal blocks cooled in liquid nitrogen (Ohno and Fujii, 1990, 1991; Naramoto *et al.*, 1991a, 1991b). The quickly frozen kidney tissues are then freeze-substituted in absolute acetone containing osmium tetroxide, by the following procedure. The samples are placed in the acetone solution containing osmium tetroxide and kept at about $-80°C$ for 20 hr (Furukawa *et al.*, 1991). The temperature is then raised, first to $-20°C$ for 2 hr, to $4°C$ for 2 hr, and finally to room temperature. The samples are washed twice in pure acetone and embedded in epoxy resin. Ultrathin sections are routinely prepared and examined in electron microscopes.

1. Basement Membrane

In the GBM of fresh kidneys prepared by the QF–FS method, the obscure lamina densa is directly attached to the cell membrane of foot processes and endothelial cells (Fig. 4). The distance between foot processes has a mean value of about 32 nm. A detailed description of the morphometric data to determine the width of slit diaphragms is discussed in Section II,C,3 (Furukawa *et al.*, 1991). In contrast, conventionally fixed specimens show artificial ultrastructures of the GBM and the foot processes, which are probably caused by the routine fixation and dehydration steps (Furukawa *et al.*, 1991; Ohno *et al.*, 1992).

2. Anionic Sites

The ultrastructure of anionic sites was examined by using PEI with two different molecular weights (MW = 1800 or 70,000), which were prepared

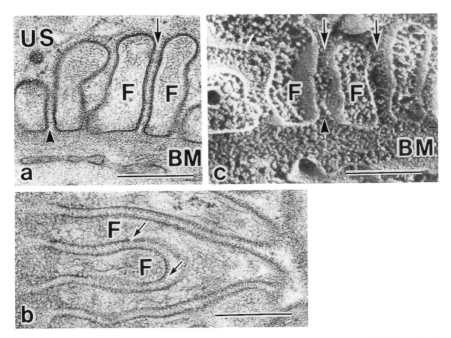

FIG. 4   Electron micrographs of glomerular capillary loops prepared by the QF–FS method (a,b) and the QF–DE method (c). Bar = 0.5 μm. (a) Flocculent materials are localized in the basement membrane (BM) and urinary space (US). The epithelial slits (arrows) run relatively straight, with narrow spaces between foot processes (Fs). A slit diaphragm is seen spanning the filtration cleft (arrowhead). The lamina densa, LRE, and LRI are less discernible in these specimens than in Fig. 2. (b) Slit diaphragms (arrows) are seen between foot processes (Fs) in ultrathin sections obliquely cut to the BM. They appear to be networks bridging adjacent foot processes (Fs). Neither central filament nor periodic cross-bridge is seen. (c) A replica electron micrograph of capillary loop freeze-fractured transversely to the BM. The slit diaphragm (arrowhead) is observed between foot processes (Fs). The epithelial clefts run with relatively straight courses (arrows).

for the QF–FS method (Yoshimura *et al.*, 1992). A 0.5% PEI solution was injected into rats, followed by perfusion fixation with paraformaldehyde. The left kidneys were then removed and processed by the previously described QF–FS method. Numerous PEI-labeled anionic sites are identified as tiny particles that are distributed in the LRE and LRI (Fig. 5a). Fine fibrils are also identified in both LRE and LRI or meshwork structures of lamina densa. In contrast, the tiny PEI particles disappear and a regular arrangement of large particles is often observed in the LRE of insufficiently frozen glomeruli, where large ice crystals are easily formed (Figs. 5b, 5c). In the badly frozen areas the visualized size of PEI particles is far larger than that in well-frozen areas, and the filamentous structures of the GBM are mostly destroyed (Fig. 5c). PEI particles can be more precisely localized by using small PEI molecules (MW = 1800) as opposed to large ones (MW = 70,000), probably because an aggregation of large PEI particles is too big for identification of tiny anionic sites. Thus, small molecular cationic probes are better for observing ultrastructures of anionic sites in the GBM, as shown also in Fig. 8e. Thus, the regular arrangement of anionic sites observed in conventional ultrathin sections (Fig. 3) is artificially produced by the preparation steps, which are usually necessary for visualization of PEI particles. As shown in Fig. 5c, the localization of PEI particles similar to that in conventional sections is demonstrated in the GBM concomitant with ice crystals because of insufficient freezing time, indicating that the

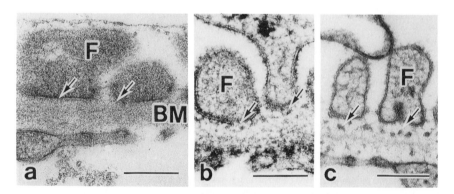

FIG. 5 Electron micrographs of PEI localization in the glomerular basement membrane (BM) prepared by the QF–FS method. Bar = 0.25 μm. (a) Many PEI-decorated anionic sites (arrows) are identified as pinpoint-sized grains, which are distributed in well-frozen areas. (b, c) The number of PEI particles (arrows) is decreased, and their regular arrangement in the LRE is observed in badly frozen areas, where large ice crystals are formed. The size of particles is far larger, and the filamentous structure of the GBM is largely destroyed in (c). F, foot process.

regular arrangement of large PEI particles is artificially produced by some damage to the GBM during conventional tissue preparations.

3. Slit Diaphragm and Foot Process

The glomerular slit diaphragm had been reported to consist of zipperlike substructures in normal kidneys that were treated with TGO (Rodewald and Karnovsky, 1974). However, it was recently demonstrated that the three-dimensional ultrastructures of slit diaphragms in unfixed rat glomeruli were originally sheetlike, as revealed by the QF–DE method (Hora et al., 1990; Ohno et al., 1992). In contrast, the slit diaphragms in TGO-fixed glomeruli appeared to be zipperlike and more widely open than those in the unfixed glomeruli. It is also known that ultrastructural images obtained by the QF–FS method represent the in vivo structures of kidneys more closely than do those of specimens prepared by the conventional fixation method. This section clarifies whether the width of the freeze-substituted slit diaphragm is narrower than that fixed by the conventional fixation method.

Fresh kidney tissues are quickly frozen and processed by the FS method as already described (Furukawa et al., 1991). The distances between the neighboring foot processes and the widths of foot processes are measured at 50 points in the cross-sections from several glomeruli for each fixation group, as schematically described in Fig. 6. Data are expressed as the mean ± standard deviation (SD). The morphometric data are shown in Table I. The width of slit diaphragms is significantly narrower after the FS than after the conventional TGO or GO fixation ($P < 0.001$): After FS with 2% osmium tetroxide, the slit diaphragms have a mean width of 33.8 ± 3.6 nm, whereas after the TGO and GO fixations, they have mean widths of 47.0 ± 4.3 nm and 48.7 ± 5.5 nm, respectively. Similarly, the space between neighboring foot processes at 50 nm from the slit diaphragms is significantly narrower with the FS method ($P < 0.001$): 54.7 ± 18.7 nm versus 98.6 ±

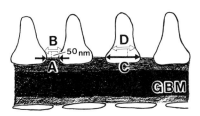

FIG. 6   Schematic representation of measured areas in the glomerular capillary loop. A, width of slit diaphragm; B, distance between the neighboring foot processes at 50 nm from the slit diaphragm; C, width of foot process at the level of the slit diaphragm; D, width of foot process at 50 nm from the slit diaphragm. GBM, glomerular basement membrane.

TABLE I

Morphometric Data Obtained from the Measured Areas as Shown in Fig. 6

	FS fixation (mean ± SD, $N = 50$)	TGO fixation (mean ± SD, $N = 50$)	GO fixation (mean ± SD, $N = 50$)
A	33.8 ± 3.6 nm	47.0 ± 4.3 nm	48.7 ± 5.5 nm
B	54.7 ± 18.7 nm	98.6 ± 17.1 nm	122.0 ± 29.3 nm
C	253.7 ± 89.5 nm	228.4 ± 61.0 nm	246.8 ± 86.9 nm
D	235.1 ± 87.2 nm	179.2 ± 59.4 nm	188.8 ± 78.2 nm

* $P < 0.01$.
** $P < 0.001$.

17.1 nm with the TGO fixation and 122.0 ± 29.3 nm with the GO fixation. Foot processes at 50 nm from slit diaphragms are significantly broader with the FS method: 235.1 ± 87.2 nm versus 179.2 ± 59.4 nm with the TGO fixation ($P < 0.001$) and 188.8 ± 78.2 nm with the GO fixation ($P < 0.01$). Because the tissue shrinkage and pulling apart of the foot processes are probably minimized with the FS method, the epithelial clefts run with relatively straight and narrow courses. This is due to the close apposition of the glycocalyx on adjacent foot processes *in vivo*. Where the plane of an ultrathin section passes almost parallel to the basement membrane the slit diaphragms are observed as consistently heterogeneous structures bridging adjacent foot processes (Fig. 4b). The foot processes are probably contracted during the conventional TGO or GO fixation, except for the attachment of their bases to the GBM. Unlike the foot processes prepared by the QF–FS method, the TGO-fixed glomeruli show their shrunken surface contours.

It is well known that TGO-fixed kidney specimens show an increased contrast in the extracellular structures. Although the ultrastructure of slit diaphragms is clearly revealed in such specimens, it can be modified during the conventional fixation, dehydration, and embedding steps. It is probable that the broadening of slit diaphragms is due to a shrinkage artifact that causes adjacent foot processes to be pulled apart. Latta (1970) also reported that slit diaphragms would be pulled apart by the shrinkage of foot processes after chemical fixation. In contrast, the QF–FS method causes less tissue shrinkage, because the osmium tetroxide fixation occurs below the freezing point.

Rodewald and Karnovsky (1974) demonstrated that the slit diaphragm was a continuous junctional band, 39.4 ± 2.9 nm wide after TGO fixation and had zipperlike substructures. However, it was doubtful that such morphology was an accurate description of the slit diaphragm. It is probable that the slit diaphragm between neighboring foot processes acts as a cell junction apparatus (Schnabel *et al.*, 1990). Because the TGO fixation has a high osmolarity, the zipperlike substructure is likely pulled by the shrinkage of foot processes. It has been demonstrated that the width of slit diaphragms is narrower after FS fixation. Therefore, it is probable that the popular zipperlike substructure of slit diaphragms seen in the TGO-fixed glomeruli is an artifact caused by the conventional fixation procedure, which is discussed in Section II,D,4.

D. Deep-Etching Method

To obtain three-dimensional images of glomerular structures, the previously frozen kidney tissues are freeze-fractured with a scalpel in liquid nitrogen (Takami *et al.*, 1990, 1991). They are deeply etched under vacuum conditions of $2-6 \times 10^{-7}$ torr at a temperature of $-95°C$ for 20–30 min. After being deep-etched, the specimens are placed on a rotary stage and first shadowed with platinum at angles of 25–35° for several seconds and then rotary shadowed up to the total thickness of about 2 nm (Ohno *et al.*, 1992). The samples are additionally coated with carbon at an angle of 90°. A drop of 2% collodion in amylacetate is placed on the replicas as soon as they are taken out of the freeze-etching machine to prevent their breaking into pieces during the subsequent digestion procedure. The replica membranes coated with dried collodion are treated with household bleach to dissolve the tissue components. They are washed in distilled water, cut into small pieces with scissors, picked up on Formvar-filmed copper grids, and immersed in amylacetate solution to dissolve the collodion film. The sections are then observed in electron microscopes. Some stereo pictures are taken at tilting angles of ±5°. Electron micrographs are printed from the inverted negative films.

1. Basement Membrane

The three-dimensional structure of fibrils in the GBM is not well defined in conventional ultrathin sections. In contrast, the DE method is highly versatile for visualizing three-dimensional intracellular structures (Ohno, 1985; Takasu *et al.*, 1988; Naramoto *et al.*, 1990; Ohno and Fujii, 1990) and extracellular matrices (Larabell and Chandler, 1988). The three-dimensional ultrastructure of the GBM was clarified by the QF–DE method (Takami *et al.*, 1990, 1991). Male rats were perfused through the aorta with 2% paraformaldehyde in 0.1 *M* phosphate buffer (PB). The renal cortices were cut into small pieces and washed in PB to remove soluble proteins from the glomeruli. They were postfixed with 0.25% glutaraldehyde, immersed in 10% methanol, and quickly frozen, as described in Section II,B (Jehl *et al.*, 1981; Ohno, 1985). The frozen tissue surfaces were freeze-fractured in liquid nitrogen and processed for replica preparations as described in Section II,D. The long and short dimensions of meshwork pores and the diameters of fibrils were measured on the replica electron micrographs. These data were expressed as the mean ± SD.

The replica electron micrographs show three layers in the structure of the GBM with widths of 250 to 280 nm (Fig. 7). However, it is not clear whether these layers are completely identical with those commonly ob-

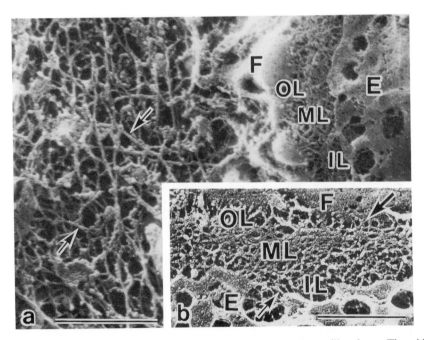

FIG. 7   Replica electron micrographs of freeze-fractured glomerular capillary loops. The middle layer (ML) of the GBM is composed of fibrils, which form the polygonal meshwork structure. At the outer layer (OL), fibrils connect the meshwork structure of the ML with the true cell surface of foot processes (b, large arrow). At the inner layer (IL), fibrils also connect the ML with the cell surface of endothelial cells (b, small arrow). In the cytoplasm of podocytes, intermediate filaments (a, arrows) form main cytoskeletal networks. F, foot process; E, endothelial cell. Bar = 0.5 μm.

served in conventional ultrathin sections, such as the LRE, LRI, and lamina densa. To avoid confusion with the terminology used in the case of ultrathin sections, we temporarily designate the three layers observed with the QF–DE method as outer, inner, and middle layers, respectively. The outer layer is 30 to 40 nm wide and consists of fibrils that connect cell membranes of foot processes perpendicularly with meshworks of the middle layer (Fig. 7b). The middle layer is 160 to 200 nm wide and has polygonal meshwork structures that are composed of fibrils. The spaces between fibrils of the meshwork have a long diameter of 16.8 ± 8.7 nm and a short one of 12.0 ± 6.2 nm. Stereo pictures demonstrate that the meshwork has a three-dimensional organization. The inner layer is 30 to 40 nm wide and consists of perpendicular fibrils (9.0 ± 2.2 nm in diameter) that connect cell membranes of endothelial cells with the middle layer.

The GBM has been assumed to contain type IV collagen, laminin, entactin, heparan sulfate proteoglycan, and fibronectin (Laurie et al., 1984). These components have been demonstrated to undergo spontaneous aggregation and to form orderly meshworks (Yurchenco and Furthmayr, 1984; Yurchenco and Rubin, 1987; Grant et al., 1989), which may provide the framework of the middle layer. The GBM has been identified as a size-barrier structure, and its structural determinants of permeability for macromolecules have been also analyzed in some detail (Caulfield and Farquhar, 1974; Farquhar, 1975). It was assumed initially that the GBM acts only as a molecular sieve, filtering various molecules on the basis of their size and shape. Pappenheimer (1953) first proposed the hydrodynamic model of solute transport through the GBM. This model envisions transport of solutes as taking place through many cylindrical pores. However, more recent data do not support the cylindrical pore model (Yurchenco and Rubin, 1987) and instead indicate that the GBM has a polygonal meshwork structure. The space between meshwork fibrils appears to produce an insufficiently tight sieve to exclude albumin protein. It is also known that a charge barrier provided by heparan sulfate proteoglycan acts as another molecular sieve (Kanwar and Farquhar, 1979a). Moreover, the hydrostatic pressure and hemodynamics may play other roles in GBM filtration (Simpson, 1986). It is not yet known whether the meshwork structure of the GBM is altered by high blood pressure or increased blood viscosity. The meshwork structure is seemingly not a rigid framework but rather a flexible structure imparted by superhelix formation of type IV collagen.

The LRE and LRI are seen as electron-lucent layers in conventional ultrathin sections. It is known that anionic sites, composed mainly of HSPG, are distributed throughout the LRE and LRI, as revealed by using cationized ferritins, ruthenium red (Kanwar and Farquhar, 1979c), and high-iron diamine (Washizawa *et al.*, 1989). It was demonstrated by alcian blue staining that 9-nm-thick fibrils run perpendicular to the cell membrane of foot processes. The same perpendicular fibrils are also found in the LRI (Caulfield, 1979) and could be made of protein backbones of heparan sulfate proteoglycans (Reale *et al.*, 1983). Functionally, the fibrils seem to anchor foot processes with lamina densa in the LRE, because the podocytes receive the glomerular transcapillary hydraulic pressure. We also discuss partial interruption of perpendicular fibrils under some pathological conditions in Section III,A,1.

2. Anionic Sites

The anionic sites present in the GBM are important in regulating the permeability of GBM to various proteins. Many studies with conventional ultrathin sections have demonstrated that cationized PEI particles are regularly spaced along the LRE. Such anionic sites seem to match the location of HSPGs (Kanwar and Farquhar, 1979a, 1979b). The three-dimensional ultrastructure of the anionic sites was examined by the QF–DE method (Yoshimura *et al.*, 1991; Duan *et al.*, 1993). Male rats were injected via the left renal artery with 0.5% PEI solution under ether anesthesia. At 15 min after the injection, they were perfused with 2% paraformaldehyde in PB. The kidneys were taken out and cut into small pieces ($2 \times 2 \times 4$ mm) with razor blades. They were washed in PB to remove soluble proteins from the tissue surface and fixed again with 0.25% GA. The specimens in group 1 were treated with phosphotungstic acid (PTA) and directly processed by the QF–DE method (Fig. 8a). To assess the contrasting effect of PEI, the fixed specimens in group 2 were washed in PB and then incubated in PTA–GA and osmium tetroxide (Fig. 8b, c). They were then processed by the QF–DE method.

Figure 8a shows many PEI particles around fibrils in the LRE of group 1. The arrangement of such particles is not as regular as that observed in conventional ultrathin sections (see Fig. 3). The PEI particles are about 20 nm in diameter. In contrast, the glomerular tissues, which are postfixed with PTA–GA and osmium tetroxide, show a shrunken appearance on replica membranes (Fig. 8b, c). The number and arrangement of PEI particles in group 2 is almost the same as observed in the conventional preparation. The number of PEI particles in the LRE is markedly decreased in group 2 as compared with group 1. Interruption of fibrils is often observed in some parts of the LRE.

In group 1, no contrasting with osmium tetroxide is necessary for visualization of anionic sites, and numerous PEI particles are detected around fibrils in the LRE. This finding supports a previous report that filamentous structures are present in the LRE and might be heparan sulfate proteogly-

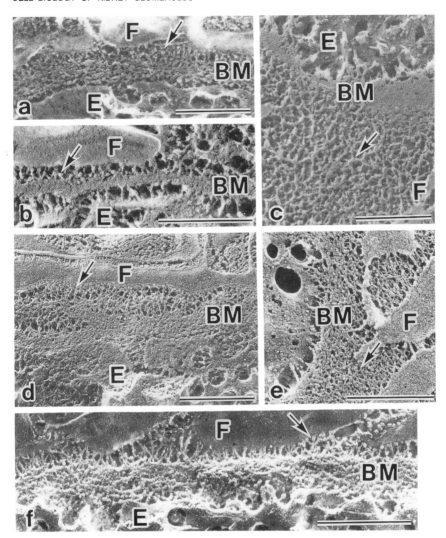

FIG. 8   Ultrastructural localization of PEI particles in the basement membrane (BM) under various experimental conditions. Bar = 0.5 μm. (a) Many PEI particles (arrow), about 20 nm in diameter, are localized around fibrils after the perfusion of PEI *in vivo*. (b) The number of PEI particles (arrow) in the LRE is decreased after osmium tetroxide fixation. Interruption of fibrils is also prominent in the LRE. (c) LRE freeze-fractured horizontally to the BM. The fine fibrils are mostly disturbed, and the number of PEI particles (arrow) is decreased. (d) Many PEI particles (arrow), about 10 nm in diameter, are located around fibrils in the LRE by perfusion of PEI after the paraformaldehyde fixation. (e) Small PEI particles with molecular weight of 1800 decorate fibrils of networks in LRE (arrow). (f) The perfused cationized ferritins are also localized in the LRE (arrow), LRI, and lamina densa. F, foot process; E, endothelial cell.

cans (Reale *et al.*, 1983). However, the ultrastructure of anionic sites in group 1 is completely different from that in group 2, in which the replicas examined were prepared after the contrasting and fixation necessary for conventional visualization of PEI particles. Therefore, the decreased density of PEI particles and destruction of fibrils seen in group 2 were caused by the chemical procedures, such as contrasting and fixation with PTA–GA and osmium tetroxide (Yoshimura *et al.*, 1991). Although anionic sites visualized with PEI are located regularly in conventional ultrathin sections and also in group 2, their structural localizations are disarranged by the chemical treatments. In contrast, the QF–DE method produced much less modification of the anionic sites (Figs. 8d, 8e, 8f), as revealed in group 1.

3. Mesangial Matrix

The three-dimensional organization of the mesangial matrix (MM) has been poorly defined in conventional ultrathin sections. This section describes how the three-dimensional ultrastructure of MM was clarified by the QF–DE method (Heuser and Kirschner, 1980; Ohno, 1985). The mesangium is composed of mesangial cells and the MM (Fig. 1). The MM has a loose meshwork structure made up of fibrils 6 to 10 nm thick and pores 32 ± 16 nm in diameter (Fig. 9). The meshwork structures of the GBM and MM are gradually transformed into each other. Mesangial fibrils are also associated with cell membranes of both endothelial and mesangial cells. Stereo pictures show the three-dimensional relationship between mesangial cells and the MM. In paramesangium the MM is located between the paramesangial GBM and the mesangial cell (Fig. 9). The mesangial cell is usually irregular in shape and has numerous cytoskeletal networks that are associated with cell organelles. The endothelial cell is also connected with the meshwork of the MM by fibrillar structures.

By employing the QF–DE method, we demonstrated the three-dimensional meshwork structure of the MM (Takami *et al.*, 1990), which is rarely observed in conventional ultrathin sections, because fixation with osmium tetroxide may have considerable distorting and destructive effects on various tissue components (Jost and Griffith, 1973; McMillan and Luftig, 1973; Maupin-Szamier and Pollard, 1978). Fibrils 6 to 10 nm in diameter are the main components of meshworks in the MM. The localization of collagen type IV, fibronectin, laminin, entactin, and HSPG was demonstrated in the MM by immunohistochemistry (Courtoy *et al.*, 1980; Madri *et al.*, 1980; Roll *et al.*, 1980; Bender *et al.*, 1981). Moreover, the meshwork structure has been reported to be organized by an *in vivo* self-assembly mechanism (Yurchenco and Rubin, 1987; Grant *et al.*, 1989). The microfibrillar connections between the GBM and mesangial cells have recently been clarified at mesangial angles

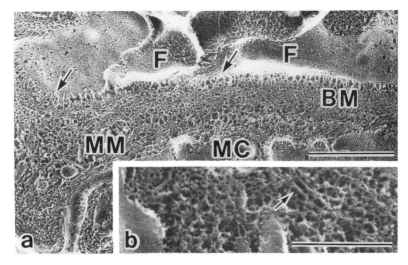

FIG. 9   Replica electron micrographs of the paramesangial region. Bar = 0.5 μm. (a) The mesangial matrix (MM) fills the space between paramesangial basement membranes (BMs) and mesangial cells (MCs). The BM and MM have meshwork structures. Note the looser meshwork structure of the MM in contrast with the rather compact BM. Arrows indicate fibrils that connect the foot processes (Fs) with the meshwork of the BM. (b) Mesangiolysis induced with anti-thymocyte serum. The destruction of fibrils and appearance of thicker fibrils are often identified in the MM.

(Sakai and Kriz, 1987; Mundel *et al.*, 1988). The thicker fibrils could be microfibrils, probably composed of amyloid P, which is associated with fibronectin (Schwartz *et al.*, 1985; Rostango *et al.*, 1986).

It is important to clarify some of the complex mechanisms involved in the movement of macromolecules into and out of the mesangium. Some ultrastructural studies with a variety of tracers support the concept of plasma flow through the MM. The permeability of the MM is dependent on a wide variety of determinants, such as blood pressure, contraction of mesangial cells, and changes of anionic sites. Moreover, pore sizes of the mesangial meshwork could be one of the determinants of permeability change. It is well known that macromolecules can pass more easily through the MM than through the GBM (Leiper *et al.*, 1977; Keane and Raij, 1980; Latta and Fligiel, 1985). This functional permeability of the MM is compatible with the present data that the MM has larger pores than the GBM. The contractility of mesangial cells has been shown by experiments *in vitro* (Ausiello *et al.*, 1980; Singhal *et al.*, 1986). The mesangial contraction could be transmitted to the GBM by mesangial processes or by the three-dimensional meshwork structure of the MM. Because the GBM has a compact meshwork structure, the transmitted contractile force would alter the diameter of capillary loops to regulate glomerular blood flow. Thus, the QF–DE method has elucidated the morphological background for some functions in the mesangium, such as the mesangial pathway and glomerular contraction (Takami *et al.,* 1990). The QF–DE method could provide a variety of new information about the mesangium in the future.

4. Slit Diaphragm

The glomerular slit diaphragms have been reported to show zipperlike or weblike substructures by a conventional freeze-fracture replica method (Karnovsky and Ryan, 1975) and irregular zipperlike substructures by the QF–DE method (Kubosawa and Kondo, 1985). Despite these reports, it can still be argued that these substructures simply reflect the artificial appearance of the glycocalyx within epithelial slits, because adjacent foot processes are pulled apart during conventional tissue preparations. Hora *et al.* (1990) reported that the slit diaphragms in isolated glomeruli were easily modified to form zipperlike, ladderlike, and sheetlike substructures by various fixation methods (Fig. 11c). This study suggests that substructure rearrangement might easily occur during the specimen preparation. The QF–DE method was used to reevaluate the zipperlike substructure of slit diaphragms (Ohno *et al.,* 1992).

Rat kidneys were conventionally perfused with tannic acid and glutaraldehyde in PB (Rodewald and Karnovsky, 1974; Furukawa *et al.,* 1991). They were cut into small pieces and postfixed with osmium tetroxide. The fixed samples were quickly frozen and processed for replica preparation. Fresh, unfixed kidneys were also processed for the QF–DE method.

In the replica membranes the slit diaphragms are observed between the shrunken foot processes and consist of zipperlike substructures in the conventionally prepared specimens (Fig. 10a). The cross-bridges appear to extend between the two sides of a central filament, thus giving the slit diaphragm a zipperlike appearance. Conversely, the width of slit diaphragms, as measured from the outer surface of one foot process to the next, averages 25 nm in the fresh, unfixed specimens processed for the QF–DE method (Fig. 10b). The replica membranes, which were freeze-fractured parallel to the basement membrane, demonstrate the widely viewed ultrastructure of the slit diaphragms in the fresh, unfixed ones (Fig. 10c). Interdigitating foot processes are in close apposition, and sheetlike substructures are detected beween them. Figure 10d illustrates that various substructures are identified in some parts of horizontally freeze-fractured capillary loops under different experimental conditions. The irregular zipperlike substructures are sometimes observed

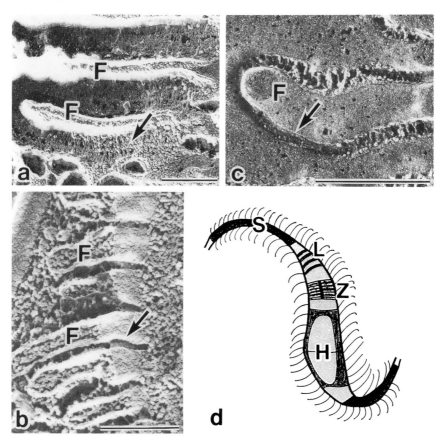

FIG. 10 Replica electron micrographs of slit diaphragms under various experimental conditions. Bar = 0.5 μm. (a) The freeze-fractured capillary loop, which was fixed with TGO, is observed from the urinary space. Shrunken foot processes (Fs) are separated from each other by large spaces. The slit diaphragm (arrow) consists of a single line in the middle and cross-bridging filaments to form zipperlike substructures. (b) A capillary loop of fresh, unfixed tissues is obliquely freeze-fractured. The slit diaphragms between foot processes (Fs) show sheetlike substructures (arrow). (c) A capillary loop is horizontally freeze-fractured along the basement membrane, giving an *en face* view of sheetlike substructures (arrow) between foot processes (Fs). (d) Schematic representation of various substructures in slit diaphragms of differently prepared specimens. S, sheetlike; L, ladderlike; Z, zipperlike; H, destroyed structure.

in specimens fixed with paraformaldehyde, which are continuously transformed to ladderlike substructures (Figs. 11a, 11b). Isolated glomeruli also show two types of substructures, ladderlike and sheetlike ones (Fig. 11c).

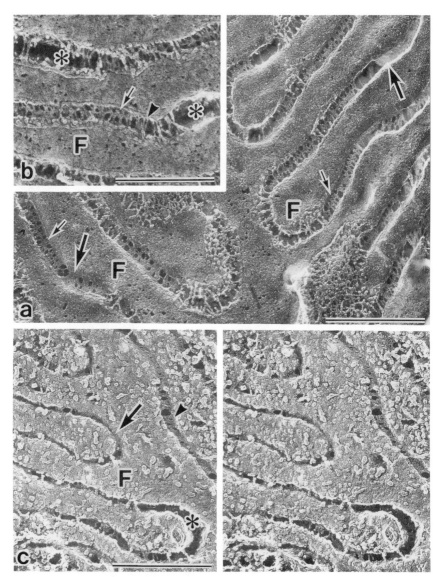

FIG. 11   Replica electron micrographs of various substructures in slit diaphragms. Bar = 0.5 μm. (a, b) The slit diaphragms show sheetlike (large arrows), ladderlike (arrowhead), and irregular, zipperlike (small arrows) substructures, which were prepared by paraformaldehyde fixation. Some structures are destroyed, forming large holes (asterisks in b and c). (c) Stereo pictures of slit diaphragms in isolated glomeruli. Ladderlike (arrowhead) and sheetlike (arrow) substructures are identified between foot processes (Fs).

The slit diaphragms in the fresh, unfixed glomeruli actually show sheetlike substructures, although they are composed of regular zipperlike substructures in glomeruli fixed with TGO (Ohno *et al.*, 1992), which suggests that the zipperlike appearance is an artifact caused by the TGO fixation. These findings indicate that the slit diaphragms really exist in the glomeruli *in vivo.* It has been reported that they have a regular pattern when observed by transmission electron microscopy after the TGO fixation method and the conventional freeze-fracture replica method (Rodewald and Karnovsky, 1974; Karnovsky and Ryan, 1975; Ryan *et al.*, 1975; Schneeberger *et al.*, 1975). They have periodic cross-bridges extending from the foot processes to a central filament to form homogeneous pores (about 4×14 nm) and are continuous between all foot processes. Moreover, physiological data indicate that the glomerular filter functions as an isoporous membrane that excludes proteins larger than serum albumin (Caulfield and Farquhar, 1974; Purtell *et al.*, 1979).

There has been considerable controversy concerning the main filtration barrier of the glomerulus, principally between the basement membrane theory (Caulfield and Farquhar, 1974; Shikata *et al.*, 1990) and the slit diaphragm theory (Graham and Karnovsky, 1966; Rennke and Venkata-chalam, 1977). Proteinuria has been reported to result primarily from an abnormality in the GBM or epithelial detachment (Kanwar, 1984; Messina *et al.*, 1987). The central lamina densa is known to restrict the passage of large protein molecules (Ryan and Karnovsky, 1976). However, the morphological findings support the notion that the slit diaphragm represents an additional filtration barrier (Graham and Karnovsky, 1966), as suggested by a study using a monoclonal antibody against slit diaphragms (Orikasa *et al.*, 1988). During conventional TGO fixation, tiny holes are easily produced in the fragile sheetlike substructures of the slit diaphragms. Physiologically, the dimensions of rectangular pores in the zipperlike substructures are almost compatible with the size of albumin molecules (Brenner *et al.*, 1978), supporting their role as a filtration barrier against plasma proteins. However, the present evidence suggests a different view of physiological functions of the slit diaphragm, because of its sheetlike substructure (Shea and Morrison, 1975). It is still unclear how small proteins, such as horseradish peroxidase and small amounts of albumin, filter through the glomerular capillary walls (Graham and Karnovsky, 1966; Venkatachalam *et al.*, 1970b; Bariety *et al.*, 1978). It is probable that tiny porous structures exist in the slit diaphragms and are masked by shadowing with platinum metal. They are discussed in Sections III,B,1 and III,B,2. However, another possibility is that the sheetlike substructures are hemodynamically transformed into porous structures to allow the passage of protein molecules.

III. Experimental Alteration of Glomerulus

A. Basement Membrane and Mesangial Matrix

1. Serum Sickness Nephritis

The development of glomerulonephritis is due mainly to immunological reactions. The category includes immunocomplex glomerulonephritis such as IgA nephropathy, membranous glomerulonephritis, and membranoproliferative or poststreptococcal glomerulonephritis. Three types of immune mechanisms, including anti-glomerular antibody, *in situ* immune complex formation, and circulating immune complexes, participate in the initiation and development of glomerulonephritis. Experimental serum sickness nephritis has been a suitable model for the immunocomplex glomerulonephritis induced by circulating immune complex (Bolton and Sturgill, 1978; Yamamoto *et al.*, 1978; Shigematsu and Yano, 1986). In this model, however, the three-dimensional relationship between immune deposits and extracellular matrices of the GBM has not been analyzed. The QF–DE method has been used to reveal how the immune deposits appear in the extracellular matrices of the GBM and mesangium (Naramoto *et al.*, 1991b; Nakazawa *et al.*, 1992). Particular attention was also paid to cytoskeletal changes in podocytes under the experimental conditions.

Granular immune deposits are localized in filamentous networks of the GBM (Fig. 12). These constitutional fibrils with diameters of 8–15 nm are directly attached to the immune deposits. The filamentous networks become markedly loosened around some deposits (Fig. 12a). In epithelial podocytes, reticular microfilaments with positive decoration by myosin subfragment 1 (S1) are increased in flattened foot processes and directly attached to the cell membrane (Fig. 12b). Fine filaments with diameters of 4–7 nm are undecorated with S1 and connect with actin filaments as cross-bridges. Intermediate filaments are also increased in the cell bodies and primary processes of podocytes. The immune deposits are primarily detected in the networks of lamina densa and actually destroy the size barrier composed of filamentous networks. Moreover, the mesangial deposits also disorganize mesangial networks and probably alter mesangial flow through the matrices. Increased actin filaments in the foot processes seemingly reinforce the binding between the cell membrane and the connecting fibrils in LRE, which prevents the initial detachment of podocytes from the GBM.

The alteration of anionic sites in the GBM of serum sickness nephritis was studied with the QF–DE method using PEI as a cationic probe (Duan *et al.*, 1993). The fibrils around small immune deposits radiate into the lamina densa, and anionic sites are well preserved. These results suggest that some of the small deposits simply pass through the GBM and decorate

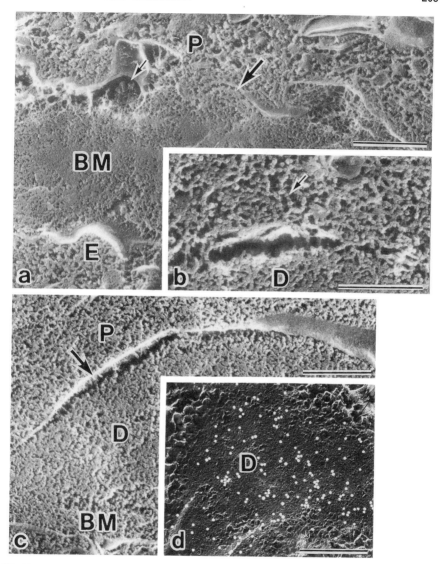

FIG. 12   Replica electron micrographs of peripheral glomerular basement membranes (BM) and podocytes (P) in serum sickness nephritis. Bar = 0.5 μm. (a) A small deposit (large arrow) is covered with densely packed microfilaments in the flattened foot process (P). Some structures in the LRE are destroyed (small arrow). E, endothelial cell. (b) The S1-decorated actin filaments (arrow) are localized under the cell membrane near immune deposits (D). (c) A humplike immune deposit (D) is shown in the LRE. Many fibrils organizing the BM are directly attached to the deposits, but connecting fibrils in the LRE are partially destroyed along the cell membrane (arrow). (d) Immunolocalization of egg albumin on replica membranes by freeze-fracture immunohistochemistry. The humplike deposit (D) is positively labeled by secondary immunogold antibody.

the fibril structures, but other large deposits probably destroy many fibrils, including anionic sites.

Ultrastructural changes in the GBM and MM are induced by immune deposits in serum sickness nephritis. In the initial phase, immune deposits appear to be trapped within fibrillar networks of lamina densa, because the connecting fibrils organizing the LRE are well preserved in the areas with little deposition. The immune deposits grow and translocate toward the LRE and finally localize as subepithelial deposits. The fibrillar networks of the lamina densa become irregular due to the dilatation of meshes around the immune deposits at later stages. The lamina densa in intact glomeruli, however, has a size barrier, which is composed of regular fibrillar networks (Laurie *et al.,* 1984). Therefore, immune deposition may cause the disruption and disorganization of this size barrier, resulting in the development of proteinuria. It was reported that the deposits in the extracellular matrix were positive for antigen, antibody, and complement factors but negative for type IV collagen, laminin, fibronectin, and heparan sulfate proteoglycan. The observation implies that the immune deposits destroy the normal fibrillar networks. It was also reported that larger immune deposits were preferentially precipitated in the MM (Elema *et al.,* 1976; Koyama *et al.,* 1978). Thus, the mesangial immune deposits might also destroy the fibrillar networks of the MM, resulting in abnormal mesangial flow.

Podocytes contain three major cytoskeletal elements—microfilaments, intermediate filaments, and microtubules (Vasmant *et al.,* 1984). Drenkhahn and Franke (1988) reported the regional difference in the cytoskeletal organization using postembedding immunoelectron microscopy. Actin microfilaments are localized in foot processes and under the cell membrane of cell bodies. However, vimentin and tubulin, the unit proteins of intermediate filaments and microtubules, respectively, are localized in the cell bodies and their major processes. In the previous experiments (Naramoto *et al.,* 1991b), S1-decorated actin filaments increased in the flattened foot processes, and intermediate filaments increased in the cell bodies and primary processes. The increased intermediate filaments probably work to maintain the structural integrity of podocytes under the pathological conditions (Drenckhahn and Franke, 1988). Moreover, S1-undecorated fine filaments between the actin microfilaments are recognized in the flattened foot processes. Some proteins of fine filaments with diameters of 2–5 nm have been identified as new cytoskeletal components (Roberts, 1987), which suggests that their function may contribute to the cell mobility and cytoarchitecture.

The detachment of podocytes from GBM is a serious tissue injury in relation to increasing proteinuria. Their focal detachment is known to enhance the permeability of the GBM, so large molecular proteins could escape into the urinary space (Kanwar and Rosenzweig, 1982; Whiteside *et al.,* 1989). Moreover, detachment is correlated to the development of focal glomerular sclerosis (Kanwar and Farquhar, 1980). In some areas, the only partial detachment of podocytes is seen, and the connecting fibrils in LRE are disrupted focally in such areas (Fig. 12). This finding may represent the initiation stage of the podocyte detachment. It was reported that anionic sites in LRE were necessary for the attachment of podocytes to GBM. In fact, anti–heparan sulfate proteoglycan antibodies could not bind with the areas of subepithelial immune deposition. Moreover, actin microfilaments in foot processes might be connected with the fibrils in LRE via the binding proteins (Drenckhahn and Franke, 1988). Therefore, it is suggested that the increased actin microfilaments reinforce the connecting fibrils to prevent the detachment of podocytes from the GBM.

2. Masugi Nephritis

Mesangial sclerosis, or a significant increase of mesangial matrix, is the characteristic feature of chronic glomerulonephritis. Analysis of the developing matriceal changes in chronic glomerulonephritis is important, because in some cases progressive mesangial sclerosis results in global glomerular sclerosis (Striker *et al.*, 1984). Moreover, an increased number in obsolescent glomeruli correlates with deterioration of renal function, which is often accompanied by persistent proteinuria (Shigematsu *et al.*, 1985). The chronic phase of accelerated Masugi nephritis is a suitable model for analyzing the organization and development of mesangial sclerosis (Shigematsu and Kobayashi, 1973). The three-dimensional changes of the GBM as well as the MM in Masugi nephritis were examined by the QF–DE method (Naramoto *et al.*, 1991a).

Newly formed MM, which is increased axially, is composed of fibrillar networks similar to those in the lamina densa of the GBM (Fig. 13a). These fibrils are 10–20 nm in diameter and directly attached to the cell membrane of mesangial cells and endothelial cells. The fibrillar networks are also seen in the areas of mesangial interposition of the glomerular capillary wall. The fibrils organizing the networks of lamina densa are partially thickened, accompanied by some decorations. The connecting fibrils are preserved along the areas of foot processes.

The three-dimensional localization of laminin was also investigated during the chronic phase of Masugi nephritis by the QF–DE method combined with immunohistochemistry (Naramoto *et al.*, 1992). The laminin immunolocalization was revealed with diaminobenzidine (DAB) reaction products of secondary peroxidase-labeled antibody following anti-laminin antibody (Fig. 13b). These immunoreaction products are not uniformly distributed in the newly formed glomerular matrix. Although the fibrils organizing lamina densa are immunostained with anti-laminin antibody, the fibrils associated with mesangial cells, foot processes, and endothelial

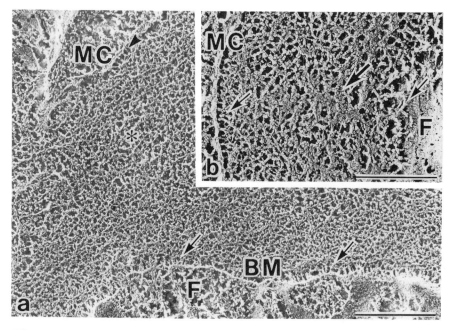

FIG. 13   Replica electron micrographs of mesangial areas in Masugi nephritis. Bar = 0.5 μm.
(a) The increased mesangial matrix is composed of fine fibrillar networks (asterisk). They
spread around mesangial cells (MC) and are directly attached to the cell membranes (arrowhead). (b) The proliferated mesangial matrix with anti-laminin immunostaining. The fibrils
organizing the proliferated mesangial matrix are heterogeneously decorated (large arrows).
The connecting fibrils around mesangial cells (MCs) and foot processes (Fs) are not decorated
(a,b, small arrows).

cells have small amounts of DAB immunoreaction products. It is suggested
that one of the components of proliferated fibrils is the laminin protein,
which is heterogeneously distributed in the newly formed matrix.

The backbone structure of mesangial sclerosis is composed of fibrillar
networks resembling the lamina densa. Laminin is one of the components
organizing the fibrils. Other components include type IV collagen, fibronectin, entactin, and glycosaminoglycan (Ishimura *et al.,* 1989). The fibrillar
networks have been reported to be formed by a self-assembly mechanism
(Yurchenco and Furthmayr, 1984). Moreover, mesangial sclerosis results
from an increased synthesis of the components and disturbance of their
degradation in the mesangial compartment (Romen and Morath, 1979).
Functionally, newly formed mesangial networks seem to interfere with the
contractile ability of the mesangial cells. It is known that contraction and
relaxation of normal mesangial cells play an important role in the regulation
of glomerular filtration through the interconnecting fibrillar networks (Sa-

kai and Kriz, 1987; Mundel *et al.,* 1988). Therefore, mesangial sclerosis probably impairs the regulation of filtration.

Newly formed networks are also seen in the areas of mesangial interposition, which is one of the specific changes in progressive glomerulonephritis. The mesangial cells do not extend into the peripheral capillary wall under normal physiological conditions, possibly because endothelial cells or podocytes secrete heparin-like materials that inhibit mesangial cell proliferation (Castellot and Hoover, 1986). In Masugi nephritis, however, severe damage to endothelial cells may allow the proliferation or elongation of mesangial cells into the edematous subendothelial space. These interposed mesangial cells also produce the matrix of fibrillar networks. Thus, the newly formed fibrillar networks in the mesangial interposition may result from impaired recovery from lysis of extracellular matrix.

Only partial detachment of endothelial cells from the GBM is seen in glomerular capillary walls, where their connecting fibrils in LRI are usually disrupted (Naramoto *et al.,* 1991a). This finding presumably explains how the detachment of endothelial cells occurs in the GBM. The connecting fibrils around endothelial cells are easily disrupted under some pathological conditions. It has also been reported that other connecting fibrils are disrupted in the LRE of serum sickness nephritis and that detachment of podocytes is then induced (Naramoto *et al.,* 1991b). Prolonged injury with detachment of endothelial cells seemingly results in failure of the repair of the glomerular structure, often followed by progressive sclerosis.

In accelerated Masugi nephritis, nephrotoxic rabbit globulin and anti–rabbit IgG are known to be linearly localized along the peripheral basement membrane, as revealed by an immunofluorescent method (Masugi, 1980). The QF–DE method demonstrates that the fibrils that organize networks of lamina densa are diffusely thickened with some modification. It has been also reported from immunoelectron microscopy that autologous antibody against nephrotoxin is located throughout the GBM (Hoedemaeker *et al.,* 1972). Therefore, nephrotoxin and anti–rabbit IgG may be associated with the fibrils of the GBM. The matriceal fibrils are known to consist of type IV collagen, laminin, entactin, and fibronectin (Laurie *et al.,* 1984), although the injection of monoclonal antibodies against some proteins could not induce the proliferative glomerulonephritis of Masugi nephritis (Abrahamson and Caulfield, 1982; Yaar *et al.,* 1982). The principal distribution of antigens in Masugi nephritis has not yet been elucidated.

3. Diabetes Mellitus Nephropathy

Typical morphological changes of human advanced diabetic nephropathy are nodular mesangial expansion, thickening of the GBM, and hyalinosis of arteries. However, the detailed morphological changes in early diabetic nephropathy have not yet been clarified. Treatment at an early stage of diabetic nephropathy is clinically very important, because it is most likely reversible. Therefore, it is necessary to elucidate initial morphological changes in diabetic nephropathy. Although streptozotocin (STZ)-induced diabetic rats have been commonly used for the study of diabetic nephropathy, initial morphological changes have not yet been clarified. The three-dimensional ultrastructure of GBM in STZ-induced diabetic rats has been examined by the QF–DE method (Moriya *et al.,* 1993).

In diabetic rats the inner layer corresponding to the LRI is diffusely enlarged, and the meshwork structure of the lamina densa—the middle layer—becomes markedly irregular due to the rupture of fibrils and thickened fibrils with adherent materials (Fig. 14). These changes probably correspond to subendothelial edema and fluffy materials in the GBM, as revealed in conventional ultrathin sections. Thus, the initial morphological change in STZ-induced diabetic nephropathy is induced by disruption of

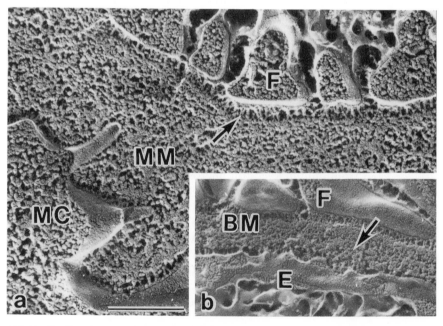

FIG. 14   Replica electron micrographs of glomerular capillary loops in diabetic rats at an early stage. Bar = 0.5 μm. (a) Tiny materials are deposited in the MM, and irregularity of meshwork structures is seen. An arrow indicates perpendicular fibrils of the LRE, which are well preserved. (b) The meshwork structures of the lamina densa have tiny deposits (BM), and enlargement of the LRI (arrows) with disruption of fibrils is often observed. F, foot process; E, endothelial cell; MC, mesangial cell.

matricial fibrils in the GBM, which seems to indicate disturbance of size or charge barriers. Only a few reports have described the glomerular morphology at the stage of microalbuminuria in early diabetic nephropathy (Inomata *et al.*, 1987; Chavers *et al.*, 1989). It is necessary to evaluate the mechanism of early diabetic nephropathy, because clinically advanced nephropathy is already at an irreversible stage and results sooner or later in chronic renal failure or uremia (Boushey *et al.*, 1986; Rosenstock and Raskin, 1986). Some reports clarified morphological changes in STZ-induced diabetic rats at late diabetic stages (Weil *et al.*, 1976; Rasch, 1979; Hirose *et al.*, 1982; Østerby *et al.*, 1984; Templeton, 1989), but it was not fully understood what early morphological changes are. These changes could be clarified by the QF–DE method.

It has been reported that albuminuria occurs within 4 to 8 weeks in STZ-induced diabetic rats (Cooper *et al.*, 1988; Anderson *et al.*, 1989) and that their blood pressure is higher than in control rats after 8 weeks (Anderson *et al.*, 1989). Focal subendothelial edema in diabetic rats is identified earlier than 8 weeks, as revealed in conventional ultrathin sections. The edema becomes more prominent at later stages and extends into mesangial areas or peripheral capillary loops, followed by mesangial interposition. There has been no report that subendothelial edema is detected as an early diabetic change in animals. In human diabetes some authors reported subendothelial edema (Koda *et al.*, 1991) or mesangiolysis (Saito *et al.*, 1988) as an early morphological change. It is reasonable to consider that the initial morphological change in diabetic animals is also glomerular subendothelial edema. It is also worth noting that as early as 4 weeks tiny materials are seen focally in the subendothelial space (Moriya *et al.*, 1993). They are usually attached to meshwork structures of the middle layer, which have some irregularity. Thus, we conclude that the tiny materials, probably originating from the bloodstream, adhere to fibrils in the meshwork structures of GBM at an early stage in STZ-induced nephropathy.

It is generally accepted that charge and size barriers against plasma macromolecules are formed in the GBM. Considering the mechanism of proteinuria, some authors have indicated the importance of the charge barrier (Vehaskari *et al.*, 1982), and others have stressed the size barrier (Friedman *et al.*, 1983; Nakamura and Myers, 1988; Myers, 1990). Proteinuria in diabetic nephropathy usually results from the disturbance of the charge barrier at an early stage, followed by damage to the size barrier at later stages (Viberti and Wiseman, 1986; Deckert *et al.*, 1988). The charge barrier disturbance was demonstrated by a decrease in the number of cationic tracers bound to anionic sites of the GBM (Chakrabarti *et al.*, 1989, 1991). Another way to detect its disturbance is to examine the main structure of the anionic sites, such as HSPG. In diabetic humans and animals the HSPG has been reported to be decreased (Shimomura and Spiro, 1987).

The HSPG is mainly localized at the LRE, and the decreased content of HSPG may result in disturbance of the charge barrier. By the QF–DE method, the outer layer of the GBM corresponding to the LRE shows a slight irregularity of fibril organization, which may be related to the charge barrier disturbance (Moriya *et al.*, 1993). The disturbance of the size barrier has also been indicated as the second mechanism of proteinuria in diabetic nephropathy (Deckert *et al.*, 1988; Myers, 1990). However, there has been no report that the size barrier disturbance was detected by a morphological method. The three-dimensional ultrastructure of normal GBM in Fischer rats was described by the QF–DE method (Takami *et al.*, 1991; Nakazawa *et al.*, 1992). The GBM in normal Wistar rats is also composed of three layers, as reported before (Moriya *et al.*, 1993). The size of GBM meshworks in normal Fischer rats is almost the same as in Wistar rats. Considering the previous reports, the meshwork structures of the lamina densa in diabetic nephropathy seem to act as the size barrier against plasma macromolecules.

In diabetic rats the enlarged space in the inner layer, as revealed by the QF–DE method, corresponds to the subendothelial edema observed in conventional ultrathin sections (Moriya *et al.*, 1993). Tiny materials around fibrils are about 16–32 nm in size on the replica membranes. The conventional splitting of the lamina densa also corresponds to the distortion of the fibrils in the middle layer, which provides the lesion for the easy escape of plasma macromolecules. However, what causes these morphological changes remains to be determined. The initial change is adhesion of tiny materials to the fibrils, and then the meshwork structures show their progressive irregularity (Moriya *et al.*, 1993). These morphological findings might be due to metabolic disorders of endothelial cells and podocytes, resulting in later mesangial proliferation and sclerosis.

4. Lupus Nephritis Model

Murine chronic graft-versus-host reaction (GVHR) was induced by injecting DBA/2 spleen cells into F1(B10XDBA/2) mice (Lewis *et al.*, 1968; Gleichmann *et al.*, 1982, 1984). In this model system, host B cells are stimulated by donor T cells and produce various kinds of autoantibodies, which result in immune complex glomerulonephritis. This GVHR is considered to be suitable for clarifying the mechanism of human lupus nephritis (Bruijin *et al.*, 1988, 1990). Immune complex deposits are noted mainly in the MM and subepithelial portions of the GBM. IgG antibodies tend to be localized in the GBM and IgM in the MM. Moreover, focal mesangiolytic changes are seen at stages later than the seventh week. These morphological findings suggest that the deposition of the immune complex might appear at an early stage, which is compatible with the biochemical data. The three-dimensional ultrastructures of glomerular lesions in the GVHR were examined by the QF–DE method (Otani *et al.*, 1993) and compared with those in other animal models of glomerulonephritis.

In the GVHR the immune complex composed of autoantibodies has been considered to induce animal models of human lupus nephritis. The replica membranes show various sizes of the immune deposits in the GBM (Fig. 15a). Moreover, fibrils around large deposits are strongly disrupted (Fig. 15b). Microfilaments are increased in the cytoplasm of podocytes, and endothelial cells are slightly detached from the GBM. As reported previously, the deposition of immune complexes is noticed in the glomeruli of this animal model system (Otani *et al.*, 1993). Immunofluorescent examination revealed that IgM deposition is dominant in the mesangial area from an early stage. Various immune deposits are also seen in the mesangial area by the QF–DE method (Fig. 15b). Fibrillar meshworks around the deposits are irregularly disrupted, and their pore sizes become enlarged.

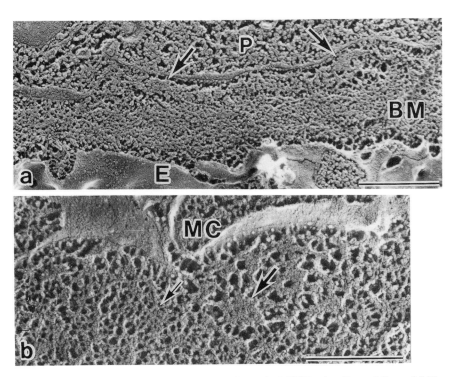

FIG. 15   Ultrastructures of glomerular capillary loops in GVHR mice. Bar = 0.5 μm. (a) The GBM is thickened with some small deposits (arrows). P, disappearance of foot process; E, endothelial cell. (b) Immune deposits in the mesangial area. Some meshwork structures are destroyed (large arrow), and others are preserved without deposits (small arrows). MC, mesangial cell.

It was reported that the MM in normal glomeruli have meshwork structures and that both foot processes and endothelial cells are attached to the lamina densa with perpendicular fibrils. In serum sickness nephritis, immune deposits are also seen in the MM, and meshworks are disrupted around them (Naramoto *et al.,* 1991b). In Masugi nephritis, glomerular sclerosis consists of condensed fibrillar structures, in which more thickened fibrils are connected with each other (Naramoto *et al.,* 1991a). It was reported that mesangial immune deposits might induce the secretion of IL-1 or cytokines and result in glomerular sclerosis (Iskandar *et al.,* 1988; Potanova *et al.,* 1988). In the GVHR the MM contains thick fibrils resembling collagen fibers, which might be related to the development of glomerular sclerosis.

Immune deposition induces irregularity and thickening of the fibrillar structures in the peripheral GBM as well as in the MM (Otani *et al.,* 1993). In the GVHR the morphological changes of the lamina densa corresponded to the splitting and distortion of the GBM in conventional ultrathin sections. Under light microscopy most of the glomeruli show crescent formation, which results from destruction of the GBM and exuded fibrin with infiltration of inflammatory cells. The immune deposits are localized mainly in the LRE, and the endothelial cells are partly detached from the lamina densa. Moreover, epithelial podocytes show the disappearance of foot processes and increased microfilaments in their cytoplasm. These epithelial and endothelial damages may accelerate the glomerular sclerosis, although heparin-like substances are reported to be beneficial for glomerular repair (Castellot *et al.,* 1982, 1985). It is suggested that immune complex deposition and the subsequent matricial damage may play an important role in the development of glomerular lesions.

B. Anionic Site and Slit Diaphragm

1. Polyethyleneimine Administration

We have reported that zipperlike substructures are easily formed by TGO treatment and that natural structures of slit diaphragms are sheetlike ones, as revealed in fresh, unfixed kidneys (Hora *et al.,* 1990; Ohno *et al.,* 1992). In this section we discuss the ultrastructural changes in rats treated with PEI *in vivo.* Rats were injected with PEI solution via the renal artery (Schurer *et al.,* 1978). At 15 min after the injection, they were perfused with 2% paraformaldehyde in PB, and the left kidneys were taken out and processed for the QF–DE method. In PEI-treated rats, many tiny holes form on the sheetlike substructures of slit diaphragms (Fig. 16a). Moreover, aggregated PEI particles are identified in the LRE. In some parts, cross-bridging fibrils are disrupted by the association with PEI particles.

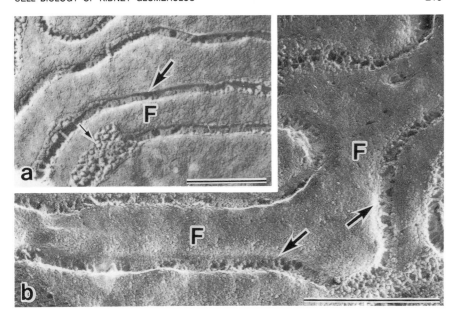

FIG. 16   (a) Replica electron micrograph of PEI-treated rats *in vivo*. Many tiny holes (large arrow) are formed on sheetlike substructures of slit diaphragms between foot processes (Fs). A small arrow indicates aggregated PEI particles. (b) Ultrastructural changes of slit diaphragms in rats treated with anti–slit diaphragm antibody *in vivo*. Many tiny holes (arrows) are also formed on sheetlike substructures between foot processes (Fs). Bar = 0.5 μm.

It is generally accepted that the negative charge barrier of the GBM is important for regulating the glomerular permeability. The anionic sites in GBM have been often examined with PEI as a cationic tracer. It has also been reported that the slit diaphragms are composed mainly of sheetlike substructures (Hora *et al.*, 1990; Ohno *et al.*, 1992). As shown in Fig. 16a, the PEI injection induces the formation of tiny holes on the sheetlike substructures and the disruption of cross-bridging fibrils in the LRE. However, the zipperlike substructures are not observed under these experimental conditions. Therefore, it is concluded that the sheetlike substructures are modified by PEI treatment *in vivo*, which might correlate with abnormal functions of the GBM in proteinuria.

2. Anti–Slit Diaphragm Antibody

It was reported that proteinuria was also induced by injecting monoclonal anti–slit diaphragm antibody into rats (Orikasa *et al.*, 1988). At 24 hr after the injection, cross-bridging fibrils in LRE become thick, and the cell surface of foot processes is modified with granular structures. The slit diaphragm has tiny porous substructures, as revealed on horizontally freeze-fractured GBM (Fig. 16b). Moreover, severe proteinuria occurs at later stages, and then the slit diaphragm shows many tiny holes on sheetlike substructures. This means that the sheetlike substructures are functionally important for the filtration barrier of proteins.

IV. Glomerular Morphology by an "*in Vivo* Cryotechnique"

The glomerulus consists of intricate networks of capillaries through which the blood passes under the influence of blood pressure. Therefore, it is reasonable that hemodynamic factors, such as the blood pressure and flow, exert an important influence on glomerular ultrastructures (Griffith *et al.*, 1967; Kanwar, 1984; Kriz *et al.*, 1994). The changes in glomerular hemodynamics may also affect the driving force that modulates the permeability property of the filtration barrier (Ryan and Karnovsky, 1976; Ryan *et al.*, 1976; Bohrer *et al.*, 1977; Brenner *et al.*, 1977; Olivetti *et al.*, 1981). The molecular sieving against plasma proteins usually occurs across the GBM under normal hydraulic pressures. The distribution of albumin proteins in rat glomeruli was reported to be altered due to different hemodynamic conditions (Ryan and Karnovsky, 1976; Olivetti *et al.*, 1981). So the functional role of normal blood circulation is definitely important in maintaining the glomerular barrier funciton. However, information about the significance of the glomerular ultrastructures, which are revealed by conventional preparation methods, has been limited (Yoshimura *et al.*, 1991; Chan *et al.*, 1993; Chan and Inoue, 1994), because the chemical fixation of tissues takes considerable time, during which their morphology could change. It is generally accepted that renal ischemia easily produces ultrastructural changes in glomeruli (Griffith *et al.*, 1967). The ultimate goal of morphological study is that all features to be examined should reflect the physiology under investigation and that no artifacts should be introduced by the preparation procedure.

The preservation of tissues in functioning organs is necessary in studies to define their ultrastructures (Griffith *et al.*, 1967; Harreveld *et al.*, 1974). Small specimens are commonly frozen within several seconds following the excision of tissues (Furukawa *et al.*, 1991; Nicolas, 1991; Chan *et al.*, 1993). Most cryotechniques have been based on the use of prior excised tissues. When kidney tissues are resected from the blood circulation system, they are immediately exposed to severe ischemic condition. Even the QF method has some limitations with such specimens. It is well known that the ultrastructures of kidneys are easily changed by stopping the blood supply (Griffith *et al.*, 1967). Thus, it has been difficult to elucidate the natural morphology of kidneys by using conventional cryotechniques. There has been a need for a new method capable of freezing tissues *in vivo* and obtaining acceptable morphology in functioning organs (Chang *et al.*, 1980; Zglinicki *et al.*, 1986). In this section we discuss the morphological aspects of functioning kidneys, whose ultrastructures are reassessed, particularly with regard to hemodynamic conditions operating at the time of cryofixation.

A. Ultrastructure under Normal Blood Circulation

In the "*in vivo* cryotechnique," male mice are anesthetized by injecting sodium pentobarbital via the peritoneal cavity. The abdomen is exposed through pararectus incision, and the left kidney is detected. The kidney *in situ* is carefully put on a wax plate without disturbing vessels or ureter. Then the *in vivo* cryotechnique is performed under the normal blood flow condition. First, the cryoknife edge is precooled in liquid nitrogen and allowed to reach liquid nitrogen temperature ($-196°C$) and is then positioned over the kidney organ *in vivo*. While the heart is beating normally the renal cortical tissues are cut and simultaneously frozen with the cryoknife edge. In practice the cryoknife is manually pushed into the kidney organ as fast as possible, and liquid isopentane–propane cryogen ($-193°C$) is immediately poured over it. Moreover, liquid nitrogen is poured on the frozen tissues to avoid their rewarming. The metal knife edge is withdrawn after the whole kidney is completely frozen with liquid nitrogen. The frozen kidney is then excised from the remaining abdominal tissues and transferred into liquid nitrogen, in which the well-frozen parts are identified and trimmed off with cooled nippers. They are then submitted to the DE method or the FS method.

The ultrastructures of areas of glomeruli that are judged to be relatively free of ice crystals are remarkably different from those revealed by the conventional preparation method, as shown in Fig. 17. The GBM consists of thin lamina densa and intervening lamina lucida in contact with endothelial cells or foot processes (Fig. 17a), whose shapes are more round. Under normal hemodynamic conditions the spaces between adjacent foot processes are heterogeneously narrow at the level of slit diaphragms. At higher magnification, the lamina densa is composed of irregular cords with high electron dense line (Fig. 17a). The fibrillar network is also observed at both sides of the thin lamina densa. The replica membrane of slit diaphragms, which are freeze-fractured parallel to the GBM, reveals that interdigitating foot processes are heterogeneously in close apposition (Fig. 17b). The width

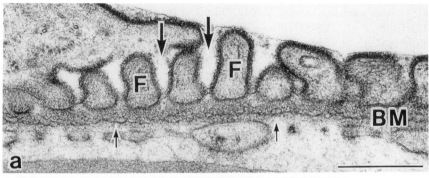

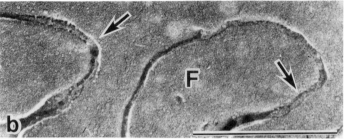

FIG. 17   Electron micrographs of glomerular capillary loops under normal blood flow, prepared by the FS method after the "*in vivo* cryotechnique." Bar = 0.5 μm. (a) The thin layer resembling lamina densa (small arrows) can be identified. The spaces between foot processes (Fs) appear a little dilated (large arrows). (b) Replica electron micrograph of horizontally freeze-fractured capillary loop. The distance between foot processes (Fs) is narrow and variable in some parts (arrows).

of slit diaphragms, as measured from the outer surface of one foot process to the next, is 20–45 nm ($N = 50$).

A common technique of cryofixation like slam freezing is characterized by a sudden exposure of excised tissues to copper or silver metals at a low temperature (Harreveld and Crowell, 1964; Harreveld *et al.*, 1974; Dempsey and Bullivant, 1976; Nicolas, 1991). However, in the present study, isopentane–propane was selected as the liquid cryogen because it remains liquid even at liquid nitrogen temperature and possesses good thermal conductivity (Jehl *et al.*, 1981; Murray *et al.*, 1989; Cole *et al.*, 1990; Furukawa *et al.*, 1991; Takami *et al.*, 1991). All living processes in kidney organs are stopped instantly, and their components are maintained *in situ*. There is no need to isolate the kidney tissue from the blood circulation system as reported for hearts and livers (Zglinicki *et al.*, 1986). Thus, ischemic influence on kidney tissues can be minimized by using the described cryofixation technique. The GBM in living kidneys is continuously stretched by the hydraulic pressure in glomerular capillaries, whereas it does not receive such stretching forces in excised kidneys, as summarized in Fig. 20. The space between foot processes of podocytes is maintained wide open to facilitate passage of the glomerular filtrate by the contractile proteins, such as actin, myosin, and α-actinin, and the foot processes are capable of such actin-mediated movement (Andrews, 1988). The podocytes may regulate the glomerular filtration rate through this active movement, thereby influencing hydraulic pressure across the GBM, as reported by Kriz *et al.* (1994).

In the functioning glomerulus the filtration slits are probably maintained open by the hydraulic pressure and the polyanionic charge repulsion, so blood pressure should be maintained steady before exposure to the QF method. The filtration slits are found to be narrow under normal blood flow. It has been reported that glomerular hydraulic pressure may be regulated by the relative width of the filtration slits. The total width of the epithelial slits plays an important role in providing a filtration function that controls hydraulic conductivity and water flow (Drumond and Deen, 1994; Kriz *et al.*, 1994). The present study provides the ultrastructural confirmation, indicating that the passage of solutes is possibly affected by glomerular hemodynamics under normal blood flow.

It has been generally accepted that kidney morphology and function are extremely dependent on maintenance of normal blood pressure (Griffith *et al.*, 1967; Kriz *et al.*, 1994). It has been shown that stopping the blood flow into kidneys results in notable changes of glomeruli, probably due to rapid loss of the fluid volume after interruption of the blood supply. Under normal blood flow, structural pores in the GBM may be too small to allow proteins to penetrate through them (Ryan and Karnovsky, 1976). Dilation of the pore size may occur as the blood flow stops, by loosening a compact network in the GBM, so normal hemodynamic conditions are definitely required for optimal retention of protein molecules (Ryan and Karnovsky, 1976; Ryan *et al.*, 1976; Olivetti *et al.*, 1981). Using the specially designed cryoknife to perform cryofixation *in situ* solves the morphological problem.

B. Glomerular Changes by Abnormal Hemodynamics

In another experimental protocol, under sodium pentobarbital anesthesia, the abdominal aorta is ligated at the region caudal to the point where the renal arteries branch. Under ligation of the aorta the kidneys gain a temporary increase in blood supply and take on balloon appearance. The *in vivo* cryotechnique is then performed under this experimental condition. The glomerular ultrastructure changes, depending on the renal hemodynamic state at the time of freezing *in vivo* (Fig. 18). An acutely elevated

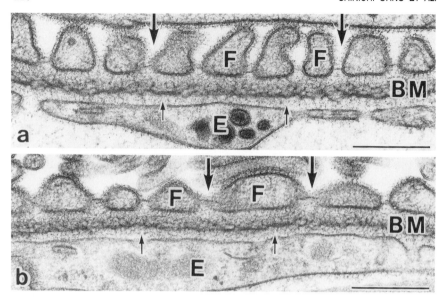

FIG. 18   Glomerular capillary loops under higher blood supply into the kidneys after ligation of the aorta, prepared by the FS method following the *"in vivo* cryotechnique." Bar = 0.5 μm. (a) The spaces between foot processes (Fs) are more widely dilated (large arrows). The lamina densa (small arrows) is clearly identified in the basement membrane (BM). (b) Some foot processes (Fs) in another glomerulus are flattened over the BM. The spaces between foot processes are more widely dilated (large arrows). The lamina densa is also clearly identified (small arrows). E, endothelial cell.

blood supply into the kidneys increases the width of slit diaphragms (Fig. 18a) to a mean value of 41.3 nm ($N = 50$; Table II). Moreover, a transient increase in intravascular pressure may cause stretching of some epithelial foot processes (Fig. 18b). Figure 20 is a schematic representation of glomerular capillary loops prepared by various electron microscopic procedures,

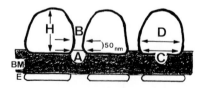

FIG. 19   Schematic representation of various measured parts in the glomerular capillary loop, which correspond to data in Table II. A, width of slit diaphragms; B, distance between neighboring foot processes at 50 nm from the slit diaphragm; C, width of foot process at the level of the slit diaphragm; D, width of foot process at 50 nm from the slit diaphragm; H, height of foot process.

TABLE II

Morphometric Data Obtained from the Measured Areas As Shown in Fig. 19

	Conventional quick-freezing of resected fresh specimen (mean ± SD, $N = 50$)	Freezing under normal blood flow *in vivo* (mean ± SD, $N = 50$)	Freezing after ligation of aorta *in vivo* (mean ± SD, $N = 50$)
A	32.2 ± 3.3 nm	36.5 ± 5.7 nm	41.3 ± 7.2 nm
B	35.7 ± 7.2 nm	50.7 ± 13.3 nm	84.2 ± 25.9 nm
C	257.3 ± 92.5 nm	289.2 ± 84.1 nm	297.4 ± 105.4 nm
D	246.1 ± 90.2 nm	240.7 ± 62.5 nm	175.9 ± 94.5 nm
H	460.9 ± 115.0 nm	294.3 ± 91.5 nm	224.8 ± 44.6 nm

* $P < 0.1\%$.
** $P < 0.05\%$.

such as the conventional double fixation (Fig. 20a), QF method with fresh, unfixed kidneys (Fig. 20b), *in vivo* cryotechnique under normal blood pressure (Fig. 20c), and *in vivo* cryotechnique under higher blood supply into the kidneys (Fig. 20d). In summary, the shapes of foot processes and the ultrastructures of basement membranes are clearly altered, depending on the various techniques (Figs. 20a, 20b) and the experimental conditions *in vivo* (Figs. 20c, 20d).

The *in vivo* cryotechnique, which is directly performed in the animal body, is quick enough to arrest transient physiological processes for morphological

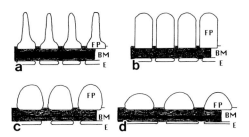

FIG. 20   Schematic representation of glomerular capillary loops prepared by various electron microscopic procedures. (a) Conventional double fixation method. (b) Quick-freezing method with fresh, unfixed kidneys. (c) The *in vivo* cryotechnique under normal blood pressure. (d) The *in vivo* cryotechnique under higher blood supply into the kidneys. FP, foot process; BM, basement membrane; E, endothelium.

study of kidneys. Therefore, the ultrastructural changes created by stopping blood circulation are avoided in the outermost tissues due to the simultaneity of cutting and freezing. The time interval between excising and freezing is dramatically shorter than that used in the conventional preparation procedure. It is tempting to propose that glomerular tissues have no typical basement membrane in their living state, as shown in Fig. 17a. During normal blood flow the glomerular ultrastructures are different from those in conventionally fixed kidneys, as summarized in Fig. 20. The functional meaning of glomerular morphology is valid only if hemodynamic factors are taken into consideration.

The morphological dependence of kidneys on physiological blood supply has been demonstrated by the present study, in which glomeruli in living states were observed before and after ligation of the aorta. The ultrastructural changes can be explained, in large part, by hemodynamic factors. The greater hydrostatic pressure at the glomerular level leads to an acute increase in filtration rate. It has been reported that some change in glomerular passage is a consequence of hormonal alteration in the permeability of the capillary wall (Bohrer *et al.,* 1977; Brenner *et al.,* 1977; Olivetti *et al.,* 1981). However, with ligation of the aorta, the glomerular filtration rate may temporarily increase, depending on the acute changes of renal blood flow. The ultrastructure with wider filtration slits probably reflects an increase of the collective slit areas, which represent the exit pathway for the glomerular filtrate (Kriz *et al.,* 1994). It is concluded that the transcapillary passage of materials is dependent on hemodynamic factors and also structural changes in the capillary loop.

V. Concluding Remarks

Ultrastructural differences in the kidney glomerulus have been demonstrated under various preparation conditions. More experiments are needed to clarify other functions of the glomerulus in animals. This review on the glomerular ultrastructure of kidneys indicates the type of information *in vivo* that may contribute to the elucidation of other glomerular physiological functions in the future.

Acknowledgments

The authors wish to express their thanks to many co-workers involved in this kidney research since the mid-1980s, particularly Dr. H. Shigematsu and Dr. H. Oguchi, Shinshu University School of Medicine; Dr. H. Takami, Tokyo Medical and Dental University; Dr. A. Yoshimura,

Showa University Fujigaoka Hospital; Dr. K. Otani, Toho University School of Medicine; Dr. F. Shimizu, Niigata University School of Medicine; and Dr. T. Moriya, Kitasato University School of Medicine. In addition, the authors acknowledge the skillful technical assistance of Miss Y. Kato and Miss K. Ariizumi in our laboratory.

References

Abrahamson, D. R., and Caulfield, J. P. (1982). Proteinuria and structural alteration in rat glomerular basement membranes induced by intravenously injected anti-laminin immuno-globulin G. *J. Exp. Med.* **156,** 128–145.

Anderson, S., Rennke, H. G., Garcia, D. L., and Brenner, B. M. (1989). Short and long term effects on antihypertensive therapy in the diabetic rat. *Kidney Int.* **36,** 526–536.

Andrews, P. (1988). Morphological alterations of the glomerular (visceral) epithelium in response to pathological and experimental situations. *J. Electron Microsc. Tech.* **9,** 115–144.

Ausiello, D. A., Kreisberg, J. I., Roy, C., and Karnovsky, M. J. (1980). Contraction of cultured rat glomerular cells of apparent mesangial origin after stimulation with angiotensin II and arginine vasopressin. *J. Clin. Invest.* **65,** 754–760.

Bariety, J., Druet, P., Laliberte, F., Sapin, C., Belair, M. F., and Paing, M. (1978). Ultrastructural evidence, by immunoperoxidase technique, for a tubular reabsorption of endogenous albu-min in normal rat. *Lab. Invest.* **38,** 175–180.

Barnes, J. L., and Venkatachalam, M. A. (1985). The role of platelets and polycationic mediators in glomerular vascular injury. *Semin. Nephrol.* **5,** 57–68.

Bastford, S. R., Rohrbach, R., and Vogt, A. (1987). Size restriction in the glomerular capillary wall: Importance of lamina densa. *Kidney Int.* **31,** 710–717.

Bender, B. L., Jaffe, R., Carlin, B., and Chung, A. E. (1981). Immunolocalization of entactin, a sulfated basement membrane component, in rodent tissues, and comparison with GP-2 (laminin). *Am. J. Pathol.* **103,** 419–426.

Bohrer, M. P., Deen, W. M., Robertson, C. R., and Brenner, B. M. (1977). Mechanism of angiotensin II–induced proteinuria in the rat. *Am. J. Physiol.* **233,** F13–F21.

Bolton, W. K., and Sturgill, B. C. (1978). Bovine serum albumin chronic serum sickness nephropathy. *Br. J. Exp. Pathol.* **59,** 167–177.

Boushey, H. A., Warnock, D. G., and Smith, L. H. (1986). The pathogenesis and prevention of diabetic nephropathy. *West. J. Med.* **145,** 222–227.

Brenner, B. M., Baylis, C., and Deen, W. M. (1976). Transport of molecules across renal glomerular capillaries. *Physiol. Rev.* **56,** 502–534.

Brenner, B. M., Bohrer, M. P., Baylis, C., and Deen, W. M. (1977). Determinants of glomerular permselectivity: Insights derived from observations *in vivo. Kidney Int.* **12,** 229–237.

Brenner, B. M., Hostetter, T. H., and Humes, H. D. (1978). Molecular basis of proteinuria of glomerular origin. *N. Engl. J. Med.* **298,** 826–833.

Bruijin, J. A., Van Elven, E. H., Hogendoorn, P. C. W., Corver, W. E., Hoedenmaeker, P. J., and Flueren, G. J. (1988). Murine chronic graft-versus-host disease as a model for lupus nephritis. *Am. J. Pathol.* **130,** 639–641.

Bruijin, J. A., Van Leer, E. H. G., Baelde, H. J. J., Corver, W. E., Hogendoorn, P. C. W., and Fleuren, G. J. (1990). Characterization and in vitro transfer of nephritogenic autoanti-bodies directed against dipeptidyl peptidase IV and laminin in experimental lupus nephritis. *Lab. Invest.* **63,** 350–359.

Castellot, J. J., Favreau, L. V., Karnovsky, M. J., and Rosenberg, R. D. (1982). Inhibition of vascular smooth muscle cell growth by endothelial cell-derived heparin. *J. Biol. Chem.* **257,** 11256–11260.

Castellot, J. J., and Hoover, R. L. (1986). Glomerular endothelial cells secrete a heparin-like inhibitor and a peptide stimulator of mesangial cell proliferation. *Am. J. Pathol.* **125,** 493–500.

Castellot, J. J., Hoover, R. L., Harper, P. A., and Karnovsky, M. J. (1985). Heparin and glomerular epithelial cell–secreted heparin-like species inhibit mesangial cell proliferation. *Am. J. Pathol.* **120,** 427–435.

Caulfield, J. P. (1979). Alterations in the distribution of alcian blue–staining fibrillar anionic sites in the glomerular basement membrane in aminonucleoside nephrosis. *Lab. Invest.* **40,** 503–511.

Caulfield, J. P., and Farquhar, M. G. (1974). The permeability of glomerular capillaries to graded dextrans. Identification of the basement membrane as the primary filtration barrier. *J. Cell Biol.* **63,** 883–903.

Caulfield, J. P., and Farquhar, M. G. (1976). Distribution of anionic sites in glomerular basement membranes: Their possible role in filtration and attachment. *Proc. Natl. Acad. Sci. USA* **73,** 1646–1650.

Caulfield, J. P., and Farquhar, M. G. (1978). Loss of anionic sites from the glomerular basement membrane in aminonucleoside nephrosis. *Lab. Invest.* **39,** 505–512.

Chakrabarti, S., Ma, N., and Sima, A. A. F. (1989). Reduced number of anionic sites is associated with glomerular basement membrane thickening in the diabetic BB-rat. *Diabetologia* **32,** 826–828.

Chakrabarti, S., Ma, N., and Sima, A. A. F. (1991). Anionic sites in diabetic basement membranes and their possible role in diffusion barrier abnormalities in the BB-rat. *Diabetologia* **34,** 301–306.

Chan, F. L., and Inoue, S. (1994). Lamina lucida of basement membrane: An artefact. *Microsc. Res. Tech.* **28,** 48–59.

Chan, F. L., Inoue, S., and Leblond, C. P. (1993). The basement membranes of cryofixed or aldehyde-fixed, freeze-substituted tissues are composed of a lamina densa and do not contain a lamina lucida. *Cell Tissue Res.* **273,** 41–52.

Chang, R. L. S., Deen, W. M., Robertson, C. R., and Brenner, B. M. (1975). Permselectivity of the glomerular capillary wall: III. Restricted transport of polyanions. *Kidney Int.* **8,** 212–218.

Chang, S. H., Mergner, W. J., Pendergrass, R. E., Bulger, R. E., Berezesky, I. K., and Trump, B. F. (1980). A rapid method of cryofixation of tissues *in situ* for ultracryomicrotomy. *J. Histochem. Cytochem.* **28,** 47–51.

Chavers, B. M., Bilous, R. W., Ellis, E. N., Steffes, M. W., and Mauer, S. M. (1989). Glomerular lesions and urinary albumin excretion in type I diabetes without overt proteinuria. *N. Engl. J. Med.* **320,** 966–970.

Cole, R., Matuszek, G., See, C., and Rieder, C. L. (1990). A simple pneumatic device for plunge-freezing cells grown on electron microscopy grids. *J. Electron Microsc. Tech.* **16,** 167–173.

Cooper, M. E., Allen, T. J., Macmillan, P., Bach, L., Jermus, G., and Doyle, A. E. (1988). Genetic hypertension accelerates nephropathy in the streptozotocin diabetic rat. *Am. J. Hypertens.* **1,** 5–10.

Courtoy, P. J., Kanwar, Y. S., Hynes, R. O., and Farquhar, M. G. (1980). Fibronectin localization in the rat glomerulus. *J. Cell Biol.* **87,** 691–696.

Deckert, T., Feldt-Rasmussen, B., Djurup, R., and Deckert, M. (1988). Glomerular size and charge selectivity in insulin-dependent diabetes mellitus. *Kidney Int.* **33,** 100–106.

Dempsey, G. P., and Bullivant, S. (1976). A copper block method for freezing non-cryoprotected tissue to produce ice-crystal-free regions for electron microscopy. *J. Microsc.* **106,** 251–260.

Drenckhahn, D., and Franke, R. P. (1988). Ultrastructural organization of contractile and cytoskeletal proteins in glomerular podocytes of chicken, rat, and man. *Lab. Invest.* **59,** 673–682.

Drumond, M. C., and Deen, W. M. (1994). Structural determinants of glomerular hydraulic permeability. *Am. J. Physiol.* **266,** F1–F12.

Duan, H. J., Ohno, S., and Shigematsu, H. (1993). Anionic sites in glomerular basement membrane of rats with serum sickness nephritis: Quick-freezing and deep-etching study. *Histochem. J.* **25**, 367–375.

Elema, J. D., Hoyer, J. R., and Vernier, R. L. (1976). The glomerular mesangium. Uptake and transport of intravenously injected colloidal carbon in rats. *Kidney Int.* **9**, 395–406.

Farquhar, M. G. (1975). The primary glomerular filtration barrier—basement membrane or epithelial slits? *Kidney Int.* **8**, 197–211.

Farquhar, M. G., Wissig, S. L., and Palade, G. E. (1961). Glomerular permeability. I. Ferritin transfer across the normal glomerular capillary wall. *J. Exp. Med.* **113**, 47–66.

Friedman, S., Jones, H. W., III, Golbetz, H. V., Lee., J. A., Little, H. L., and Myers, B. D. (1983). Mechanisms of proteinuria in diabetic nephropathy. II. A study of the size-selective glomerular filtration barrier. *Diabetes* **32**, 40–46.

Furukawa, T., Ohno, S., Oguchi, H., Hora, K., Tokunaga, S., and Furuta, S. (1991). Morphometric study of glomerular slit diaphragms fixed by rapid-freezing and freeze-substitution. *Kidney Int.* **40**, 621–624.

Gleichmann, E., Pals, S. T., Rolink, A. G., Radaszkiewicz, T., and Gleichmann, H. (1984). Graft-versus-host reaction: Clues to the etiopathology of a spectrum of immunological diseases. *Immunol. Today* **5**, 324–332.

Gleichmann, H., Van Elven, E. H., and Van der Veen, J. P. W. (1982). A systemic lupus erythematosus (SLE)-like disease in mice induced by abnormal T–B cell corporation: Preferential formation of autoantibodies characteristic of SLE. *Eur. J. Immunol.* **12**, 152–158.

Graham, R. C., and Karnovsky, M. J. (1966). Glomerular permeability: Ultrastructural cytochemical studies using peroxidases as protein tracers. *J. Exp. Med.* **124**, 1123–1134.

Grant, H. K., Leblond, C. P., Kleinman, H. K., Inoue, S., and Hassel, J. R. (1989). The incubation of laminin, collagen IV, and heparan sulfate proteoglycan at 35°C yields basement membrane–like structures. *J. Cell Biol.* **108**, 1567–1574.

Greene, W. B., and Walsh, L. G. (1992). An improved cryo-jet freezing method. *J. Microsc.* **166**, 207–218.

Griffith, L. D., Bulger, R. E., and Trump, B. F. (1967). The ultrastructure of the functioning kidney. *Lab. Invest.* **16**, 220–246.

Harreveld, A. V., Trubatch, J., and Steiner, J. (1974). Rapid freezing and electron microscopy for the arrest of physiological processes. *J. Microsc.* **100**, 189–198.

Harreveld, A. V., and Crowell, J. (1964). Electron microscopy after rapid freezing on a metal surface and substitution fixation. *Anat. Rec.* **149**, 381–386.

Heuser, J. E., and Kirschner, M. W. (1980). Filament organization revealed in platinum replicas of freeze-dried cytoskeletons. *J. Cell Biol.* **86**, 212–234.

Hippe-Sanwald, S. (1993). Impact of freeze substitution on biological electron microscopy. *Microsc. Res. Tech.* **24**, 400–422.

Hirose, K., Østerby, R., Nozawa, M., and Gundersen, H. J. G. (1982). Development of glomerular lesions in experimental long-term diabetes in the rat. *Kidney Int.* **21**, 689–695.

Hoedemaeker, P. J., Feenstra, K., Nijkeuter, A., and Arends, A. (1992). Ultrastructural localization of heterologous nephrotoxic antibody in the glomerular basement membrane of the rat. *Lab. Invest.* **26**, 610–613.

Hora, K., Ohno, S., Oguchi, H., Furukawa, T., and Furuta, S. (1990). Three-dimensional study of glomerular slit diaphragm by the quick-freezing and deep-etching replica method. *Eur. J. Cell Biol.* **53**, 402–406.

Inomata, S., Oosawa, Y., Itoh, M., Inoue, M., and Masamune, O. (1987). Analysis of urinary proteins in diabetes mellitus—with reference to the relationship between microalbuminuria and diabetic renal lesions. *J. Jpn. Diabetes Soc.* **30**, 429–435.

Ishimura, E., Sterzel, B., Budde, K., and Kashgarian, M. (1989). Formation of extracellular matrix by cultured rat mesangial cells. *Am. J. Pathol.* **134**, 843–855.

Iskandar, S. S., Giffod, D. R., and Emancipator, S. N. (1988). Immune complex acute necrotizing glomerulonephritis with progression to diffuse glomerulosclerosis: A murine model. *Lab. Invest.* **59,** 772–779.

Jehl, B., Bauer, R., Dörge, A., and Rick, R. (1981). The use of propane/isopentane mixtures for rapid freezing of biological specimens. *J. Microsc.* **123,** 307–309.

Jost, P. C., and Griffith, O. H. (1973). The molecular reorganization of lipid bilayers by osmium tetroxide. A spin-label study of orientation and restricted y-axis anisotropic motion in model membrane system. *Arch. Biochem. Biophys.* **159,** 70–81.

Kanwar, Y. S. (1984). Biophysiology of glomerular filtration and proteinuria. *Lab. Invest.* **51,** 7–21.

Kanwar, Y. S., and Farquhar, M. G. (1979a). Presence of heparan sulfate in the glomerular basement membrane. *Proc. Natl. Acad. Sci. USA* **76,** 1303–1307.

Kanwar, Y. S., and Farquhar, M. G. (1979b). Isolation of glycosaminoglycans (heparan sulfate) from glomerular basement membrane. *Proc. Natl. Acad. Sci. USA* **76,** 4493–4497.

Kanwar, Y. S., and Farquhar M. G. (1979c). Anionic sites in the glomerular basement membrane. *In vivo* and *in vitro* localization to the laminae rarae by cationic probes. *J. Cell Biol.* **81,** 137–153.

Kanwar, Y. S., and Farquhar, M. G. (1980). Detachment of endothelium and epithelium from the glomerular basement membrane produced by kidney perfusion with neuraminidase. *Lab. Invest.* **42,** 375–384.

Kanwar, Y. S., and Rosenzweig, L. G. (1982). Altered glomerular permeability as a result of focal detachment of the visceral epithelium. *Kidney Int.* **21,** 565–574.

Karnovsky, M. J. (1979). The ultrastructure of glomerular filtration. *Ann. Rev. Med.* **30,** 213–224.

Karnovsky, M. J., and Ainsworth, S. K. (1972). The structural basis of glomerular filtration. *Adv. Nephrol.* **2,** 35–60.

Karnovsky, M. J., and Ryan, G. B. (1975). Substructure of the glomerular slit diaphragm in freeze-fractured normal rat kidney. *J. Cell Biol.* **65,** 233–236.

Kashgarian, M. (1985). Mesangium and glomerular disease. *Lab. Invest.* **52,** 569–571.

Kaysen, G. A., Meyers, B. D., Couser, W. G., Robkin, R., and Felts, J. (1986). Mechanisms and consequences of proteinuria. *Lab. Invest.* **34,** 479–498.

Keane, W. F., and Raij, L. (1980). Impaired mesangial clearance of macromolecules in rats with chronic mesangial ferritin–antiferritin immune complex deposition. *Lab. Invest.* **43,** 500–507.

Koda, Y., Maruyama, Y., Ogino, S., Hayashi., N., Nishi, S., Saito, T., Ueno, H., and Arakawa, M. (1991). Early mesangiolysis and monocyte influx observed in diabetic glomerulosclerosis: Relation to nodule formation. *Jpn. J. Nephrol.* **33,** 33–42.

Koyama, A., Niwa, Y., and Shigematsu, H. (1978). Studies on passive serum sickness. II. Factors determining the localization on antigen–antibody complexes in the murine renal glomerulus. *Lab. Invest.* **41,** 253–262.

Kriz, W., Hackenthal, E., Nobiling, R., Sakai, T., and Elger, M. (1994). A role for podocytes to counteract capillary wall distension. *Kidney Int.* **45,** 369–376.

Kubosawa, H., and Kondo, Y. (1985). Ultrastructural organization of the glomerular basement membrane as revealed by a deep-etch replica method. *Cell Tissue Res.* **242,** 33–39.

Larabell, C. A., and Chandler, D. E. (1988). The extracellular matrix of *Xenopus laevis* eggs: A quick-freeze, deep-etch analysis of its modification at fertilization. *J. Cell Biol.* **107,** 731–741.

Latta, H. (1970). The glomerular capillary wall. *J. Ultrastruc. Res.* **32,** 526–544.

Latta, H. and Fligiel, S. (1985). Mesangial fenestrations, sieving, filtration, and flow. *Lab. Invest.* **52,** 591–604.

Laurie, G. W., Leblond, C. P., Inoue, S., Martin, G. R., and Chung, A. (1984). Fine structure of the glomerular basement membrane and immunolocalization of five basement membrane components to the lamina densa (basal lamina) and its extensions in both glomeruli and tubules of the rat kidney. *Am. J. Anat.* **169**, 463–481.

Leiper, J. M., Thompson, D., and Macdonald, M. K. (1977). Uptake and transport of imposil by the glomerular mesangium in the mouse. *Lab. Invest.* **37**, 526–533.   .

Lewis, R. M., Armstrong, M. Y. K., Andre-Schwarz, J., Muftuoglu, A., Beldotti, L., and Schwarz, R. S. (1968). Chronic allogeneic disease. I. Development of glomerulonephritis. *J. Exp. Med.* **128**, 653–679.

Madri, J. A., Roll, F. J., Furthmayer, H., and Foidart, J. M. (1980). Ultrastructural localization of fibronectin and laminin in the basement membranes of the murine kidney. *J. Cell Biol.* **86**, 682–687.

Masugi, Y. (1980). Immunohistochemistry of nephrotoxic nephritis. *In* "Masugi nephritis and its immunopathologic implications" (A. Okabayashi and Y. Kondo, eds.), 1st ed., pp. 58–82. Igaku-Shoin, Tokyo.

Maupin-Szamier, P., and Pollard, T. D. (1978). Actin filament destruction by osmium tetroxide. *J. Cell Biol.* **77**, 837–852.

McMillan, P. N., and Luftig, R. B. (1973). Preservation of erythrocyte ghost ultrastructure achieved by various fixatives. *Proc. Natl. Acad. Sci. USA* **70**, 3060–3064.

Menco, B. P. M. (1986). A survey of ultra-rapid cryofixation methods with particular emphasis on applications to freeze-fracturing, freeze-etching, and freeze-substitution. *J. Electron Microsc. Tech.* **4**, 177–240.

Messina, A., Davies, D. J., Dillane, P. C., and Ryan, G. B. (1987). Glomerular epithelial abnormalities associated with the onset of proteinuria in aminonucleoside nephrosis. *Am. J. Pathol.* **126**, 220–229.

Michael, A. F., Keane, W. F., Raij, L., Vernier, R. L., and Mauer, S. M. (1980). The glomerular mesangium. *Kidney Int.* **17**, 141–154.

Moriya, T., Ohno, S., Nakazawa, K., Shigematsu, H., and Yajima, Y. (1993). Ultrastructural study of glomerular basement membrane in diabetic rats by quick-freezing and deep-etching method. *Virchows Arch. B* **64**, 107–114.

Mundel, P., Elger, M., Sakai, T., and Kriz, W. (1988). Microfibrils are major component of the mesangial matrix in the glomerulus of the rat kidney. *Cell Tissue Res.* **254**, 183–187.

Murray, P. W., Robards, A. W., and Waites, P. R. (1989). Countercurrent plunge cooling: A new approach to increase reproducibility in the quick freezing biological tissue. *J. Microsc.* **156**, 173–182.

Myers, B. D. (1990). Pathophysiology of proteinuria in diabetic glomerular disease. *J. Hypertens.* **8**, S41–46.

Nakamura, Y., and Myers, B. D. (1988). Charge selectivity of proteinuria in diabetic glomerulopathy. *Diabetes* **37**, 1202–1211.

Nakazawa, K., Ohno, S., Naramoto, A., Takami, H., Duan, H. J., Itoh, N., and Shigematsu, H. (1992). Immune deposits in the glomerular extracellular matrix detected by the quick-freezing and deep-etching method. *Nephron* **62**, 203–212.

Naramoto, A., Ohno, S., Itoh, N., Shibata, N., Nakazawa, K., Takami, H., Duan, H. J., Kasahara, H., and Shigematsu, H. (1991a). Ultrastructure of matriceal changes in chronic phase of Masugi nephritis by quick-freezing and deep-etching method. *Virchows Arch. A* **418**, 51–59.

Naramoto, A., Ohno, S., Itoh, N., Shibata, N., and Shigematsu, H. (1992). Localization of laminin in nephritic glomeruli as revealed by a quick-freezing and deep-etching method with immunohistochemistry. *Histochem. J.* **24**, 717–726.

Naramoto, A., Ohno, S., Itoh, N., Takami, H., Nakazawa, K., and Shigematsu, H. (1990). Three-dimensional identification of actin filaments in phalloidin-treated rat livers by quick-freezing and deep-etching method. *Virchows Arch. A* **417**, 15–20.

Naramoto, A., Ohno, S., Nakazawa, K., Takami, H., Itoh, N., and Shigematsu, H. (1991b). Three-dimensional ultrastructure of glomerular injury in serum sickness nephritis using the quick-freezing and deep-etching method. *Virchows Arch. A* **418**, 185–192.

Nicolas, G. (1991). Advantages of fast-freeze fixation followed by freeze-substitution for the preservation of cell integrity. *J. Electron Microsc. Tech.* **18**, 395–405.

Ohno, S. (1985). Immunocytochemical study on the cytoplasmic side of cell membranes infected with vesicular stomatitis virus by quick-freezing and deep-etching replica method. *Histochem.* **82**, 565–575.

Ohno, S., and Fujii, Y. (1990). Three-dimensional and histochemical studies of peroxisomes in cultured hepatocytes by quick-freezing and deep-etching method. *Histochem. J.* **22,** 143–154.

Ohno, S., and Fujii, Y. (1991). Three-dimensional studies of the cytoskeleton of cultured hepatocytes: A quick-freezing and deep-etching study. *Virchows Arch. A* **418**, 61–70.

Ohno, S., Hora, K., Furukawa, T., and Oguchi, H. (1992). Ultrastructural study of the glomerular slit diaphragm in fresh unfixed kidneys by a quick-freezing method. *Virchows Arch. B.* **61**, 351–358.

Olivetti, G., Kithier, K., Giacomelli, F., and Wiener, J. (1981). Glomerular permeability to endogenous proteins in the rat. Effects of acute hypertension. *Lab. Invest.* **44**, 127–137.

Orikasa, M., Matsui, K., Oite, T., and Shimizu, F. (1988). Massive proteinuria induced in rats by a single intravenous injection of a monoclonal antibody. *J. Immunol.* **141**, 807–814.

Østerby, R., Brekke, I. B., Gunderson, H. J. G., Jorgensen, H. E., Lokkegaard, H., Mogensen, G. E., Nyberg, G., Parving, H. H., and Westberg, G. (1984). Quantitative studies of glomerular ultrastructure in human and experimental diabetes. *Appl. Pathol.* **2**, 205–211.

Otani, M., Ohno, S., Aoki, I., Misugi, K., Nakazawa, K., and Shigematsu, H. (1993). Three-dimensional study of glomerular lesions in murine chronic graft-versus-host reaction by the quick-freezing and deep-etching method. *Nephron* **64**, 268–274.

Pappenheimer, J. R. (1953). Passage of molecules through capillary walls. *Physiol. Rev.* **33**, 387–423.

Plattner, H., and Bachmann, L. (1982). Cryofixation: A tool in biological ultrastructural research. *Int. Rev. Cytol.* **79**, 237–304.

Potanova, J. P., Ebling, F. M., Hammond, W. S., Hahn, B. H., and Kotzin, B. L. (1988). Allogeneic MHC antigen requirements for lupus-like autoantibody production and nephritis in murine graft-versus-host disease. *J. Immunol.* **141**, 3370–3376.

Purtell, J. N., Pesce, A. J., Clyne, D. H., Miller, W. C., and Pollak, V. E. (1979). Isoelectric point of albumin: Effect on renal handling of albumin. *Kidney Int.* **16**, 366–376.

Rasch, R. (1979). Prevention of diabetic glomerulopathy in streptozotocin diabetic rats by insulin treatment. Glomerular basement membrane thickness. *Diabetologia* **16**, 319–324.

Reale, E., and Luciano, L. (1993). Further observations on the morphological alterations of the glomerular capillary wall of the rat kidney caused by chemical and physical agents: Standard procedures versus quick-freezing and freeze-substitution. *Histochem. J.* **25**, 357–366.

Reale, E., Luciano, L., and Kühn, K. W. (1983). Ultrastructural architecture of proteoglycans in the glomerular basement membrane: A cytochemical approach. *J. Histochem. Cytochem.* **31**, 662–668.

Rennke, H. G., Cotran, R. S., and Venkatachalam, M. A. (1975). Role of macromolecular charge in glomerular permeability: Tracer studies with cationized ferritins. *J. Cell Biol.* **67**, 638-646.

Rennke, H. G., Patel, Y., and Venkatachalam, M. A. (1978). Glomerular filtration of proteins: Clearance of anionic, neutral, and cationic horseradish peroxidase in the rat. *Kidney Int.* **13**, 324–328.

Rennke, H. G., and Venkatachalam, M. A. (1977). Glomerular permeability: *In vivo* tracer studies with polyanionic and polycationic ferritins. *Kidney Int.* **11**, 44–53.

Roberts, T. (1987). Fine (2–5 nm) filaments: New types of cytoskeletal structures. *Cell Motil. Cytoskeleton* **8**, 130–142.

Rodewald, R., and Karnovsky, M. J. (1974). Porous substructure of the glomerular slit diaphragm in the rat and mouse. *J. Cell Biol.* **60**, 423–433.

Roll, F. J., Madri, J. A., Albert, J., and Furmayr, H. (1980). Codistribution of collagen type IV and AB2 in basement membrane and mesangium of the kidney. An immunoferritin study of ultrathin frozen sections. *J. Cell Biol.* **85**, 597–616.

Romen, W., and Morath, R. (1979). Diffuse glomerulonephritis—A dysfunction of the mesangium? *Virchows Arch. B* **31**, 205–210.

Rosenstock, J., and Raskin, P. (1986). Early diabetic nephropathy: Assessment and potential therapeutic interventions. *Diabetes Care* **9**, 529–545.

Rostango, A., Frangione, B., Pearlstein, E., and Garcia-Pardo, A. (1986). Fibronectin binds to amyloid P component. Localization of the binding site to the 31,000 dalton C-terminal domain. *Biochem. Biophys. Res. Commun.* **140**, 12–20.

Ryan, G. B., Hein, S. J., and Karnovsky, M. J. (1976). Glomerular permeability to proteins. Effects of hemodynamic factors on the distribution of endogenous immunoglobulin G and exogenous catalase in the rat glomerulus. *Lab. Invest.* **34**, 415–427.

Ryan, G. B., and Karnovsky, M. J. (1976). Distribution of endogenous albumin in the rat glomerulus: Role of hemodynamic factors in glomerular barrier function. *Kidney Int.* **9**, 36–45.

Ryan, G. B., Rodewald, R., and Karnovsky, M. J. (1975). An ultrastructural study of the glomerular slit diaphragm in aminonucleoside nephrosis. *Lab. Invest.* **33**, 461–468.

Saito, Y., Kida, H., Takeda, S., Yoshimura, M., Yokoyama, H., Koshino, Y., and Hattori, N. (1988). Mesangiolysis in diabetic glomeruli: Its role in the formation of nodular lesions. *Kidney Int.* **34**, 389–396.

Sakai, T., and Kriz, W. (1987). The structural relationship between mesangial cells and basement membrane of the renal glomerulus. *Anat. Embryol.* **176**, 373–386.

Schnabel, E., Anderson, J. M., and Farquhar, M. G. (1990). The tight junction protein zo-1 is concentrated along slit diaphragms of the glomerular epithelium. *J. Cell Biol.* **111**, 1255–1263.

Schneeberger, E. E., Levey, R. H., McCluskey, R. T., and Karnovsky, M. J. (1975). The isoporous substructure of the human glomerular slit diaphragm. *Kidney Int.* **8**, 48–52.

Schurer, J. W., Hoedemaeker, P. H. J., and Molenaar, I. (1977). Polyethyleneimine as tracer particle for (immuno) electron microscopy. *J. Histochem. Cytochem.* **25**, 384–387.

Schurer, J. W., Kalicharan, D., Hoedemaeker, P. H. J., and Molenaar, I. (1978). The use of polyethyleneimine for demonstration of anionic sites in basement membrane and collagen fibrils. *J. Histochem. Cytochem.* **26**, 688–689.

Schwartz, E., Goldfischer, S., Coltoff-Schiller, B., and Blumenfeld, O. (1985). Extracellular matrix microfibrils are composed of core proteins coated with fibronectin. *J. Histochem. Cytochem.* **33**, 268–274.

Shea, S. M. and Morrison, A. B. (1975). A stereological study of the glomerular filter in the rat. Morphometry of the slit diaphragm and basement membrane. *J. Cell Biol.* **67**, 436–443.

Shigematsu, H., and Kobayashi, Y. (1973). The distortion and disorganization of the glomerulus in progressive Masugi nephritis in the rat. *Virchows Arch. B* **14**, 318–328.

Shigematsu, H., Kobayashi, Y., Tateno, S., and Hiki, Y. (1985). Prognostic significance of mesangial sclerosis in IgA nephropathy. *Jpn. J. Nephrol.* **27**, 303–309.

Shigematsu, H., and Yano, A. (1986). Participation of antigen presenting cells in glomerulonephritis. *Acta Pathol. Jpn.* **36**, 489–497.

Shikata, K., Makino, H., Ichiyasu, A., and Ota, Z. (1990). Three-dimensional meshwork structure of glomerular basement membrane revealed by chemical treatment. *J. Electron Microsc.* **39**, 182–185.

Shimomura, H., and Spiro, R. G. (1987). Studies on macromolecular component of human glomerular basement membrane and alterations in diabetes. Decreased levels of heparan sulfate proteoglycan and laminin. *Diabetes* **36**, 374–381.

Simpson, L. O. (1986). Is current research into basement membrane chemistry and ultrastructure providing any new insights into the way the glomerular basement membrane functions? *Nephron* **43**, 1–4.

Singhal, P. C., Schrschmidt, L. A., Gibbons, N., and Hays, R. M. (1986). Contraction and relaxation of cultured mesangial cells on a silicon rubber surface. *Kidney Int.* **30**, 862–873.

Striker, L. M. M., Killen, P. D., Chi, E., and Striker, G. S. (1984). The composition of glomerular sclerosis. I. Studies in focal sclerosis, crescentic glomerulonephritis, and membranoproliferative glomerulonephritis. *Lab. Invest.* **51**, 181–192.

Takami H., Naramoto, A., Nakazawa, K., Shigematsu, H., and Ohno, S. (1990). Ultrastructure of glomerular mesangial matrix by quick-freeze and deep-etch methods. *Kidney Int.* **38**, 1211–1215.

Takami, H., Naramoto, A., Shigematsu, H., and Ohno, S. (1991). Ultrastructure of glomerular basement membrane by quick-freeze and deep-etch methods. *Kidney Int.* **39**, 659–664.

Takasu, N., Ohno, S., Takasu, M., and Yamada, T. (1988). Polarized thyroid cells in monolayers cultured on collagen gel: Their cytoskeleton organization, iodine uptake, and resting membrane potentials. *Endocrinology* **122**, 1021–1026.

Templeton, D. M. (1989). Retention of glomerular basement membrane–proteoglycans accompanying loss of anionic site staining in experimental diabetes. *Lab. Invest.* **61**, 202–211.

Vasmant, D., Maurice, M., and Feldmann, G. (1984). Cytoskeleton ultrastructure of podocytes and glomerular endothelial cells in man and in the rat. *Anat. Rec.* **210**, 17–24.

Vehaskari, V. M., Root, E. R., Germuth, F. G., and Robson, A. M. (1982). Glomerular charge and urinary protein excretion: Effects of systemic and intrarenal polycation infusion in the rat. *Kidney Int.* **22**, 127–135.

Venkatachalam, M. A., Cotran, R. S., and Karnovsky, M. J. (1970a). An ultrastructural study of glomerular permeability in aminonucleoside nephrosis using catalase as a tracer protein. *J. Exp. Med.* **132**, 1168–1180.

Venkatachalam, M. A., Karnovsky, M. J., Fahimi, H. D., and Cotran, R. S. (1970b). An ultrastructural study of glomerular permeability using catalase and peroxidase as tracer proteins. *J. Exp. Med.* **132**, 1153–1167.

Viberti, G. C., and Wiseman, M. J. (1986). The kidney in diabetes: Significance of the early abnormalities. *Clin. Endocrinol. Metab.* **15**, 753–782.

Washizawa, K., Ishii, K., Itoh, N., Mori, T., Akabane, T., and Shigematsu, H. (1989). Morphometric change in glomerular anionic sites during aminonucleoside nephrosis. *Acta Pathol. Jpn.* **39**, 558–565.

Weil, R., Nozawa, M., Koss, M., Weber, C., Reemts, K., and McIntosh, R. (1976). The kidney in streptozotocin diabetic rats. *Arch. Pathol. Lab. Med.* **100**, 37–49.

Whiteside, C., Prutis, K. Cameron, R., and Thompson, J. (1989). Glomerular epithelial detachment, not reduced charge density, correlates with proteinuria in adriamycin and puromycin nephrosis. *Lab. Invest.* **61**, 650–660.

Yaar, M., Foidart, J. M., Brown, K. S., Rennard, S. L., Martin, G. R., and Liotta, L. (1982). The Goodpasture-like syndrome in mice induced by intravenous injections of anti–type IV collagen and anti-laminin antibody. *Am. J. Pathol.* **107**, 79–91.

Yamamoto, T., Kihara, I., Morita, T., and Oite, T. (1978). Bovine serum albumin (BSA) nephritis in rats. I. Experimental model. *Acta Pathol. Jpn.* **28**, 859–866.

Yoshimura, A., Nakano, K., Oniki, H., Ideura, T., Koshikawa, S., and Ohno, S. (1992). Ultrastructure of anionic sites of glomerular basement membrane by quick-freezing and deep-etching method: Studies of two different molecular cationic probes and freeze-substitution fixation method. *J. Clin. Electron Microsc.* **25**, 763–764.

Yoshimura, A., Ohno, S., Nakano, K., Oniki, H., Inui, K., Ideura, T., and Koshikawa, S. (1991). Three-dimensional ultrastructure of anionic sites of the glomerular basement membrane by a quick-freezing and deep-etching method using a cationic tracer. *Histochem.* **96**, 107–113.

Yurchenco, P. D., and Furthmayr, H. (1984). Self-assembly of basement membrane collagen. *Biochemistry* **23**, 1839–1850.

Yurchenco, P. D., and Rubin, G. C. (1987). Basement membrane structure *in situ:* Evidence for lateral association in the type IV collagen network. *J. Cell Biol.* **105**, 2259–2268.

Zglinicki, T. V., Rimmler, M., and Purz, H. J. (1986). Fast cryofixation technique for X-ray microanalysis. *J. Microsc.* **141**, 79–90.

Distribution and Functions of Parathyroid Hormone–Related Protein in Vertebrate Cells

Patricia M. Ingleton* and Janine A. Danks†
*Institutes of Endocrinology and Cancer Studies, Sheffield University Medical
School, Sheffield S10 2RX, United Kingdom
†St. Vincent's Institute of Medical Research, Fitzroy, Victoria 3065,
Australia

Parathyroid hormone–related protein (PTHrP) was isolated from tumors and identified as the agent of humoral hypercalcemia of malignancy (HHM) in 1987. Since then its gene structure in several mammalian and an avian species has been analyzed and its gene expression demonstrated in many adult and embryonic tissues derived from all three germ layers. The composition and structure of PTHrP peptide depends on both differential gene splicing and posttranslational processing, which result in a range of peptides of potentially diverse functions.

This chapter describes the distribution of PTHrP in both normal and neoplastic adult and embryonic tissues. PTHrP is of fundamental importance to cell survival because the absence of the gene is fatal; this aspect of PTHrP function in cell physiology becomes overwhelmingly important in neoplasia. Intracrine or paracrine actions for PTHrP seem to be most likely in mammalian and avian physiology, but in fishes high circulating levels suggest classic endocrine functions as well. Much remains to be learned of the biology of this fascinating protein.

KEY WORDS: Parathyroid hormone–related protein, Humoral hypercalcemia, Neoplasia.

I. Introduction

For many years the syndrome of humoral hypercalcemia of malignancy (HHM) was believed to be due to ectopic secretion of parathyroid hormone

231

(PTH), since the biochemical features are similar to those of hyperparathy-roidism (Martin and Mundy, 1987), as described by Albright (1941). How-ever, increased levels of circulating PTH could not be found unequivocally, and it was only the preparation of tumor extracts, with PTH-like bioactivity but lacking PTH immunological characteristics, that established the exis-tence of a distinct factor. The factor was isolated and cloned from tumor tissues in 1987 (Burtis *et al.*, 1987; Moseley *et al.*,1987; Strewler *et al.*, 1987; Suva *et al.*, 1987). Analysis of the peptide showed homology with PTH only at the N-terminus, where 8 of the first 13 amino acids are identical. This limited sequence homology, together with structural homology in the 14–34 region (Kemp *et al.*, 1987; Caulfield *et al.*, 1990), is sufficient to allow interactions at the PTH receptors, principally in bone and kidney, which lead to development of HHM (Kukreja *et al.*, 1988; Insogna, 1989; Martin, 1990) and also explains the cross-reactivity with some antibodies to PTH. The factor was named *parathyroid hormone–related protein* because of its limited homology with PTH (it has also been called PTH-like peptide). Since the initial identification and characterization of PTHrP in tumors a wealth of information has been gained about its gene structure, amino acid composition and tissue distribution in normal as well as malignant tissues and in vertebrate groups and species in addition to humans.

1. Chemistry of PTHrP

Human PTHrP, the first to be analyzed, includes a 36–amino acid "pre–pro" sequence that is cleaved to produce mature forms that are 139, 141, and 173 amino acids (aa) long, respectively, by differential splicing between exons VI, VIII, and IX of a nine-exon gene (Martin *et al.*, 1991) (Fig. 1). Rat (Thiede and Rodan, 1988; Yasuda *et al.*, 1989a), mouse (Mangin *et al.*, 1990), and chicken PTHrP (Thiede and Rutledge, 1990; Schermer *et al.*, 1991) are apparently single molecular weight forms of 141 and 139 amino acids, whereas chicken PTHrP (Thiede and Rutledge, 1990; Schermer *et al.*, 1991) may occur in two forms, approximately 139 aa long, resulting from differential splicing at the 3′ terminus of the gene. Their compositions have been deduced from cDNA analysis, which has shown there is marked sequence conservation in the first 111 N-terminal residues (Fig. 1); indeed, rat and mouse differ from human PTHrP in only 2 of these residues, but thereafter there is little homology. PTHrP has not been isolated from any other vertebrates, but immunoreactive PTHrP in sea bream (a marine teleost) pituitary appears in two principal molecular weight forms of 29 and 26 kDa and a minor one of 14.3 kDa (Danks *et al.*, 1993), suggesting possible differential splicing of the gene. The mammalian and avian PTHrP molecules show interesting composition and structural features that offer many possibilities for tissue processing by proteolytic cleavage, amidation,

Human PTHrP Translation Products

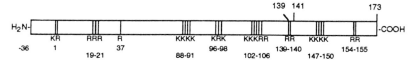

Linear composition

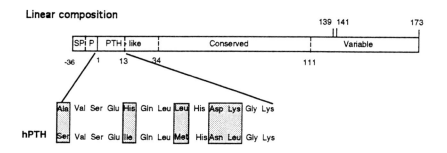

hPTH

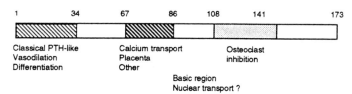

Functional domains

FIG. 1   Chemical structure of human PTHrP showing gene splice products of 139, 141, and 173 aa, and possible processing sites by proteolysis (aa 19, 88, 96, 102), by O-glycosylation between aa 84 and 108 and by amidation after aa 86 and 94. Only the N-terminus is similar to PTH, the midmolecule region, aa 34 to 110, is highly conserved, and the C-terminus is variable. Definitive functions have been assigned to the N-terminus and C-terminus but only tentatively to the midmolecule sequence.

and glycosylation, which result in a variety of molecular forms with potentially different biological functions. Within the conserved region there are four potential proteolytic processing sites, at amino acids 19, 88, 96, and 102 (Fig. 1), where multiple or dibasic amino acids (Arg-Lys; Arg-Arg or Lys-Lys) are located. Two potential amidation sites also are found in this domain following proline residues at 86 and 94, as well as multiple basic serine- and threonine-rich motifs creating possible O-glycosylation sites between residues 84 and 108 (Soifer et al., 1992). Residues 107–111 form a highly conserved, basic pentapeptide (-Thr-Arg-Ser-Ala-Trp-), which has been found to be essential for the inihibitory action of PTHrP on osteoclast resorbing activity (Fenton *et al.*, 1991a,b, 1993) and may possibly take part with other basic regions in nuclear transport functions (Dingwall and Laskey, 1991).

PTHrP is synthesized by many phenotypically distinct cell types, as is discussed later, and its pluripotent structure lends itself to processing in individual tissues to create a form specific to that tissue. Convertases, a group of enzymes, have been identified as associated with processing of precursor molecules, such as pro-opiomelanocortin (POMC) by cleavage of intramolecular dibasic sites, to produce several hormones and neuro-transmitters (Seidah *et al.*, 1993). Thus, the function of PTHrP produced in or acting on any particular tissue may depend on posttranslational pro-cessing and the spectrum of convertases present in the cells of that tissue. The potential diversity of molecular forms also predicts problems in assay systems, particularly for circulating forms, and emphasizes the need for combined bioassays and immunoassays to allow accurate interpretation of the biological functions of PTHrP isoforms.

2. Gene Structure

The human gene for PTHrP is large and complex: a single copy gene composed of 1.5 kb organized into nine exons (Thiede *et al.*, 1988; Mangin *et al.*, 1989; Suva *et al.*, 1989; Yasuda *et al.*, 1989b, Martin *et al.*, 1991) (Fig. 2). It is located on the short arm of chromosome 12 at p11.2–p12.1, whereas the parathyroid hormone (PTH) gene, composed of only three exons, is on chromosome 11, the evolutionary precursor of chromosome 12. Thus, the two genes appear to share a common origin. Analysis of their genomic structures has shown that they share homologies only in the exons coding for the pre–pro sequences. The gene structures for PTHrP of the rat (Yasuda *et al.*, 1989b; Karaplis *et al.*, 1990) and mouse (Mangin *et al.*, 1990) (Fig. 2) have been determined and found to be simpler than the human gene, consisting of only five exons encoding a single translation product. The gene for chicken PTHrP (Thiede and Rutledge, 1990; Schermer *et al.*, 1991) also has only five exons but apparently includes a splice donor site allowing for alternate splicing at the 3' end of the primary mRNA to be translated into two different proteins (Fig. 2). The principal coding sequence for human PTHrP(1–139) is located in exon VI, with the corresponding region of mouse and rat PTHrP in exon IV and in the chicken gene in exon III. Alternative splicing regions of human PTHrP gene are contained in exons VIII and IX, with the latter corresponding to exon V of the other species' genes (Fig. 2).

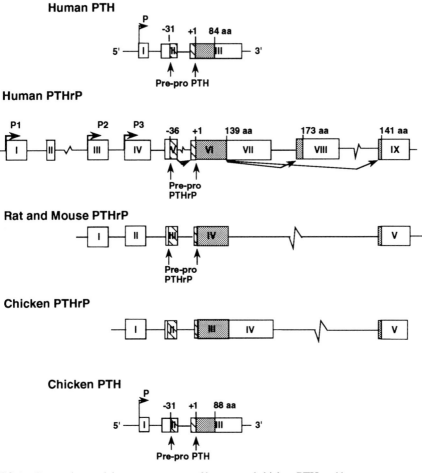

FIG. 2   Comparisons of the gene structure of human and chicken PTH and human, rat, mouse, and chicken PTHrP.

The human gene has four 5′ promoter regions, including two TATA boxes and a GC-rich region, which suggest a variety of controlling factors for gene expression. In normal human amnion, for example, PTHrP(1–139) is the principal isoform, and the preferred promoter is the GC region (Brandt *et al.*, 1992), which may therefore be linked to the specific isoform. Analysis by exon-specific primers with reverse transcriptase–polymerase chain reaction of nonneoplastic human tissues has shown that transcripts encoding PTHrP (1–139) and (1–141) are much more abundant than the long form (1–173) transcript (Campos *et al.*, 1994). However, the mouse and rat genes have only two promoter regions, and the chicken gene has only one. It is notable that the structures of the PTHrP genes of the four species, three mammalian and one avian, so far analyzed are so variable while that for PTH is highly conserved (Fig. 2). Such differences point to the roles the gene products PTH and PTHrP play in vertebrate biology: PTH governs a more restricted range of biological functions, principally elevation of serum calcium, whereas PTHrP is apparently involved in diverse activities, including tissue development and differentiation and control of vascular tone as well as calcium metabolism.

3. Parathyroid Gland Evolution

The parathyroid gland first evolved as a distinct organ in the amphibians, and it has been established that it is important for calcium regulation in adult anurans (Cortelyou *et al.*, 1960). However, not all amphibians depend on the parathyroid for control of hypercalcemia, a function that is instead associated with the pituitary (Pang and Yee, 1980). The urodeles are considered to be more primitive than their fellow amphibians, the anurans, and it is in this group that the parathyroid gland first appears, with the larval stages in some species lacking the gland but developing it in the adult. Fishes do not have parathyroid glands at any stage of development and depend on factors from other organs for hypercalcemic control. It was suggested by Parsons *et al.* (1979), from results of studies on the cod, that the pituitary gland of fishes contains a hypercalcemic factor related to but distinct from PTH. This factor may be PTHrP because it has been identified in the pituitary and in pituitary incubation media of the sea bream, a marine teleost (Danks *et al.*, 1993). Thus, although PTHrP has been detected in embryonic and pathological parathyroid glands, it is not a significant secretory product, and it seems likely that PTH and PTHrP coexisted before there was a specialized gland producing only PTH.

II. Parathyroid Hormone–Related Protein in Tissues

A. Methods of Detection and Assay

1. Immunohistochemistry

Immunohistochemistry has been extremely useful in identifying the tumors and normal tissues containing PTHrP. This visualization technique in conjunction with Western and Northern blotting allowed identification of the protein in tissues before radioimmunoassay for PTHrP was developed.

Initially the tumors associated with HHM were examined for the PTHrP antigen: all squamous cell carcinomas (SCC) were found to be positive for PTHrP, regardless of their tissue origin (Danks et al., 1989). Examination of SCC of the skin for PTHrP by this technique also revealed that the surrounding normal skin contained PTHrP in the spinous keratinocyte layer.

Antisera for immunohistochemistry have been both polyclonal and monoclonal and have been directed to various portions of the human PTHrP molecule; in particular, antibodies to the N-terminus of the molecule have been used extensively in immunohistochemistry. Kitazawa et al. (1991, 1992a, 1992b) raised and used a monoclonal antibody to human PTHrP (1–34) in studies looking for PTHrP in normal skin, bronchial epithelium, uterine cervix, and neoplastic conditions of the tissues. Polyclonal antisera raised in rabbits to human PTHrP(1–34) and PTHrP(1–16) as well as sheep antibody to midmolecule human PTHrP(50–69) have been used in studies of normal human tissues and tumors (Danks et al., 1989, 1990) as well as tissues from fish (Danks et al., 1993; Ingleton et al., 1995) and amphibians. Antibodies to the C-terminus of the PTHrP molecule have been used by Iwamura et al. (1993) in studies of prostatic tumors that demonstrated that all neoplasms were positive with an antibody to PTHrP(109–141), but none of the surrounding stroma or inflammatory cells reacted. These results contrasted with their own findings in a subsequent study using an antibody to the N-terminus of PTHrP (Iwamura et al., 1994), in which they showed that only the neuroendocrine cells were positive for PTHrP. This difference in staining patterns indicates possible alternative tissue processing of the PTHrP molecule and warrants further investigation for possible convertase activity and posttranslational processing. N-terminus-directed antibodies whether polyclonal (Southby et al., 1990) or monoclonal (Kohno et al., 1994) detected PTHrP in about 60% breast cancers, and polycolonal antibody to PTHrP(37–67) (Bundred et al., 1991) or PTHrP(38–64) (Liapis et al., 1993a) produced similar results; however, the samples were not examined with both region-specific antibodies.

Validation of antisera is extremely important regardless of whether they are to be used in immunohistochemistry, Western blotting, radioimmunoassay or biological assay. All new antisera raised for immunohistochemistry should be subjected to vigorous testing and validation by the following protocols that have been used in all our studies. After initial testing by ELISA, the antisera are tested in a PTHrP immunohistochemical assay against known positive and negative tissues with positive and negative controls (see next paragraph). If the antisera look promising, they are then tested for positive and negative reactions by Western blotting (Danks et al., 1990) against recombinant human PTHrP and human PTH peptides, because of the potential cross-reactivity with N-terminus PTH. The antisera we use do not recognize human PTH(1–84) but do recognize region-specific epitopes within PTHrP(1–141). For example, anti-PTHrP(50–69) antiserum recognizes PTHrP(1–141), PTHrP(1–108), PTHrP(1–84) but not PTHrP(1–34), whereas anti-PTHrP(1–34) antiserum recognizes all the recombinant PTHrP peptides. Then the antisera are tested for specificity and sensitivity by standard radioimmunoassay, and their ability to block the biological activity of PTHrP(1–34) is determined (Danks et al., 1989). In this assay the antisera are incubated with human PTHrP(1–34) and human PTH(1–34) for 18 hours at 4°C before the biological activity is measured by the generation of cAMP in UMR106-01 osteogenic sarcoma cells (Forrest et al., 1985).

Validation of the immunohistochemical technique is also fundamental. It cannot be overly stressed that the panel of controls included in each assay must be complete. Immunohistochemistry fell into some disrepute in the 1970s and 1980s because some studies were not properly controlled, and the work was seen as "quick and dirty." Results from such studies are now in doubt, and the harm done to the field of molecular histology can be overcome only by adhering to strict guidelines. In each experiment there must be at least duplicates of each variable that is run, and there must be antibody-specificity controls consisting of tissues that are known to be positive and those that are known to be negative for the antigen. Affinity purification of antibodies on specific antigen columns allows for comparison with the total antiserum and elimination of false-positive results. Preabsorption of antisera with specific antigen provides the best negative control: We have used these methods in our studies with antisera to human PTHrP(1–34) and (1–16) (Danks et al., 1989, 1990). Method-specificity controls consist of alternating deletion of antibody layers (primary antiserum, secondary antibody, and peroxidase–antiperoxidase) and the substitution of nonimmune serum (rabbit) and an unrelated immune serum (in our case rabbit anti-HIV antiserum) for the primary antibody. If any of the preceding controls do not give the expected result, the whole assay and its results must be discarded. Two antibody dilutions may be used to give semiquantitative results. The lower dilution is chosen so that it will stain all the tissues that contain PTHrP, whereas the higher dilution will stain only those elements that are strongly positive for PTHrP. This panel of controls has been included in all the textbooks written about immunohistochemistry but sometimes seems to be sadly lacking in papers describing immunohistochemical studies, including some that examine the localization of PTHrP.

2. Radioimmunoassay

With the value of hindsight it is now clear that before PTHrP was identified and analyzed, a number of antisera used for PTH RIA could also detect

PTHrP. This lack of specificity led to confusion and problems in interpreting results. Thus, since 1987 considerable effort has been devoted to developing a sensitive and specific method for measuring the PTHrP in the circulation. Previously, cytochemical and adenylate cyclase assays were the only ways of measuring the circulating PTH-like activity.

The first radioimmunoassay capable of detecting PTHrP in the human circulation (Budayr et al., 1989) was based on a polyclonal antiserum to the N-terminus of PTHrP, using ^{125}I-labeled PTHrP(1–34) as ligand. Serum was found to interfere with the RIA, and PTHrP had to be extracted from plasma samples and concentrated using an affinity column. Two similar assays were developed using N-terminus polyclonal antisera and ^{125}I-labeled PTHrP(1–34). That of Kao et al. (1990) still requires an extraction and concentration step and has a detection limit of 2 pmol/liter, whereas that of Henderson et al. (1990) does not require extraction of the samples but has a high sensitivity limit of 50 pmol/liter.

Grill et al. (1991) developed an N-terminus RIA that does not require extraction of the PTHrP from the patient samples and has a sensitivity 2 pmol/liter. Patient plasma samples are collected in tubes containing EDTA and aprotinin to stabilize any PTHrP. The assay utilizes polyclonal antiserum raised to human PTHrP(1–40) and ^{125}I-labeled Tyr^0-PTHrP(1–34). There is no cross-reactivity with hPTH(1–34) and hPTHrP(1–20). The method does detect hPTHrP(1–25),hPTHrP(1–29), hPTHrP(15–34), and hPTHrP(1–141), and it was concluded that the antibody-binding epitopes are between PTHrP 20 and 29. This RIA has been extremely useful on human plasma samples from cancer patients, pregnant and lactating women (Grill et al., 1992), and in measuring circulating immunoreactive PTHrP in teleosts (Danks et al., 1993) and elasmobranchs (Ingleton et al., 1995).

Two site immunoassays have been developed that demonstrate, at least partially, the molecular size of PTHrP in tissue and plasma samples. Immunoradiometric assays (IRMA) utilize two antisera, one that captures the protein by affinity chromatography and a second, to a distant epitope, that reacts with the radiolabeled ligand. Ratcliffe et al. (1991b) developed an IRMA based on a monoclonal antibody to PTHrP(1–34) as the capture antibody and a polyclonal antibody to PTHrP(37–67) as the signal antibody. This assay detected PTHrP(1–86) but not PTHrP(1–34) or (37–67) or PTH. By this assay patients with HHM had detectable PTHrP(1–86), but normocalcemic patients with solid tumors and those with hyper-and hypoparathyroidism and chronic renal failure generally did not have detectable plasma levels. A similar assay (Burtis et al., 1990) used a polyclonal antibody to PTHrP(37–74) as the capture antibody and an iodinated antiserum to PTHrP(1–34) as the signal antibody. This assay detected all PTHrP isoforms containing the N-terminus (1–34) (i.e., PTHrP(1–74; 1–108; 1–141) but not PTHrP(37–74). This assay, too, showed that only hypercalcemic patients had detectable plasma PTHrP levels. A refinement of the technique by Pandian *et al.*, (1992) increased the sensitivity of the assay to allow detection of PTHrP in approximately half their samples of normal plasmas as well as those from hypercalcemic patients with cancers. The first commercial IRMA was developed by Fraser *et al.* (1993). It has a detection limit of <0.7–2.6 pmol/liter of PTHrP.

Ratcliffe *et al.* (1991a) also developed an RIA for PTHrP using the polyclonal antibody to midmolecule PTHrP(37–67) and [^{125}I]PTHrP(37–67) as radioligand. This antiserum recognized the epitope PTHrP(52–61) and was relatively insensitive, with a lower limit of 57pmol/liter. the mid-molecule PTHrP RIA developed by Burtis *et al.* (1994) with antiserum to PTHrP(1–74) and [^{125}I]PTHrP(37–74) allowed comparisons to be made between assays for C-terminus PTHrP by RIA and N-terminus by IRMA. All patients with HHM had elevated plasma PTHrP by all three assays. C-terminus PTHrP levels were elevated in chronic renal failure, but N-terminus levels were undetectable, and only 1 of 14 patients had elevated PTHrP by the mid-molecule assay. Plasma levels of N-terminus PTHrP were not elevated in patients with osteolytic hypercalcemia, but three out of six samples had detectable C-terminus PTHrP, and one out of six had detectable mid-molecule PTHrP. The RIA results for the HHM patients were characterized by extraction on an affinity column using antibody to PTHrP(37–74) and HPLC. Fractions were assayed by the N-terminus IRMA, and midmolecule and C-terminus RIAs. These methods showed that the midmolecule region was more abundant than either terminus in HHM patients. It is possible that the use of recombinant mid-molecule standards may not allow the same competition for antibody binding as does the native fragment, which may lead to misleading results.

Two further mid-molecule RIAs have been reported; Blind *et al.* (1992) showed that 22 of 27 patients with HHM had raised plasma PTHrP, but no normocalcemic patients had elevated levels. A more sensitive mid-molecule assay was developed by Bucht *et al.* (1992) using extraction of plasma by silica cartridges or immune serum. The silica cartridge extraction method improved assay sensitivity from 8.4 pmol/liter to 0.8 pmol/liter, so that PTHrP could be detected in normal plasma. This assay also showed that hypercalcemic patients or those with impaired renal function all had significantly raised plasma PTHrP concentrations.

Radioimmunoassay with C-terminus PTHrP antisera have been developed by Burtis *et al.* (1990) and Imamura *et al.*(1991). The immunogen used to raise polyclonal antibodies Tyr^{109} PTHrP(109–138) was also used as the iodinated ligand for RIA that had a detection limit of 2 pmol/liter. Patients with chronic renal failure had elevated levels of C-terminus PTHrP, which suggests that impaired renal clearance leads to accumulation of C-terminus fragments in the circulation. This possibility was supported by subsequent evidence showing that poor glomerular filtration correlated with elevated C-terminal PTHrP plasma levels (Orloff *et al.*, 1993).

The circulating forms of PTHrP have not been characterized, so the significance of the results of region-specific assays must be considered with care. Assays continue to be developed and refined. The theoretically ideal assay for any specific region of PTHrP would be an IRMA using two antibodies to epitopes at opposite ends of the fragments, for example, for the N-terminus, PTHrP(1–14) and PTHrP(20–34). The antibodies would be very specific and the assay sensitivity sufficient to detect normal plasma levels, which would be lower than the levels in normocalcemic patients with SCC, which would, in turn, be lower than the levels in similar patients with hypercalcemia.

3. Gene Expression

Standard Northern blot analysis was one of the first techniques used to identify the tissues and tumors expressing PTHrP (Ikeda *et al.*, 1988a, 1988b); intracellular localization of PTHrP expression by *in situ* hybridization (ISH) was developed later. Thiede *et al.* (1990) looked at rat uterine tissues and used a PTHrP 26-base oligonucleotide probe corresponding to amino acids 62–70. The probe was 3′ end–labeled with deoxyadenosine 5′-[a-[^{35}S]thio]triphosphate. The images were visualized with X-ray film, and the slides were dipped in emulsion and exposed for 2–4 weeks at 4°C.

Radioactive *in situ* hybridization was used in most of the studies in the period 1988–1992. Heath *et al.*(1990) examined human tumors with sense and antisense PTHrP riboprobes labeled with ^{32}P. Vargas *et al.* (1992) used ^{35}S-labeled riboprobes to visualize PTHrP mRNA in human breast cancer metastases. Riboprobes were selected because PTHrP mRNA appeared to be of a very small copy number, and this choice of probe and lable gave the sensitivity required to localize PTHrP mRNA. However, there was high nonspecific background reaction and the specificity of radioactive *in situ* hybridization was not sufficient to localize PTHrP mRNA in normal skin.

With an improvement in the sensitivity of nonradioactive *in situ* hybridization Kitazawa *et al.* (1992a) used a cDNA probe labeled with bromodeoxyuridine (BrdU) and visualized with fluorescein isothiocyanate on parathyroid adenoma sections. By this method PTHrP mRNA was localized to the acidophil cells, which reflected the findings of immunohistochemistry. Iwamura *et al.* (1994) also used a PTHrP cDNA probe, but labeled with digoxygenin, to probe sections of normal and neoplastic prostate. Again the PTHrP message relfected the protein distribution detected by immunohistochemistry in the neuroendocrine cells in both normal and neoplastic tissues. These studies did not indicate the sensitivity of the method and referred only to negative controls.

The most sensitive and specific method for ISH was developed by Danks *et al.* (1995), using antibody enhancement. Oligonucleotide probes offer high specificity but demonstrating hybridization, particularly where mRNA copy numbers are low, requires increased sensitivity. Most nonradioactive *in situ* methods used a single enzyme-linked antibody—for example, fluorescein isothiocyanate (FITC)–conjugated anti-BrdU antibody or alkaline phosphatase–conjugated anti-digoxygenin. To increase the sensitivity we used an unlabeled anti-digoxygenin antibody and a second species-specific enzyme-linked antibody. We chose 33-mer oligonucleotide probes to three regions of the human PTHrP molecule, equivalent to the pre–pro, N-terminus, and midmolecule regions of the amino acid sequence of PTHrP. The probes were 3′ end–labeled with digoxygenin, and after hybridization the second antibody was applied. Sheep anti-digoxygenin antiserum was applied for 1 hour, and then alkaline phosphatase–linked anti-sheep immunoglobulins were applied for another hour. The PTHrP mRNA was localized in gynecological and skin tumors and in normal skin. A panel of controls was used in the course of this study. Controls in hybridization histochemistry are as important as in immunohistochemical studies. Normal human skin was used as a positive control, and there were several negative controls, including sections hybridized in the absence of probe and sections subjected to RNase A treatment and hybridized with unlabeled probe in excess prior to hybridization with labeled probe. A general positive control was hybridization with poly d(T), and similar negative controls were hybridization with poly d(A) probes as well as a "nonsense" probe (reverse antisense probe to the constant region of the κ-immunoglobulin light chain).

This technique has allowed PTHrP mRNA to be localized in normal skin as well as squamous cell carcinoma of the skin (see Fig. 5B) and gynecological tract. We have also used this method to visualize PTHrP mRNA in amphibian tissues (see Fig. 6) using oligonucleotide probes to a chicken PTHrP sequence.

B. Distribution and Functions

1. The Cardiovascular System

The principal cell types of the cardiovascular system—endothelial, smooth, and cardiac muscle—all appear to be involved in production and/or response to PTHrP. The effect of PTHrP on smooth muscle of the major blood vessels is to cause muscle relaxation and hence vasodilation (Folkman and Singh, 1992), an action similar to that on bladder, chicken oviduct, and rat uterus smooth muscle. Receptors for PTH/PTHrP have been detected in major blood vessels, for example, in bovine and rat dorsal aorta (Orloff *et al.,* 1989), but the vasodilatory effects of PTHrP are also evident in rat coronary vessels (Nickols *et al.,* 1989), rabbit renal arteries (Trizna and Edwards, 1991), and rat renal vascular bed (Musso *et al.,* 1989). In addition to responding to PTHrP, vascular smooth muscle cells also produce the protein constitutively (Hongo *et al.,* 1991), which suggests that it may act as an autocrine factor. In rat aortic smooth muscle cells, maintained *in vitro,* production of PTHrP can be stimulated by angiotensin II as well as by other vasoconstrictive drugs (Pirola *et al.,* 1993), whereas a vasodilatory agent, atrial natriuretic peptide, has no effect. In this study angiotensin II acted both by increasing PTHrP gene transcription and by prolonging the mRNA half-life in the smooth muscle cells. The action of PTHrP in rat aorta smooth muscle cells appears to be principally via the PTH/PTHrP receptor, since PTHrP(1–34) peptide induces the relaxation response, and the consequent rise in cAMP suggests that the adenyl cyclase system is instrumental in the vasodilatory process (Crass and Scarpace, 1993), acting via cyclic-nucleotide-dependent protein kinases (Hino *et al.,* 1994). Systemic hypertension in rats induced by angiotensin II and salt loading increases cardiac and aorta PTHrP mRNA content (Takahashi *et al.,* 1995), correlating with the preceding *in vitro* results. Factors present in fetal calf serum, of unknown identity, also stimulate PTHrP synthesis by cultured rat aortic smooth muscle cells, an effect that can be partially inhibited by dexamethasone (Hongo *et al.,* 1991). In this system cell replication was involved in the increased PTHrP synthesis and release, but vasoactive agents—endothelin, noradrenaline, and thrombin—were also stimulatory. PTHrP in rat aorta smooth muscle cells may play a role in controlling the cell cycle and cell division (Okano *et al.,* 1995), as it appears to do in human skin keratinocytes (Kaiser *et al.,* 1992). However, in the vascular system, endothelial cells themselves also produce PTHrP, which, due to the cells' association with smooth muscle cells, means that PTHrP may act as a paracrine factor. Endothelial cells from bovine aortas (Ishikawa, 1994) and human umbilical veins (Rian *et al.,* 1994) express immunodetectable PTHrP when cultured *in vitro,* but these latter cells have no PTH/PTHrP receptors, which suggests the absence of an extracellular autocrine feedback system. Phorbol esters stimulate endothelial cell differentiation and angiogenesis *in vitro* and also up-regulate PTHrP expression, suggesting a causal relationship between the two (Rain *et al.,* 1994). Thus, PTHrP joins numerous other factors involved in blood vessel physiology, some of which are shown in Fig. 3.

A comprehensive immunohistochemical study by Burton *et al.* (1994) using a panel of monoclonal antibodies to N-terminus, mid-molecule, and C-terminus regions of human PTHrP investigated the distribution of PTHrP in the rat heart. All three antibodies gave similar results, indicating the presence of a molecule, which may be up to 141 amino acids long, present in atria, ventricles, and associated major arteries and veins. However, RIAs

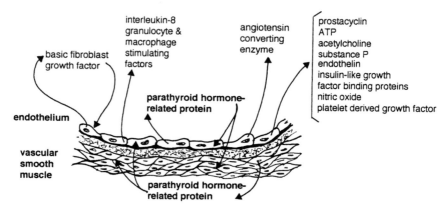

FIG. 3   PTHrP and some other vasoactive factors in the vascular wall.

with these antibodies demonstrated that the ventricles contained the lowest concentrations of PTHrP and that the levels in the atria and major vessels were similar to each other. Measurement of mRNA by competitive polymerase chain reaction (PCR) demonstrated a much lower content in the aortas and venae cavae, whereas the artia and ventricles contained similar levels of expression. Therefore, there is differential transcription and translation of the PTHrP gene in these integrated but phenotypically distinct elements of the cardiovascular system—another example of the cell-specific processing of PTHrP. PTHrP also affects the vascular bed of other tissues and organs and these aspects will be discussed in relevant sections.

2. Mammary Gland

After PTHrP was identified in tumor tissues the first normal tissue shown to contain PTHrP protein and mRNA was lactating mammary gland. Of a wide range of normal adult and embryonic rat tissues screened for expression of PTHrP gene, only lactating mammary gland contained relatively high levels of the mRNA (Thiede and Rodan, 1988). The presence of both PTHrP protein and mRNA in the gland was related to suckling; removal of the pups led to a rapid fall in both of them. There was no correlation with changes in serum calcium concentrations. Lactating mammary gland was the first source of rat PTHrP cDNA (Thiede and Rodan, 1988). Initially, PTHrP mRNA could not be detected in mammary tissue during pregnancy, but a later study (Rakapoulos et al., 1992) showed immunoreactive PTHrP in mammary epithelial nests from day 14 of gestation and in alveoli later during pregnancy, before parturition, suggesting a role for PTHrP in the development of differentiated mammary epithelium. These observations of PTHrP protein and gene expression correlate temporally with development of receptors for the pituitary hormone prolactin (Hayden *et al.*, 1979; Jahn *et al.*, 1991), which is involved in control of mammary growth and development during pregnancy and differentiated activity during lacatation. Virgin rat mammary tissue has undetectable levels of prolactin receptors by ligand binding and only very low gene expression by Northern blot analysis; levels of both increase slightly in late pregnancy, then rise significantly following parturition and induction of lactation (Jahn *et al.*, 1991). Moreover, in mice at least, only the epithelial cells express the prolactin receptor gene (Bera *et al.*, 1994). Thiede (1989) showed that PTHrP mRNA gene expression rises during lactation in response to serum prolactin elevation. The mammary gland is a complex organ in which growth and development are controlled by interacting systemic and local factors (Imagawa *et al.*, 1990), with interplay between mammary epithelium and stromal elements. PTHrP is produced specifically by the mammary epithelial cells (Ferrari *et al.*, 1992), and EGF stimulates both its gene expression and translation, possibly via tyrosine kinase–mediated mechanisms (Ferrari *et al.*, 1994); however, up-regulation of the gene may also involve protein kinase C pathways. Prolactin, too, operates via activation of JAK2 tyrosine kinase in mouse mammary gland (Campbell *et al.*, 1994). In the intact gland, it is clear from several studies that the principal control for up-regulation of PTHrP gene expression is suckling (Yamamoto *et al.*, 1992; Thiede, 1994) and that concentrations of PTHrP in milk increase during lactation (Goff *et al.*, 1991). Suckling is also the stimulus for prolactin release from the pituitary, which then stimulates the mammary gland, again demonstrating the interaction between prolactin and PTHrP production.

As in other tissues, for example the kidney, PTHrP appears to control blood flow in the lactating mammary gland (Thiede *et al.*, 1992; Davicco *et al.*, 1993). Exogenous human PTHrP increased mammary blood flow in the goat gland (Prosser *et al.*, 1994), but only pharmacological concentrations were effective. The question therefore remains whether PTHrP produced locally by epithelial cells actually interacts with smooth muscle cells of the mammary arteries and veins and the myo-epithelial cells around the alveoli, particularly in concert with neurophypohysial peptides, principally oxytocin, which is instrumental in the milk let-down process in suckling (Grosvenor and Mena, 1974). The sensitivity of these cells to PTHrP, via specific receptors, will provide important indicators of the interplay between the mammary gland epithelium and stroma.

The function of PTHrP in milk is still under debate; control of calcium uptake and differentiation of the gut in the offspring have been suggested as possible roles, but there is no evidence for these (Melton *et al.*, 1990). An interesting report by Goff *et al.* (1991) found that colostrum of Jersey cows contained high levels of PTHrP that continued to increase with lactation up to 9 months postpartum. Parallel increases in immunoreactive PTHrP, determined by using antiserum that detected an epitope of residues 20–28, were apparent in their calves' serum over the same period; however, biologically active PTHrP, tested by cAMP accumulation in ROS 17/2.8 cells, was undetectable in any of the calf sera. It appears that PTHrP from milk can enter the offspring circulation, but its biological activity, if any, is not concerned with interaction of the PTH/PTHrP receptor via the N-terminus of the molecule.

3. Connective Tissue

As appears true for other tissues, PTHrP in skeletal tissues is probably a differentiating factor that acts during fetal development and operates postpartum as a maturation and remodeling factor. The PTHrP gene is expressed in both fetal cartilage and bone (Senior *et al.*, 1991; Burton *et al.*, 1992) in the rat as early as 13 days of gestation in sclerotome mesenchyme and vertebral column cartilage. There is gradual loss of immunodetectable PTHrP from these tissues as ossification proceeds, until the differentiated chondrocytes and osteoblasts become immunopositive for PTHrP. Mice have been bred carrying the null mutation for the PTHrP gene that, as homozygotes, fail to survive the peripartum period (Karaplis *et al.*, 1994). Tissues in these mice that normally ossify postpartum—for example, perichondrium of the ribs and long bones—show accelerated calcification. There is dysplasia in the long bone growth plates, with a much reduced zone of proliferating chondrocytes but a relatively normal hypertrophic zone. The absence of the PTHrP gene therefore has profound effects on cartilage and bone development, reducing proliferation and enhancing ossification. In addition it appears to be a fatal omission for the transition from fetal to extrauterine existence, perhaps because uterine PTHrP partially supports the fetus until parturition. PTHrP may normally act as an autocrine and/or paracrine agent in bone development, and some of its actions may depend on the presence of functional receptors. Expression of the PTH/PTHrP receptor mRNA in developing rat femur is strongest in maturing chondrocytes but absent from hypertrophic chondrocytes (Lee *et al.*, 1993); however, it is possible that not all the actions of PTHrP in developing bone depend on the classical PTH/PTHrP.

Mobilization of calcium from bone was recognized as the predominant action of the hypercalcemic factor derived from some cancers before the specific molecule responsible was identified. Thus, the osteolytic action of PTHrP in adult bone, which is similar to that of PTH, operates via the common receptor and through interaction at the N-terminus of both molecules. But this action may be principally a result of the pathologically high levels of circulating PTHrP in cases of malignant tumors and may not reflect the normal activity that is derived from paracrine influence within bone tissue itself. The osteoblast-like cell line UMR 106, derived from an induced rat osteogenic sarcoma (Forrest *et al.,* 1985), has been used as a standard assay system to determine the adenylate cyclase stimulating activity of PTH and PTHrP and related molecules. However, this transformed cell and its normal counterpart do not resorb bone directly but act through osteoclasts as shown by Evely *et al.* (1991) and Fenton *et al.* (1993) in coculture systems. Therefore in normal bone, cells of osteoblast lineage act on osteoclasts to cause lysis of bone matrix. PTH is derived solely from the parathyroid gland and acts as an endocrine factor on the osteoblasts, as PTHrP may do when it is released by malignant tumors, but in bone PTHrP appears to be synthesized by osteoblast-line cells (Walsh *et al.,* 1995). Clearly, this allows the possibility for PTHrP to act as an autocrine or intracrine agent promoting the release of a factor by the osteoblasts that stimulates osteoclast bone-resorbing activity. However, this apparently relatively simple process is complicated by observations showing that the C-terminus of PTHrP inhibits bone resorption by osteoclasts and reduces the number of active osteoclasts in culture systems. Mature PTHrP (1–141) and a C-terminal fragment (107–139) act directly on osteoclasts in isolated rat cell cultures and rat long bone culture systems (Fenton *et al.,* 1991b), but the pentapeptide fragment of PTHrP, residues 107–111, is the essential region. Processing at residue 106 and the retention of a free amino terminus are necessary for full biological activity (Fenton *et al.,* 1991a). These results show the importance of the transcriptional and posttranslational processing of PTHrP that is linked to the diversity of its actions. It has yet to be determined whether the osteoblast cells producing PTHrP secrete the shorter isoform or whether other cells in normal bone possess the necessary enzyme systems responsible for this specific cleavage of PTHrP. Conflicting results using mouse bone cell cultures showed that C-terminus human PTHrP (107–130) and (107–111) stimulate the formation of osteoclast-like cells in culture, cells that appeared to have bone-resorbing activity by forming pits in dentin slices (Kaji *et al.,* 1995). These complex interacting systems in bone need more study to clearly determine all the functions of PTHrP in bone development and remodeling.

4. Urogenital System

The effects of PTHrP in the kidney became apparent initially through the development of HHM in patients with certain malignant tumors. The recognition that HHM was caused by PTHrP acting on bone to mobilize calcium and on the kidney to reduce tubular resorption of phosphate while decreasing calcium excretion focused attention on the kidney as a major site of action of PTHrP from the circulation. Indeed, the kidney subsequently became the tissue of choice for characterizing the PTH/PTHrP receptor, since it was clear that actions in the kidney were mediated by the amino terminal region of both molecules (Zhou et al., 1988). The actions of PTH and PTHrP on kidney electrolyte transport in isolated, perfused rat kidney models showed that their actions are virtually identical (Ebeling et al., 1989; Carney et al., 1991), and the signal transduction mechanisms involving formation of cAMP are equally similar (Blind et al., 1993). An action of tumor-derived PTHrP in the kidney, which appears to be mediated differently than PTH, is its action in reducing the level of 1-hydroxylase enzyme activity resulting in low levels of 1-25 dihydroxyvitamin D_3 [$1,25(OH)_2D_3$] (Stewart et al., 1990; Walker et al., 1990). A physiological condition in which PTHrP but not PTH appears to act on kidney electrolyte metabolism is during suckling (Yamamoto et al., 1991), when transient phosphaturia and elevated urinary cAMP occur. High levels of PTHrP are produced in the lactating mammary gland, and it is possible that some enters the circulation to act as an endocrine factor on the kidney. Within the kidney PTHrP has been detected immunologically in proximal and distal tubules, collecting tubules, and in the medullary rays (Kramer et al., 1991). Calcium resorption occurs principally in the distal tubule, where PTHrP may be involved in this physiological function, but it may act differently in other segments of the kidney tubular system.

Not only is the tubule the site of PTHrP action in the kidney but it may also be active in controlling renal hemodynamics by relaxing afferent arteriolar smooth muscle and hence dilating preglomerular blood vessels and constricting efferent arterioles—effects demonstrated in hydronephrotic rat kidneys (Endlich et al., 1995). The levels of PTHrP used in these studies were pharmacological, relative to normal plasma concentrations of PTHrP, but because many actions of PTHrP are paracrine, local intercellular levels may be much higher than those in the circulation. Nevertheless, the opposite actions on afferent and efferent arterioles may be pharmacologically induced, or they may be mediated by different regions of the PTHrP molecule (see Section II,B,3). Both renal tubules and microvessels possess classical PTH/PTHrP receptors for the N-terminus of both molecules (Nikols et al., 1990), which suggests that this region is responsible for both smooth muscle relaxation as well as calcium transport.

In the human urinary bladder PTHrP has been detected in the epithelium by IHC (Kramer et al., 1991), and in the rat the distended bladder smooth muscle cells contain immunodetectable PTHrP (Yamamoto et al., 1992). Expression of the PTHrP gene was specifically up-regulated by bladder distention in the rat, and exogenous PTHrP produced a muscle-relaxing effect. Whether PTHrP is found or produced in the rat urinary epithelium has not been reported, but since epithelium of human bladder contains

PTHrP, it may be present in the rat, so that the two layers of cells may interact to control bladder distention and evacuation.

5. Integument

Human skin keratinocytes were the first normal cells in which PTHrP was detected immunohistochemically (Danks *et al.*, 1989) following the observation that cultured human keratinocytes produced a parathyroid hormone–like protein (Merendino *et al.*, 1986), which was subsequently characterized as PTHrP (Insogna *et al.*, 1988). Among the population of epidermal cells only the spinous keratinocytes react with antiserum to PTHrP(1–16); there is no reaction in cells of the basal, parabasal, granular, and horny layers. The function of PTHrP in skin keratinocytes can be inferred from the results of experiments using antisense RNA technology (Kaiser *et al.*, 1992) and from the position of the keratinocytes in the differentiated epidermal layers of the skin. The spinous keratinocytes are differentiated cells generated from the basal layers, so that their position in the strata of the epidermis between the basal and horny layers suggests a relatively stable but gradually aging population of cells. The antisense RNA experiments showed that transfected human keratinocyte cell line HPK1A, which does not express PTHrP, proliferated more rapidly than normal cells, with a greater proportion of the cells in S-phase as opposed to G_0/G_1 the cell cycle. Thus, PTHrP may be concerned with inhibiting progression or entry into S-phase in keratinocytes, an action that allows stabilization of a cell population and prolongs the intermitotic stage of cell growth, a phenomenon that also occurs in tumor cells. It seems likely that this is an intracellular function of PTHrP. Indeed, human and mouse keratinocytes do not appear to have membrane receptors for PTHrP, at least none of the classic cAMP-linked receptors for the N-terminus (Orloff *et al.*, 1992). Whitfield *et al.* (1992) suggest that human keratinocytes do have receptors for PTH and PTHrP but that these are coupled to protein kinase C–activating mechanisms and do not involve cAMP activation. However, normal keratinocytes appear to synthesize and possible release a midmolecule form of PTHrP, from residue 38 to 107, that is packaged into secretory granules and hence may be released to act as a paracrine factor in the epidermis (Soifer *et al.*, 1992).

In the developing rat embryo, hair follicles, which are derived from the epidermis, express mRNA for PTHrP at 18.5 days of gestation, and in the head region at 19.5 days the outer root sheath of developing vibrissae clearly expresses the PTHrP mRNA (Senior *et al.*, 1991). The neck cells of the follicles show the strongest signal at this stage, with the keratinocytes of the epidermis developing reactive mRNA a day later, 20.5 days of gestation. In epidermis of the common frog, at the stage of emergence from water to land (Stage XXV, Taylor and Kollros, 1946) abundant mRNA for PTHrP has been detected by *in situ* hybridization using a probe to chicken PTHrP (55–65) (Danks and Ingleton, unpublished observations; Fig. 4; see color insert). The signal is in the cells of the stratum granulosum and is much stronger in the dorsal than in the ventral epidermis. Epithelium of the mucus glands, which are derived from the epidermis, also expresses the PTHrP gene, and like rat hair follicles, the neck cells express the strongest signal. Buccal epithelium in aquatic stages has abundant immuno-reactive PTHrP when it is undetectable in the dorsal epidermis (Figs. 4A, 4C; see color insert), although these cells contain low levels of mRNA (Fig. 4B; see color insert). Interestingly, there is only a very low level of gene expression in the simple epithelium of the frog tadpole (i.e., during the aquatic, premetamorphic stages), but levels increase particularly during the late metamorphic stages as the layers of epidermal cells become more numerous and mucus glands develop (Figs. 4D, 4E; see color insert).

The epidermis forms an important interface between an animal and its environment, providing a mechanical barrier for protection and a physiolog-ical barrier against dehydration in terrestrial forms. It seems likely that PTHrP is involved in aspects of these functions by providing a stable popula-tion of cells and perhaps affecting the dermal blood vessels and connective tissues by paracrine mechanisms.

6. Reproductive Tissues

Besides performing important functions in pregnancy and lactation, PTHrP has also been demonstrated in testes and ovaries, in the stromal cells responsible for producing steroids and growth factors that support germ cell maturation and development. Leydig cells of the human testis stained strongly with antisera to PTHrP(1–141) and various regions of the peptide (Asa *et al.*, 1990; Kramer *et al.*, 1991), but there was no apparent reaction in ova or spermatocytes. Kramer *et al.* (1991) also noted immunoreaction in epithelium of the prostate tubuloalveolar epithelium, but Iwamura *et al.* (1994) demonstrated both protein and gene expression only in a subpopula-tion of epithelial cells that they identified as neuroendocrine cells. These cells also contained chromogranin A, a marker for neuroendocrine cells, which suggests that PTHrP may act as a neuroendocrine or paracrine factor in the prostate; it is possible, however, by analogy with other tissues that PTHrP may influence smooth muscles involved in propelling prostatic fluid into the urethra.

At the time of ovulation in the human ovary, the follicular fluid contains high concentrations of PTHrP that originates by constitutive secretion from the granulosa–luteal cells (Gutmann *et al.*, 1993). What role this follicular PTHrP plays is still not determined, but levels in the fluid did not correlate with calcium concentrations, and secretion was not affected by either estrogen or progesterone. It is possible that in this location PTHrP allows expansion of the follicle during development and must be rapidly down-regulated at the time of ovulation. In the immature frog (*Rana temporaria*) ovary both the interstitial cells and the ova contain immunoreactive PTHrP, detected with antiserum to human PTHrP(1–16) and PTHrP mRNA hybridized *in situ* with oligonucleotide probe to chicken PTHrP(55–65) (Danks *et al.*, 1994). The function of PTHrP in any of these tissues and the factors controlling its action are so far unknown; nevertheless, the reproductive systems of males and females contain PTHrP that may be concerned in germ cell production, embryo implantation and placental development, fetal and maternal vascular reactivity, fetal calcium regulation, and lactation.

7. Brain and Central Nervous System

Gene expression for both PTHrP and the PTH/PTHrP receptor have been demonstrated in brain of the rat (Struckhoff and Turzynski, 1995). In this study the receptors were found on astrocytes, and PTHrP was detected in the meninges epithelial cells, suggesting a paracrine interaction from meningeal cells to astrocytes. However, no reactions could be detected in the neural tissue of the brain. This observation is at variance with those of Weir *et al.* (1990) and Weaver *et al.* (1995), who described widespread distribution of PTHrP bioactivity and gene expression in rat brain. The greatest bioactivity was located in the cerebral cortex and telencephalon (Weir *et al.*, 1990), with high levels of mRNA in discrete nuclei of the hypothalamus, neostriatum, hippocampus and cerebral cortex, midline thalamic nuclei, choroid plexus, and anterior pituitary gland (Weir *et al.*, 1990; Weaver *et al.*, 1995). Expression of the PTH/PTHrP receptor in rat brain was mainly in different discrete nuclei, with some in the central region of the brain. PTHrP was generated more frequently in neurons that are peripheral or in "higher" brain centers (e.g., cerebral cortex). There appears to be a system in which PTHrP acts as a neurotransmitter operating through relay centers such as the amygdala. In the brain of the emerging frog, PTHrP was detected in the olfactory lobe perikarya and in the olfactory gland epithelium with antibody to human PTHrP(1–16), and as in the epidermis, PTHrP mRNA was demonstrated more readily by *in situ* hybridization with an oligoprobe of chicken PTHrP cDNA (Danks *et al.*, 1994). The pituitary pars distalis and median eminence of the frog also contained PTHrP, but these regions hybridized only weakly with oligonucleotide probes. Abundant hybridization occurred in the pituitary pars intermedia, which reacted only very weakly with the antiserum to human PTHrP(1–16) (Figs. 4F, 4G; see color insert). These observations suggest that PTHrP may have an inductive role in the pituitary/hypothalamus system in the frog at this stage of development, but further confirmation and investigations are needed for such functions to be defined.

PTHrP may act as neurotransmitter in the dogfish, an elasmobranch, since giant neurons of the tectum, which abut the lumen of the diencephalon, contain immunoreactive PTHrP (Ingleton *et al.*, 1995). The choroid plexus of this fish is also composed of epithelial cells rich in immunoreactive PTHrP. The choroid plexus of the brain controls hydromineral balance between the cranial capillaries and the subarachnoid space (Cserr, 1971), so that PTHrP in these cells may be involved in this physiological function. In the dogfish and in sea bream and flounder, both teleosts, the saccus vasculosus epithelial cells contain immunoreactive PTHrP (Devlin *et al.*, 1996; Ingleton and Danks, 1996). The saccus vasculosus is a neurohemal organ that is found only in some cartilaginous and teleost fishes. Teleost fishes also have a distinctive neurohemal organ at the posterior end of the spinal nerve cord, the urophysis, which may receive nervous input from both the spinal cord and the brain (Kobayashi *et al.*, 1979; Kriebel *et al.*, 1985). The urophysis and spinal cord neurons of the flounder contain immunoreactive PTHrP (see Fig. 6A), which suggests that PTHrP may have function(s) in all neurohemal organs of the central nervous system. The urophysis produces two factors, urotensin I (UI), similar to corticotrophin releasing factor, and urotensin II (UII), homologous to somatostatin 14, both of which are vasoactive in several vertebrate vascular systems. The principal action is vasodilation, but in some pulmonary vessels and the dorsal aorta of rabbits both UI and UII are constrictors (Lederis, 1984). PTHrP's involvement in the physiology of this organ has not been investigated, but some of its functions appear to be similar to those of the urotensins. It will be particularly interesting to learn how these factors are integrated.

8. Immune System

In addition to its function in skeletal tissues derived from mesoderm, PTHrP is also produced by and acts on cells of the immune system, including splenic lymphocytes and T-lymphocytes. PTHrP was first associated with leukemic T-lymphocytes in patients developing humoral hypercalcemia (Honda *et al.*, 1988). Lectin-stimulated T-lymphocytes *in vitro* produce PTHrP into incubation media; PTHrP has been postulated to act as a local growth inhibitor for normal lymphocytes (Adachi *et al.*, 1990). Lymphocyte proliferation and maturation are controlled by numerous interacting factors, and PTHrP appears to be an additional cytokine that may modulate lymphocyte proliferation and function. Both gene transcription and translation for PTHrP in splenic lymphocytes is increased by endotoxin (liposaccharide,

LPS) treatment *in vivo* (Funk *et al.,* 1993). These effects appear to be mediated by local cytokines, tumor necrosis factor α, and interleukin-1β, which mediate the effects of other stimulatory and inhibitory agents. Later studies showed that there are unexpected responses to inflammatory or anti-inflammatory cytokines. The anti-inflammatory interleukin-4 (IL-4) increased LPS stimulation of PTHrP gene expression, whereas the inflammatory interferon-γ (IFN-γ) inhibited PTHrP gene expression by splenic lymphocytes (Funk *et al.,* 1994). These results should be considered in the context of the action of PTHrP in holding cells in the G_0/G_1 phase of the cell cycle: reduction of PTHrP expression would allow cells to move into proliferation and/or differentiation to enhance the inflammatory response and vice versa.

By inference, splenic lymphocytes have receptors for PTHrP, since various fragments of the molecule stimulate membrane-associated protein kinase C activity. For C-terminus fragments the conserved pentapeptide PTHrP (107–111) (TRSAW) is vital to this effect, but PTHrP(1–40) also stimulates the enzyme activity (Whitfield *et al.,* 1994). None of these interactions appear to involve increases in cAMP production, which suggests that the splenic lymphocytes do not possess classical PTH/PTHrP receptors.

C. PTHrP in the Fetus and in Pregnancy

In normal adult mammals PTHrP is found in the circulation only in low concentrations, which indicates that, unlike PTH, it acts mainly as an auto- and paracrine factor rather than as an endocrine factor. This mode of action is paramount in tissue development, so it is particularly interesting that PTHrP has been detected in many fetal cells and tissues and in cells of the placental epithelia. Thus, it seems likely that PTHrP has important functions in ontogenetic differentiation and in the fetal/maternal relationship in utero.

The earliest stage of development at which PTHrP has been detected in the mouse embryo is in the blastocyst. The first potential endoderm cells derived from the inner cell mass contain immunodetectable PTHrP (Van der Stolpe *et al.,* 1993) at a stage when the trophoblast cells are invading the uterus wall (see Section II,D,5). An earlier report of PTHrP gene expression in the rat fetus demonstrated mRNA, by *in situ* hybridization, beginning in specific areas around 15.5 days of gestation (Senior *et al.,* 1991). Cells of the dental lamina hybridized strongly at this stage and persisted, through expression in the enamel organ, to parturition. Other ectodermally derived epithelia, including hair follicles and vibrissae were positive at day 18.5, and by day 20.5 the keratinocytes were positive also. Endodermally derived lung epithelium hybridized at 15.5 days, but the widespread distribution was gradually reduced during development until only isolated patches among the bronchial ducts remained at term. Meso-dermally derived connective tissue elements gave positive results at 16.5 days. The perichondrium of the sternum, for example, hybridized strongly at this stage but reduced gradually to parturition. PTHrP peptide in the rat embryo appears to be more widespread than mRNA; at day 14, embryonic tissues, including epidermis and derived tissues, skeletal and vascular smooth muscle and cardiac myocytes, and epithelia of kidney tubules, bronchi, gut, and hepatocytes contained immunodetectable PTHrP (Campos et al., 1991). Staining intensity increased to day 18 and was more widespread, being detected also in the presumptive pars intermedia of the pituitary, choroid plexus, endocrine pancreas islets, adrenal medulla, and seminiferous tubules of the fetal testis. Postnatally most tissues retained their immunoreactive PTHrP to day 7 with the exception of the hepatocytes, in which levels were much reduced, and in the testis, where reaction was localized to the Leydig cells and was absent from the seminiferous tubules. The interplay between mesenchymal and epithelial tissues in rat embryo development is suggested by the expression of PTHrP mRNA mainly in epithelia, with the receptor mRNA more often in the mesenchyme (Lee et al., 1995). PTHrP is also found in the fetal parathyroid gland of sheep (Rodda et al., 1988) but is absent in the adult.

Not only does PTHrP appear to play a role in fetal tissue development but it is also important for implanation and activities of the placental membranes that control the internal and external ionic environment of the fetus. In the rat, PTHrP is found in the uterus wall where decidua develop at the point of implantation of the blastocyst early in gestation (Beck et al., 1993), whereas PTHrP mRNA becomes detectable in giant trophoblast cells that invade the decidua from day 7.5 (Senior et al., 1991). In human pregnancy mRNA is detectable only in the cytotrophoblast, but both the peptide and message occur in the syncytiotrophoblast and in the avascular amnion in the first trimester of pregnancy (Dunne et al., 1994a). A role for PTHrP in the amnion is likely to be associated with expansion of the amniotic cavity; at term there is a rapid down-regulation of amniotic PTHrP gene expression associated with rupture of the fetal membranes and the onset of labor (Ferguson et al., 1992). PTHrP is found in the smooth muscle of many tissues, where it induces relaxation, so that its influence has to be abrogated in the feto-placental unit for uterine contractions to take place at parturition; the signal for this is not yet known but may include prosta-glandins (White, 1989). Within the umbilical cord, PTHrP gene is expressed in both the vascular smooth muscle and the endothelial cells, whereas the receptor gene mRNA is expressed in the endothelial cells, suggesting paracrine and autocrine control of placental circulation (Ferguson et al., 1994). Rian et al. (1994) failed to detect either receptor mRNA or protein in umbilical vein cells after a period in culture, which may be due to loss of expression in the conditions used. The PCR technique also detected mRNAs for the three molecular forms of PTHrP, 1–139, 1–141, and 1–173, which were not apparent by simple Northern blast analysis (Ferguson *et al.*, 1992). Again, the physiological significance of these variations has not been investigated. In addition to its action on muscle tone PTHrP also increases epidermal growth factor receptor (EGF-R) expression in the process of conversion of cytotrophoblasts to syncytiotrophoblasts (Alsat *et al.*, 1993).

A function of PTHrP in pregnancy identified in several species is the maintenance of the differential calcium ion concentrations between the maternal and fetal circulations. The syncytiotrophoblast is instrumental in active transport of Ca^{2+} to the fetus, via a high-affinity Ca^{2+} pump (Borke *et al.*, 1989) that maintains the elevated levels in the fetal circulation. PTHrP from the placenta and from the fetal parathyroid, at least in sheep (Rodda *et al.*, 1988; Abbas *et al.*, 1989), appears to be a key factor in this process. Low external Ca^{2+} causes release of PTHrP from human cytotrophoblast cells *in vitro* (Hellman *et al.*, 1992) via a calcium-sensing mechanism similar to that on parathyroid cells. The region of the PTHrP that appears to be involved with calcium uptake in the placenta lies beyond the N-terminal 1–34, which, like PTH(1–34) and (1–84), does not stimulate placental calcium transport. Synthetic PTHrP(1–108) and (1–141) both enhance placental calcium uptake (Abbas *et al.*, 1989), which suggests that the midmolecule region may be important for this function (Care *et al.*, 1990).

The almost ubiquitous expression of the PTHrP gene during stages of mammalian fetal development suggests that it has fundamental actions with regard to differentiation and growth of tissues, functions that do not continue to operate in the adult in normal conditions except in a few tissues such as the keratinocyte layer of the skin; however, the gene can be activated in lactation and in cancers, suggesting the withdrawal of inhibitory control or reestablishment of stimulation. The widespread occurrence of PTHrP in mammalian fetal tissues provides an interesing comparison for studies on lower vertebrates that do not have an intrauterine stage of development but grow either within the amniote egg, such as reptiles and birds, or in aquatic environments, such as fish and amphibia.

D. PTHrP in Neoplastic Cells

PTHrP was first isolated and cloned from several types of neoplasms, including cell lines derived from a human lung squamous cell carcinoma (Moseley *et al.*, 1987), an adenocarcinoma of the breast (Burtis *et al.*, 1987), and a renal cell carcinoma (Strewler *et al.*, 1987). It has been detected in a diverse range of tumors and tumor cell lines associated with HHM as well as in tumors not associated with HHM. Often the presence of PTHrP relates to the normal distribution of the protein, but there are tumors that produce PTHrP derived from normal tissue that does not contain the protein. An example of the former is the presence of PTHrP in squamous cell carcinoma of the skin and in normal skin (Danks *et al.*, 1989); melanoma containing PTHrP (Danks *et al.*, 1989) and normal melanocytes, lacking the protein, exemplify the latter. Specific examples of neoplastic tumors and PTHrP production are discussed in the following sections.

1. Kidney

In common with normal kidney tubule epithelium, PTHrP antigen and mRNA have been identified in renal cell tumors that are associated with HHM. Immunoreactive PTHrP was detected in a renal carcinoma cell line (SKRC-1) (Kramer *et al.*, 1991), and immunohistochemical studies demonstrated that human renal cell carcinomas were positive for PTHrP (Danks *et al.*, 1989). Gotoh *et al.* (1993) found that granular cell subtypes tended to be more strongly positive than the clear cell subtypes, but they found no correlation between the intensity of PTHrP immunochemical staining of renal cell carcinomas and the patient's serum calcium level (Gotoh *et al.*, 1993). A possible growth stimulatory role for PTHrP in renal carcinoma was suggested by results of treatment with polyclonal antiserum to N-terminus PTHrP, which inhibited the growth of a renal cell carcinoma cell line (Burton *et al.*, 1990). The growth effects appeared to be generated through interaction at the PTH/PTHrP receptor, because growth was also significantly inhibited by a competitive PTH antagonist. Gene expression for PTHrP has been detected in human renal cell carcinomas using Northern blot analysis (Strewler *et al.*, 1987; Ikeda *et al.*, 1988a, 1998b), and *in situ* hybridization has localized PTHrP mRNA in a renal cell carcinoma in a hypercalcemic patient (Heath *et al.*, 1990). In an attempt to address the question of why some renal cell carcinoma lines produce PTHrP and some do not, Holt *et al.* (1993) examined the region-specific methylation of the PTHrP gene and found that demethylation induced the expression of PTHrP mRNA in previously nonexpressing cell lines.

2. Skin

Squamous cell carcinomas (SCCs) of the skin were found to be positive for PTHrP, as was the spinous keratinocyte layer of the adjacent normal skin, in the first immunohistochemical study of PTHrP in tumors (Danks *et al.*, 1989). Spinous keratinocytes are a differentiating feature of SCC pathologically. These findings were confirmed in similar studies by Kitazawa *et al.* (1991) and Hayman *et al.* (1989) and by *in situ* hybridization localization of PTHrP mRNA in SCC of the skin (Danks *et al.*, 1995) (see Figs. 5A and 5B). Basal cell carcinomas have been found to be negative for both the PTHrP antigen (Danks *et al.*, 1989) and mRNA (Danks *et al.*, 1995). A canine squamous cell carcinoma of the oral cavity was found to react positively with antiserum to human PTHrP(1–36) (Grone *et al.*, 1994).

3. Lung

Squamous cell carcinoma of the lung is one of the tumor types consistently associated with HHM, and a cell line (BEN) derived from a patient with lung SCC and HHM was an original source of PTHrP isolate and clone (Moseley *et al.*, 1987). Detailed analysis of the cellular location of PTHrP in lung tumors, using immunohistochemistry, has detected the protein in SCC as well as in small-cell carcinoma of the lung (Danks *et al.*, 1989), in the transitional and bronchial epithelia with squamous metaplasia, in all cases of well-differentiated and moderately differentiated SCC of the lung, and in most cases of small-cell carcinoma (Kitazawa *et al.*, 1991). In both these studies poorly differentiated tumors showed variable and inconsistent staining for PTHrP, and adenocarcinomas of the lung were negative.

Although mesothelioma is rarely associated with HHM, McAuley *et al.* (1990) found that eight out of nine malignant mesotheliomas were positive for PTHrP. A larger study (Clark *et al.*, 1995) examined malignant mesotheliomas and pleural adenocarcinomas and showed that a high proportion (84%) of the mesotheliomas stained for PTHrP, whereas only 11% of the pleural adenocarcinomas stained positively. Differential diagnosis of these two tumors is very difficult pathologically, so PTHrP staining could be useful diagnostically.

Athymic mice bearing human SCC of the lung become hypercalcemic; Kukreja *et al.* (1990) studied their bone histomorphometric parameters before and after the administration of a neutralizing PTHrP antiserum. The neutralizing antiserum proved to be as beneficially effective as tumor resection, resulting in a decrease of circulating calcium levels, decrease in bone resorption, and an increase in bone formation. Tumor cell lines that synthesize PTHrP have been established from human primary SCC—for example, BEN cells and KCC-C1 cells derived from a malignant pleural effusion (Ichinose *et al.*, 1993). These cell lines produce PTHrP that is released into the medium and hence can be used to investigate PTHrP physiology and factors controlling PTHrP production. Activators of both cAMP and protein kinase C intracellular messenger pathways stimulate production and release of PTHrP by BEN cells (Emly *et al.*, 1994; Rizzoli *et al.*, 1994), but only agents stimulating protein kinase C increase PTHrP mRNA (Emly *et al.*, 1994).

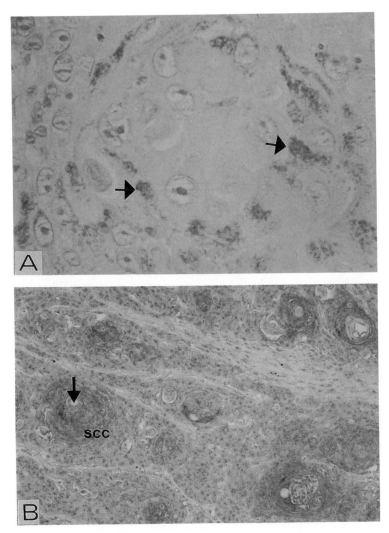

FIG. 5   (A) Immunohistochemical localization of PTHrP in a squamous cell carcinoma of
the skin using an antiserum against human PTHrP(1–16). Staining is present only in the
spinous keratinocytes (→) of the tumor and not in any of the surrounding cells. ×800
(B) Hybridization of a squamous cell carcinoma (scc) of the cervix showing expression of
PTHrP in the spinous cells of the neoplasm but not in the keratin pearls (→) or the surrounding
stroma. ×150

4. Breast

Breast cancer is the malignancy most commonly linked with hypercalcemia (Hickey *et al.*, 1981), but it was thought originally that the increase in serum calcium was due to a local lytic rather than a humoral mechanism. PTHrP was isolated and cloned from a breast tumor (Burtis *et al.*, 1987) almost simultaneously with its isolation from lung (Moseley *et al.*, 1987) and renal carcinoma (Strewler *et al.*, 1987). Several subsequent studies demonstrated immunoreactive PTHrP in approximately 60% of primary human breast cancers (Southby *et al.*, 1990; Bundred *et al.*, 1991; Liapis *et al.*, 1993a; Kohno *et al.*, 1994; Edwards *et al.*, 1995). The development of RIA for PTHrP in human plasma (Grill *et al.*, 1991) showed that humoral hypercalcemia was associated with elevated plasma PTHrP in 13 out of 20 cases. Edwards *et al.* (1995) found a similar correlation. These results provided cogent evidence that HHM is caused by tumor-derived PTHrP released into the general circulation; in these circumstances PTHrP appears to be acting like a classic hormone. However, a higher proportion of bony metastases of breast cancers have been found to contain immunoreactive PTHrP compared with primary cancers. Powell *et al.* (1991) found that 93% of metastases in bone were positive for PTHrP, which led them to suggest that PTHrP contributes to the propensity to metastasize to this site. This contention was supported by a related study (Kohno *et al.*, 1994) that showed that 83% of primary breast cancers positive for PTHrP went on to metastasize to bone; in contrast, only 38% of such primary tumors produced lung metastases. PTHrP gene expression, assessed by PCR for mRNA, gave similar results, in that higher levels of expression were found in tumors that later metastasized to bone than in those that migrated to soft tissue and those that did not recur (Bouizar *et al.*, 1993). Thus, bone may be a preferred site for some breast cancer phenotypes, which, when established, also contribute to bone matrix lysis and HHM.

Although PTHrP is associated with lactation and mammary epithelial cells during lactation (see Section II,B,2) immunostaining of normal tissue adjacent to breast cancers is inconsistent. Southby *et al.* (1990), using antiserum to the N-terminus of human PTHrP, detected no staining in normal lobular and ductal epithelium, which contrasted with the positive results of Liapis *et al.* (1993), who used an antiserum to the midmolecule portion (38–64) of PTHrP. More investigations, with a wider range of region-specific antibodies, are needed to clarify this area of study.

5. Gynecological

Hypercalcemia does not usually accompany the presence of gynecological tumors (Munday and Martin, 1982), but PTHrP is present in normal myometrium (Paspaliaris *et al.*, 1992) and in the uterus during pregnancy (Williams *et al.*, 1994) and is found in tumors of the gynecological tract. Immunohisto-chemical studies localized PTHrP in invasive squamous cell carcinomas, whereas most adenocarcinomas were negative (Kitazawa *et al.*, 1992b; Mackenzie *et al.*, 1994). These findings have been correlated with *in situ* hybridization of PTHrP mRNA (Danks *et al.*, 1995). Uterine leiomyomas overexpress the PTHrP gene (Weir *et al.*, 1994), and fibroids have greater amounts of PTHrP mRNA and protein than does normal myometrium. Similar investigations of premalignant conditions of the lower tract detected PTHrP in all the intraepithelial neoplasms of the vulva examined and in four out of five intraepithelial neoplasms of the cervix (Mackenzie *et al.*, 1994). Dunne *et al.* (1994b) found PTHrP in all 16 invasive carcinomas, using antibodies to the midmolecule region, and PTHrP mRNA in all 10 that they examined. All the tumors contained cells of squamous origin so correlating with neoplasms of other tissues. In the study by Mackenzie *et al.* (1994) the presence of human papillomavirus (HPV), a possible etiological factor in cervical cancer, did not correlate with PTHrP, which was more closely related to tumor type. In a related survey PTHrP mRNA was found in low abundance in tumors of normocalcemic patients but high in hypercalcemic patients (Liapis *et al.*, 1993b).

Small-cell carcinoma of the ovary is often associated with HHM, and Matias-Guiu *et al.* (1994) found immunoreactive PTHrP in five out of seven cases of this neoplasm. They attempted to correlate these results with circulating calcium levels and found no nexus between the two. In ovarian cancers, as in other tissue neoplasms, PTHrP is probably fulfilling intracrine and/or paracrine functions associated with growth and differentiation. Actions at distant sites, such as bone, depend on release of PTHrP from the cells by secretion or lysis.

6. Parathyroid

PTHrP is found in normal fetal human parathyroid glands; in adults the acidophil cells express PTHrP mRNA and contain immunodetectable PTHrP (Kitazawa *et al.*, 1992a). Earlier studies indicated that PTHrP was detectable only in neoplastic parathyroid glands, with very low mRNA expression in normal cow and calf parathyroids and human hyperplastic glands (Ikeda *et al.*, 1988b). In this study PTHrP mRNA was detected in human parathyroid adenomas and tertiary hyperplastic glands. Five out of six canine parathyroid adenomas contained immunodetectable PTHrP, but the normal chief cells reacted only with antiserum to PTH (Grone *et al.*, 1994). However, Danks *et al.* (1990) found PTHrP antigen in both human adenomas and hyperplastic glands from patients with chronic renal failure; primary hyperplastic glands were negative. These patients and those with chronic renal failure had undetectable plasma PTHrP levels (Grill *et al.*,

1991). Using a panel of antibodies to regions of the PTHrP molecule, to the N-terminus PTHrP(1–34), midmolecule PTHrP(50–69), and C-terminus PTHrP(106–141) for immunohistochemistry and Western blotting of tissue extracts, Danks *et al.* (1990) found that the PTHrP molecule in adenomas was truncated and appeared not to have the C-terminal region. The molecule appeared to be equivalent in size to recombinant PTHrP(1–84). Autopsy material examined by Asa *et al.* (1990) gave positive immunostaining in 8 out of 10 parathyroid adenomas and in all hyperplastic glands; however, the delay in fixation (between 6 and 36 hours) may cast doubt on the validity of the observations, since there is a dramatic increase in protease activity after death.

7. Endocrine Tumors

The endocrine pancreas is a normal tissue that expresses PTHrP; neoplasms of this organ also express PTHrP mRNA (Drucker *et al.*, 1989). There appears to be no correlation between levels of PTHrP gene expression and those of other islet cell types secreting glucagon, somatostatin, or insulin. Dunne *et al.* (1993) and Ratcliffe *et al.* (1994) localized PTHrP antigen and mRNA in an adenocarcinoma of the pancreas removed from a hypercalcemic patient who had moderate circulating PTHrP levels.

Normal bovine adrenal cortex and medulla contain PTHrP hybridizing transcripts (Ikeda *et al.*, 1988b), and whereas PTHrP immunohistochemistry has failed to locate PTHrP in normal tissues (human, Kramer *et al.*, 1991; canine, Grone *et al.*, 1994), it has detected it in pheochromocytomas (Asa *et al.*, 1990). A canine pheochromocytoma from a hypercalcemic dog stained positively for PTHrP (Grone *et al.*, 1994), and PTHrP mRNA was localized in a human malignant pheochromocytoma (Heath *et al.*, 1990). In one study, circulating PTHrP levels were assayed in patients before and after the resection of their pheochromocytoma. Basal levels that were elevated in 7 out of 10 untreated patients became undetectable after surgery. Western blot analysis and immunohistochemistry showed that the tumors contained PTHrP, which indicates that PTHrP is synthesized in pheochromocytomas and released into the circulation.

Apocrine glands of canine anal sacs produce pheromones. The epithelial cells of these glands may be transformed into adenocarcinomas that contain PTHrP antigen (Rosol *et al.*, 1990; Grone *et al.*, 1994), and mRNA (Ikeda *et al.*, 1988a).

8. Lymphoid System

PTHrP has been identified in htlv-1–induced lymphoma (Motokura *et al.*, 1988, 1989) and localized within the neoplasm (Moseley *et al.*, 1991). A carcinoid of the thymus (Yoshikawa *et al.*, 1994) as well as cultured human myeloma cells produce PTHrP. Canine lymphosarcomas express PTHrP mRNA (Ikeda *et al.,* 1988a) and immunostain for PTHrP (Grone *et al.,* 1994).

The range of neoplasms and tumors that contain PTHrP is much wider than first envisaged from the viewpoint of HHM and its possible cause. Malignant tumors of many of the tissues in which PTHrP is normally expressed also produce PTHrP. In the liver, for example, both hepatocellular carcinoma (Heath *et al.,* 1990) and cholangiocarcinoma (Roskams *et al.,* 1993) produce PTHrP, and gene expression has been demonstrated in normal liver (Ikeda *et al.,* 1988b). Moreover, a high proportion of hepatocellular carcinoma patients are hypercalcemic with elevated PTHrP levels (Yen *et al.,* 1993).

All these observations of PTHrP in neoplasias, which may or may not be associated with HHM, suggest that PTHrP has functions within the tissue that may be concerned with regulation of growth and/or differentiation. To determine whether these functions are tumor enhancing or inhibitory demands much more experimental investigation.

E. PTHrP in Nonmammalian Vertebrates

1. Birds

From the preceding sections, it is evident that PTHrP has multiple functions in many systems of adult and embryonic mammalian development and physiology. Thus, PTHrP functions are predicted in nonmammalian vertebrates, all of which are ultimately related through evolution from a common stock.

Birds are particularly interesting as regards calcium metabolism because they have to produce readily available calcium for eggshell production on a seasonal or daily basis. The most extensive studies have been made on the chicken and Japanese quail, in which calcium metabolism has been shown to be modified and amplified compared with that in mammals. (Gay, 1988) to produce rapid responses. Parathyroid hormone is the principal factor that elevates plasma calcium levels by inhibiting calcium clearance (Kenny and Dacke, 1974) through reduced bone uptake (Dacke and Shaw, 1987) and renal actions as in mammals, but PTHrP may have an influence in this and other systems in birds. The gene for chicken PTHrP as well as the protein(s) it encodes have been identified and sequenced (see Fig. 2), and the similarity to both mammalian PTHrP and the N-terminus of PTH suggested that the metabolic effects acting through the PTH/PTHrP receptor would be comparable. The binding affinity of chick PTHrP to avian and mammalian renal membranes and rat osteosarcoma cells (UMR 106 H5) is only half that of human PTHrP but is displaceable by bovine PTH,

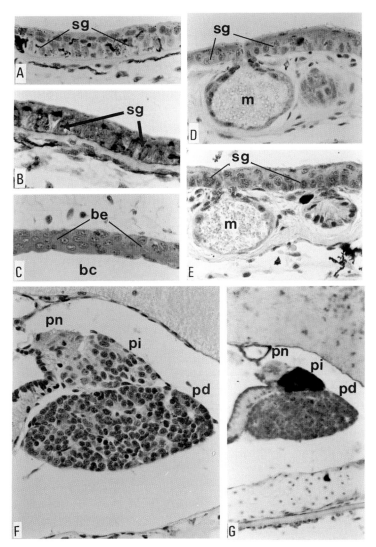

FIG. 4   Immunohistochemistry with antibody to human PTHrP(1–16) and *in situ* hybridization with an oligoprobe to chicken PTHrP(55–65) of frog (*Rana temporaria*) skin and pituitary. Immunostaining of (A) dorsal epidermis and (C) buccal epithelium (be) of the aquatic tadpole (Stage XIV); note the strong reaction in the buccal epithelium but the absence of staining in the stratum granulosum (sg) of the skin. (B) *In situ* hybridization of stage XIV dorsal skin showing hybridization in the stratum granulosum cells. (D) Immunostaining and (E) *in situ* hybridization of the dorsal skin of the stage XXIV frog immediately prior to emergence; at this stage the skin sg and mucus gland (m) epithelium contain both PTHrP protein and mRNA. (F) Immunostaining and (G) *in situ* hybridization of the stage XXIV frog pituitary; protein is detected in the neural lobe (pn) and pars distalis (pd) and slightly in the pars intermedia (pi); however, hybridization is strongest in the pi, with only low levels in the pn and pd. Magnification: A–F, ×260; G, ×140.

which indicates a common receptor in all the tissues. The osteoclast inhibitory action of the C-terminus of PTHrP is also found in the chick molecule and acts on chick osteoclasts (Fenton *et al.*, 1994). The oviduct of birds is developed for laying large megalecithal eggs, and the smooth muscle dilatory action of PTHrP has been utilized in the shell gland. PTHrP gene is expressed in serosal and smooth muscle of the shell gland of the chicken with levels of mRNA and detectable PTHrP varying with the egg-laying cycle, increasing transiently as the egg moves through the oviduct (Thiede *et al.*, 1991). The presence of PTHrP in associated arterial smooth muscle also suggests increased blood flow to the oviduct during the egg-laying cycle.

PTHrP mRNA has been found in tissues of developing chick embryos, using cloned chicken PTHrP genomic DNA and a 600-bp cDNA fragment for Northern blotting of extracted poly(A)–enriched RNA (Schermer *et al.*, 1991). In 15-day embryos tissues derived from all the germ layers—for example, brain, intestine, and smooth muscle—as well as the chorioallantoic membrane and the yolk sac hybridized with the probes. As in mammals, the PTHrP gene is widely expressed in many embryonic tissues and extraembryonic membranes, therefore appearing to be of fundamental importance to the developmental process.

2. Amphibia

There have been no studies of PTHrP in reptiles, the vertebrate group most closely related to birds, but initial studies on amphibians show its distribution in adult and larval stages to be widespread. Some of these observations have been discussed previously in relation to the development of a less permeable epidermis as the immature frog migrates from water to land. Immunohistochemical studies of developing frog (*Rana temporaria*) using antiserum to human PTHrP (1–16) and *in situ* hybridization with an oligonucleotide probe to chicken PTHrP (55–65), using protocols and controls described in the text, have shown distribution of protein and mRNA in several tissues (Danks and Ingleton, unpublished observations). The presence of PTHrP in stratum granulosa cells appears to correlate with keratinization because tadpoles with a horny "beak" for scraping algae, and the like, have a large buccal cavity lined with epithelial cells containing immunoreactive PTHrP (Fig. 4C; see color insert); however, at this stage, which is entirely aquatic (stage XIV, Taylor and Kollros, 1946), the dorsal epithelial cells do not contain immunoreactive PTHrP protein (Fig. 4A; see color insert), but mRNA can be detected in these epidermal cells (Fig. 4B; see color insert). At a later stage, immediately prior to migration to land (stage XXIV), the stratum granulosum of the dorsal epidermis contains both protein and mRNA (Fig. 4D, 4E; see color insert). Gene expression for PTHrP can also be detected by *in situ* hybridization in the cytoplasm of the ova in the developing gonad, which suggests a function in early development of the embryo. In common with other vertebrate species, larval frog kidney tubules contain immunodetectable PTHrP, which may be involved in control of ion excretion. In the emerging frog the tissue in which there is an obvious discrepancy between PTHrP protein and mRNA expression is the pituitary, in which the neural lobe and pars distalis contain the largest amount of immunoreactive PTHrP protein, but the pars interme-dia is the region containing the most mRNA, as detected by *in situ* hybridiza-tion (Figs. 4F, 4G; see color insert). These observations demonstrate that PTHrP gene transcription may occur in one cell type, but the protein is rapidly transferred to neighboring cells. Such systems have been described in developing mouse heart valves and epidermis, in which transforming growth factor β (TGF β) is the active agent (Lehnert and Akhurst, 1988; Akhurst *et al.*, 1990).

3. Fish

Parsons *et al.* (1979) first suggested that fish (cod) pituitary contained a factor similar to but distinct from PTH. Although a number of studies by Pang and colleagues (Kaneko and Pang, 1987; Fraser *et al.*, 1991) indicated hypercalcemic PTH or PTH-like activity in brain and pituitary of some teleosts, studies by Danks *et al.* (1993) were the first to identify immunoreac-tive PTHrP in pituitary and plasma of a telost fish, the sea bream. Normal plasma levels of PTHrP in this species are similar to those of humans with HHM, and the pituitary appears to be a likely principal source of the protein, because a medium of cultured pituitaries contained high concentrations of PTHrP assayed by RIA (Grill *et al.*, 1991). PTHrP was detected by immunohistochemistry in cells of the pars intermedia and also in thyro-trophs of the pituitary pars distalis. In salmonid fish the pituitary appeared to contain only few weakly reactive PTHrP-producing cells, which suggests that species and groups of teleosts express PTHrP differentially.

In teleost fishes a tissue specifically concerned with calcium metabolism is the corpuscles of Stannius (CS), which lie, in one or more pairs, on the ventral surface of the kidney, developing embryologically from the kidney duct anlagen. We have examined corpuscles from several teleostean and holostean fishes for immunoreactive PTHrP. Holostean corpuscle from *Lepisosteus* (gar pike) and *Amia* (bichir) (gift of J. Youson) did not react with antiserum to human PTHrP(1–16), which detected PTHrP in sea bream pituitary. CS of several salmonid species and the European eel also failed to react with the same PTHrP antiserum. However, cells of the CS of the euryhaline flounder (*Platichthys*) do contain PTHrP immunoreactive with antiserum to human PTHrP(1–16). Moreover, in this species the neuro-secretory neurons of the posterior nerve cord also appear to produce and transport PTHrP to the urophysis. These observations are illustrated in Fig. 6, which shows the immune reaction in the perikarya of the neurons and in their axons (Danks, Balment, Hubbard, and Ingleton, unpublished observations). In the dogfish, an elasmobranch, immunoreactive PTHrP

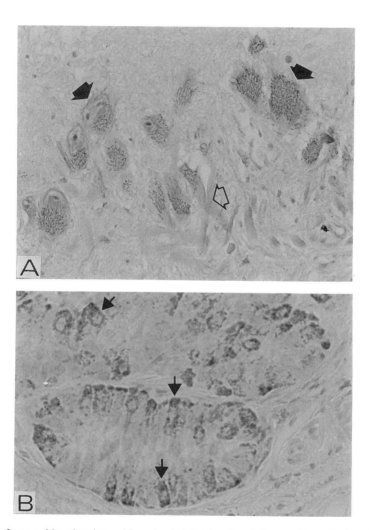

FIG. 6   Immunohistochemistry of flounder (*platichthys flesus*) tissues with antibody to human PTHrP(1–16). (A) Posterior spinal cord neurons (urophysial) showing staining in perikarya (⬧) and axons (◊); (B) staining in cells of the corpuscles of Stannius (→) ×260 (Danks, Balment, Hubbard, and Ingleton, 1995).

was detected in giant neurons of the tectum of the midbrain, which abut the lumen of the third ventricle (Ingleton *et al.*, 1995). A neurohemal organ associated with the brain of teleosts and elasmobranch fishes specifically, the saccus vasculosus, is composed of epithelial cells separating a cavity from capillaries; some of these cells contain and secrete PTHrP (Ingleton *et al.*, 1995; Devlin *et al.*, 1996). A role for PTHrP as a neurosecretory agent or neurotransmitter therefore seems likely but remains to be investigated.

III. The PTH/PTHrP Receptor

The hypercalcemic action of PTHrP, evident in HHM, results principally from interaction at kidney, bone cell, and intestine receptors to which PTH also binds. This common receptor shares at least 30% homology with receptors for other peptides, including calcitonin (Lin *et al.*, 1991), secretin (Ishihara *et al.*, 1991), vasoactive intestinal polypeptide (Ishihara *et al.*, 1992), glucagon-like peptide (Thorens, 1992), growth hormone releasing hormone (Mayo, 1992), glucagon (Jelinek *et al.* (1993), pituitary adenylate cyclase–stimulating polypeptide (Spengler *et al.*, 1993), and corticotrophin releasing hormone (Chen *et al.*, 1993), creating a related subfamily of receptors. Analysis of the cDNA sequences of renal receptors from the American opossum (Jüppner *et al.*, 1991) and from humans (Schneider *et al.*, 1993), as well as from rat (Abou-Samra *et al.*, 1992) and human osteoblast-like cells (Schipani *et al.*, 1993), showed it to be a G-protein-linked receptor with seven transmembrane domains. The human gene is located on chromosome 3 in the p21.1–p24.2 region (Gelbert *et al.*, 1994). The human gene encodes a protein of 593 amino acids, and there is considerable sequence homology between the receptor in different species, with variations occurring only in certain areas of the molecule, corresponding to amino acids 57–97, 262–276, 542–558, and 579–589 (Schipani *et al.*, 1993). Figure 7 shows a diagram of the receptor molecule representative of all the species analyzed so far. The extracellular N-terminal domain is predicted to be 200 residues, including a possible pre–pro sequence of 40 amino acids. Within this extracellular domain are six cysteine residues, with two more in extracellular loops linking transmembrane domains that are conserved in all the members of the receptor family. Mutant receptors with deletions of residues 31–47 of the N-terminal region or 431–440 in the third extracellular loop fail to bind PTH (Lee *et al.*, 1994) and probably therefore, PTHrP.

IV. Conclusion

PTHrP is a ubiquitous protein with a chemical and gene structure capable of multiple variations that may be created by processing within the cell of

PTH/PTHrP Receptor Gene

PTH/PTHrP Receptor

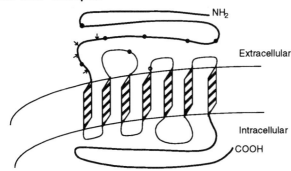

Signalling Systems

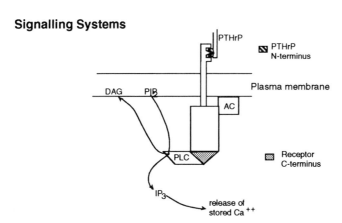

FIG. 7 Structure of the PTH/PTHrP receptor gene; U is an untranslated region, S codes for a signal peptide, E1–E3 and G code for the extracellular domain, M1–M7 and EL2 code for the transmembrane domain, and T codes for the intracellular carboxy terminal domain. The receptor is a seven-member transmembrane G-linked protein that has potential glycosylation sites in the extracellular domain (arrows); extracellular intron sites are indicated by circles. Several signaling systems have been identified with postreceptor binding mechanisms, including activation of adenylate cyclase (AC); interaction with phospholipase C (PLC) via the C-terminus of the receptor, leading to activation of the phosphoinositide pathway (PIP$_2$, PIP$_3$) and diacylglycerol (DAG) and release of ionic calcium.

origin or by target cells with which it interacts, giving this protein the potential for extraordinary versatility of function. One of these fundamental intracrine functions may be to oppose apoptosis by holding the cell in the G_0 phase of the cell cycle; this would be important for stages of tissue development in embryo and in establishing the neoplastic phenotype. Elucidating all the actions of PTHrP and how they have evolved will provide insights into the development and persistence of the neoplastic phenotype as well as into the significance of PTHrP for normal adult physiology and fetal development.

References

Abbas, S. K., Pickard, D. W., Rodda, C. P., Heath, J. A., Hammonds, R. G., Wood, W. I., Caple, I. W., Martin, T. J., and Care, A. D. (1989). Stimulation of ovine placental calcium transport by purified natural and recombinant parathyroid hormone–related protein (PTHrP) preparations. *J. Exptl. Physiol.* **74,** 549–552.

Abou-Samra, A.-B., Juppner, H., Force, T., Freeman, M. W., Kong, X.-F., Schipani, E., Urena, P., Richards, J., Bonventre, J. V., Potts, J. T., Kronenberg, H., and Segre, G. V. (1992). Expression cloning of a common receptor for parathyroid hormone and parathyroid hormone–related peptide from rat osteoblast-like cells: A single receptor stimulates intracellular accumulation of both cAMP and inositol trisphosphates and increases intracellular free calcium. *Proc. Natl. Acad. Sci. USA* **89,** 2732–2736.

Adachi, N., Yamaguchi, K., Miyake, Y., Honda, S., Nagasaki, K., Akiyama, Y., Adachi, I., and Abe, K. (1990). Parathyroid hormone–related protein is a possible autocrine growth inhibitor for lymphocytes. *Biochem. Biophys. Res. Commun.* **166,** 1088–1094.

Akhurst, R. J., Lehnert, S. A., Faissner, A., and Duffie, E. (1990). TGF β in murine morphogenetic processes: The early embryo and cardiogenesis. *Development* **108,** 645–656.

Albright, F. (1941). Case records of the Massachusetts General Hospital (case 27461). *N. Engl. J. Med.* **225,** 789–791.

Alsat, E., Haziza, J., Scioppo, M. L., Frankenne, F., and Evian-Brion, D. (1993). Increase in epidermal growth factor receptor and its mRNA levels by parathyroid hormone (1–34) and parathyroid hormone–related protein (1–34) during differentiation of human trophoblast cells in culture. *J. Cell. Biochem.* **53,** 32–42.

Asa, S. L., Henderson, J., Goltzman, D., and Drucker, D. J. (1990). Parathyroid hormone–like peptide in normal and neoplastic human endocrine tissues. *J. Clin. Endocrinol. Metab.* **71,** 1112–1118.

Beck, F., Tucci, J., and Senior, P. V. (1993). Expression of parathyroid hormone–related protein mRNA by uterine tissues and extraembryonic membranes during gestation in rats. *J. Reprod. Fertil.* **99,** 343–352.

Bera, T. K., Hwang, S. I., Swanson, S. M., Guzman, R. C., Edery, M., and Nandi, S. (1994). In situ localization of prolactin receptor message in the mammary glands of pituitary isografted mice. *Mol. Cell. Biochem.* **132,** 145–149.

Blind, E., Raue, F., Goltzman, J., Schmidt-Gayk, H., Kohl, G., and Zeigler, R. (1992). Circulating levels of midregional parathyroid hormone–related protein in hypercalcaemia of malignancy. *Clin. Endocrinol. Oxf.* **37,** 290–297.

Blind, E., Raue, F., Knappe, V., Schroth, J., and Zeigler, R. (1993). Cyclic AMP formation in rat bone and kidney cells is stimulated equally by parathyroid hormone–related protein (PTHrP) 1–34 and PTH 1–34. *Exptl. Clin. Endocrinol.* **101**, 150–155.

Borke, J. L., Caride, A., Verma, A. K., Kelly, L. K., Smith, C. H., Penniston, J. T., and Kumar, R. (1989). Calcium pump epitopes in placental trophoblast basal plasma membranes. *Am. J. Physiol.* **257**, C341–C346.

Bouizar, Z., Spyratos, F., Deytieux, S., de-Vernejoul, M. C., and Jullienne, A. (1993). Polymerase chain reaction analysis of parathyroid hormone–related protein gene expression in breast cancer patients and occurrence of bone metastases. *Cancer Res.* **53**, 5076–5078.

Brandt, D. W., Bruns, M. E., Bruns, D. E., Ferguson, J. E., Burton, D. W., and Deftos, L. J. (1992). The parathyroid hormone–related protein (PTHrP) gene preferentially utilizes a GC-rich promoter and the PTHrP 1–139 coding pathway in normal human amnion. *Biochem. Biophys. Res. Commun.* **189**, 938–943.

Bucht, E., Eklund, A., Toss, G., Lewensohn, R., Granberg, B., Sjostedt, U., Edeland, R., and Torring, O. (1992). Parathyroid hormone–related peptide measured by midmolecule radioimmunoassay, in various hypercalcaemic and normocalcaemic conditions. *Acta Endocrinol.* **127**, 294–300.

Budayr, A. A., Nissenson, R. A., Klein, R. F., Pun, K. K., Clark, O. H., Diep, D., Arnaud, C. D., and Strewler, G. J. (1989). Increased serum levels of a parathyroid hormone–like protein in malignancy-associated hypercalcemia. *Ann. Intern. Med.* **111**, 807–812.

Bundred, N. J., Ratcliffe, W. A., Walker, R. A., Coley, S., Morrison, J. M., and Ratcliffe, J. G. (1991). Parathyroid hormone related protein and hypercalcaemia in breast cancer. *Br. Med. J.* **303**, 1506–1509.

Burtis, W. J., Brady, T. G., Orloff, J. J., Ersbak, J. B., Warrell, R. P., Jr., Olson, B. R., Wu, T. L. Mitnick, M. E., Broadus, A. E., and Stewart, A. F. (1990). Immunochemical characterisation of circulating parathyroid hormone–related protein in patients with humoral hypercalcaemia of cancer. *N. Engl. J. Med.* **322**, 1106–1112.

Burtis, W. J., Dann, P., Gaich, G. A., and Soifer, N. E. (1994). A high abundance midregion species of parathyroid hormone–related protein: Immunological and chromatographic characterization in plasma. *J. Clin. Endocrinol. Metab.* **78**, 317–322.

Burtis, W. J., Wu, T., Bunch, C., Wysolmerski, J. J., Insogna, K. I., Weir, E. C., Broadus, A. E., and Stewart, A. F. (1987). Identification of a novel 17,000-dalton parathyroid hormone–like adenylate cyclase–stimulating protein from a tumor associated with humoral hypercalcaemia of malignancy. *J. Biol. Chem.* **262**, 7151–7156.

Burton, D. W., Brandt, D. W., and Deftos, L. J. (1994). Parathyroid hormone–related protein in the cardiovascular system. *Endocrinology* **135**, 253–261.

Burton, P. B. J., Moniz, C., and Knight, D. E. (1990). Parathyroid hormone related peptide can function as an autocrine growth factor in human renal cell carcinoma. *Biochem. Biophys. Res. Commun.* **167**, 1134–1138.

Burton, P. B., Moniz, C., Quirke, P., Malik, A., Bui, T. D., Juppner, H., Segre, G. V., and Knight, D. E. (1992). Parathyroid hormone–related peptide: Expression in fetal and neonatal development. *J. Pathol.* **167**, 291–296.

Campbell, G. S., Argentsinger, L. S., Ihle, J. N., Kelly, P. A., Rillema, J. A., and Carter-Su, C. (1994). Activation of JAK2 tyrosine kinase by prolactin receptors in Nb2 cells and mouse mammary explants. *Proc. Natl. Acad. Sci. USA* **91**, 5232–5236.

Campos, R. V., Asa, S. L., and Drucker, D. J. (1991). Immunocytochemical localization of parathyroid hormone–like peptide in the fetus. *Cancer Res.* **51**, 6351–6357.

Campos, R. V., Zhang, L., and Drucker, D. J. (1994). Differential expression of RNA transcripts encoding unique carboxyterminal sequences of human parathyroid hormone–related peptide. *Mol. Endocrinol.* **8**, 1656–1666.

Care, A. D., Abbas, S. K., Pickard, D. W., Barri, M., Drinkhill, M., Findlay, J. B., White, I. R., and Caple, I. W. (1990). Stimulation of ovine placental transport of calcium and magnesium by mid-molecule fragments of human parathyroid hormone–related protein. *J. Exp. Physiol.* **75,** 605–608.

Carney, S. L., Ray, C., Ebeling, P. R., Martin, T. J., and Gillies, A. H. (1991). Synthetic human parathyroid hormone–related protein and rat renal electrolyte transport. *Min. Electr. Metab.* **17,** 41–45.

Caulfield, M. P., McKee, R. L., Goldman, M. E., Duong, L., Fisher, J. E., Gay, C. T., DeHaven, P. A., Levy, J. J., Roubin, E., Nutt, R. F., Chorev, M., and Rosenblatt, M. (1990). The bovine renal parathyroid hormone (PTH) receptor has equal affinity for two different amino acid sequences: The receptor binding domains of PTH and PTH-related protein are located within the 14–34 region. *Endocrinology* **127,** 83–87.

Chen, R., Lewis, K., Perrin, M., and Vale, W. (1993). Expression cloning of a human corticotropin-releasing-factor receptor. *Proc. Natl. Acad. Sci. USA* **90,** 8967–8971.

Cheshire, I. M., Blight, A., Ratcliffe, W. A., Proops, D. W., and Heath, D. A. (1991). Production of parathyroid-hormone-related protein by cholesteatoma cells in culture. *Lancet* **338,** 1041–1043.

Clark, S. P., Chou, S. T., Martin, T. J., and Danks, J. A. (1995). Parathyroid hormone–related protein antigen localization distinguishes between mesothelioma and adenocarcinoma of the lung. *J. Pathol.* **176,** 161–165.

Cortelyou, J. R., Hibner-Owerko, A., and Mulroy, J. (1960). Blood and urine calcium changes in totally parathyroidectomized *Rana pipiens. Endocrinology* **66,** 441–450.

Crass, M. F., and Scarpace, P. J. (1993). Vasoactive properties of a parathyroid hormone–related protein in the rat aorta. *Peptides* **14,** 179–183.

Cserr, H. F. (1971). Physiology of the choroid plexus. *Physiol. Rev.* **51,** 273–311.

Dacke, C. G., and Shawe, A. J. (1987). Studies of the rapid effects of parathyroid hormone and prostaglandins on $^{45}Ca^{++}$ uptake into chick and rat bone *in vivo. J. Endocrinol.* **115,** 369–377.

Danks, J. A., Balment, R. J., Hubbard, P., and Ingleton, P. M. (1995). Parathyroid hormone–related protein in the urophysis of the flounder (*Platichthys flesus*). *J. Endocrinol.* **147** (Suppl). P61.

Danks, J. A., Devlin, A. J., Ho, P. M. W., Diefenbach-Jagger, H., Power, D. M., Canario, A., Martin, T. J., and Ingleton, P. M. (1993). Parathyroid hormone–related protein is a factor in normal fish pituitary. *Gen. Comp. Endocrinol.* **92,** 201–212.

Danks, J. A., Ebeling, P. R., Hayman, J., Chou, S. T., Moseley, J. M., Dunlop, J., Kemp, B. E., and Martin, T. J. (1989). Parathyroid hormone–related protein: Immunohistochemical localization in cancers and in normal skin. *J. Bone Min. Res.* **4,** 273–278.

Danks, J. A., Ebeling, P. R., Hayman, J. A., Diefenbach-Jagger, H., Collier, F. McL., Grill, V., Southby, J., Moseley, J. M., Chou, S. T., and Martin, T. J. (1990). Immunohistochemical localization of parathyroid hormone–related protein in parathyroid adenoma and hyperplasia. *J. Pathol.* **161,** 27–33.

Danks, J. A., McHale, J. C., Clark, S. P., Chou, S. T., Scurry, J. P., Ingleton, P. M., and Martin, T. J. (1995). *In situ* hybridization of parathyroid hormone–related protein in normal skin, skin tumors, and gynecological cancers using digoxigenin-labeled probes and antibody enhancement. *J. Histochem. Cytochem.* **43,** 5–10.

Danks, J. A., McHale, J. C., Martin, T. J., and Ingleton, P. M. (1994). Parathyroid hormone–related protein in emerging frog (*Rana temporaria*) tissues: Immunocytochemistry and *in situ* hybridisation. *J. Endocrinol.* **143,** Suppl. P25.

Davicco, M.-J., Durand, D., Lefaivre, J., and Barlet, J.-P. (1993). Parathyroid hormone–related protein might increase mammary blood flow. *J. Bone Min. Res.* **8,** 1519–1524.

Devlin, A. J., Danks, J. A., Faulkner, M. K., Power, D. M., Canario, A. V. M., Martin, T. J., and Ingleton, P. M. (1996). Immunochemical detection of parathyroid hormone–related protein in the saccus vasculosus of a teleost fish. *Gen. Comp. Endocrinol.,* in press.

Dingwall, C., and Laskey, R. A. (1991). Nuclear targeting sequences—a consensus? *TIBS* **16,** 478–481.

Drucker, D. J., Asa, S. L., Henderson, J., and Goltzman, D. (1989). The parathyroid hormone–like peptide gene is expressed in normal and human neoplastic endocrine pancreas. *Mol. Endocrinol.* **3,** 1589–1595.

Dunne, F. P., Lee, S., Ratcliffe, W. A., Hutchesson, A. C., Bundred, N. J., and Heath, D. A. (1993). Parathyroid hormone–related protein (PTHrP) gene expression in solid tumours associated with normocalcemia and hypercalcemia. *J. Pathol.* **171,** 215–221.

Dunne, F. P., Ratcliffe, W. A., Mansour, P., and Heath, D. A. (1994a). Parathyroid hormone–related protein (PTHrP) gene expression in fetal and extra-embryonic tissues of early pregnancy. *Human Reprod.* **9,** 149–156.

Dunne, F. P., Rollason, T., Ratcliffe, W. A., Marshall, T., and Heath, D. A. (1994b). Parathyroid hormone–related protein gene expression in invasive cervical tumors. *Cancer* **74,** 83–89.

Ebeling, P. R., Adam, W. R., Moseley, J. M., and Martin, T. J. (1989). Actions of synthetic parathyroid hormone–related protein(1–34) on isolated rat kidney. *J. Endocrinol.* **120,** 45–50.

Edwards, R. C., Ratcliffe, W. A., Walls, J., Morrison, J. M., Ratcliffe, J. G., Holder, R., and Bundred, N. J. (1995). Parathyroid hormone–related protein (PTHrP) in breast cancer and benign breast tissue. *Eur. J. Cancer* **31A,** 334–339.

Emly, J. F., Hughes, S., Green, E., and Ratcliffe, W. A. (1994). Expression and secretion of parathyroid hormone–related protein by a human cancer cell line. *Biochim. Biophys. Acta.* **120,** 193–198.

Endlich, K., Masfelder, T., Helwig, J. J., and Steinhausen, M. (1995). Vascular effects of parathyroid hormone and parathyroid hormone–related protein in the split hydronephrotic rat kidney. *J. Physiol.* **483,** 481–490.

Evely, R. S., Bonomo, A., Schneider, H. G., Moseley, J. M., Gallagher, J., and Martin, T. J. (1991). Structural requirements for the action of parathyroid hormone–related protein (PTHrP) on bone resorption by isolated osteoclasts. *J. Bone Min. Res.* **6,** 85–93.

Fenton, A. J., Kemp, B. E., Hammonds, R. G., Mitchelhill, K., Moseley, J. M., Martin, T. J., and Nicholson, G. (1991a). A potent inhibitor of osteoclast bone resorption within a highly conserved pentapeptide region of parathyroid–hormone related protein: PTHrP[107–111] *Endocrinology* **129,** 3424–3426.

Fenton, A. J., Kemp, B. E., Kent, G. N., Moseley, J. M., Zheng, M. H., Rowe, D. J., Britto, J. M., Martin, T. J., and Nicholson, G. C. (1991b). A carboxy terminal peptide from the parathyroid hormone–related protein inhibits bone resorption by osteoclasts. *Endocrinology* **129,** 1762–1768.

Fenton, A. J., Martin, T. J., and Nicholson, G. C. (1993). Long-term culture of disaggregated rat osteoclasts: Inhibition of bone resorption and reduction of osteoclast-like cell number by calcitonin and PTHrP[107–139]. *J. Cell. Physiol.* **155,** 1–7.

Fenton, A. J., Martin, T. J., and Nicholson, G. C. (1994). Carboxy-terminal parathyroid hormone–related protein inhibits bone resorption by isolated chicken osteoclasts. *J. Bone Min. Res.* **9,** 515–519.

Ferguson, J. E., Gormal, J. V., Bruns, D. E., Weir, E. C., Burtis, W. J., Martin, T. J., and Bruns, M. E. (1992). Abundant expression of parathyroid hormone–related protein in human amnion and its association with labor. *Proc. Nat. Acad. Sci. USA* **89,** 8384–8388.

Ferguson, J. E., Seaner, R., Bruns, D. E., Redick, J. A., Mills, S. E., Juppner, H., Segre, G. V., and Bruns, M. E. (1994). Expression of parathyroid hormone–related protein and its receptor in human umbilical cord: Evidence for a paracrine system involving umbilical vessels. *Am. J. Obstet. Gyn.* **170,** 1018–1024.

Ferrari, S. L., Rizzoli, R. and Bonjour, J. P. (1992). Parathyroid hormone–related protein production by primary cultures of mammary epithelial cells. *J. Cell. Physiol.* **150,** 304–311.

Ferrari, S. L., Rizzoli, R., and Bonjour, J. P. (1994). Effects of epidermal growth factor on parathyroid hormone–related protein production by mammary epithelial cells. *J. Bone Min. Res.* **9,** 639–644.

Folkman, J., and Singh, Y. (1992). Angiogenesis. *J. Biol. Chem.* **267,** 10931–10934.

Forrest, S. M., Ng, K. W., Findlay, D. M., Michelangeli, V. P., Livesey, S. A., Partridge, N. C., Zajac, J. D., and Martin, T. J. (1985). Characterization of an osteoblast-like clonal cell line which responds to both parathyroid hormone and calcitonin. *Calcif. Tiss. Int.* **37,** 51–56.

Fraser, R. A., Kaneko, T., Pang, P. K. T., and Harvey, S. (1991). Hypo- and hypercalcaemeic peptides in fish pituitary glands. *Am. J. Physiol.* **260,** R622–R626.

Fraser, W. D., Robinson, J., Lawton, R., Durham, B., Gallacher, S. J., Boyle, I. T., Beastall, G. H., and Logue, F. C. (1993). Clinical and laboratory studies of a new immunoradiometric assay of parathyroid hormone–related protein. *Clin. Chem.* **39,** 414–419.

Funk, J. L., Krul, E. J. T., Moser, A. H., Shigenaga, J. K., Strewler, G. J., Grunfeld, C., and Feingold, K. R. (1993). Endotoxin increases parathyroid hormone–related protein levels in mouse spleen. *J. Clin. Invest.* **92,** 2546–2552.

Funk, J. L., Shinanaga, J. K., Moser, A. H., Krul, E. J., Strewler, G. J., Feingold, K. R., and Grunfeld, C. (1994). Cytokine regulation of parathyroid hormone–related protein messenger ribonucleic acid levels in mouse spleen: Paradoxical effects of interferon-gamma and inter-leukin-4. *Endocrinology* **135,** 351–358.

Gay, C. V. (1988). Avian bone resorption at the cellular level. *CRC Crit. Rev. Poultry Biol.* **1,** 197–210.

Gelbert, L., Schipani, E., Jüppner, H., Abou-Samra, A-S., Segre, G. V., Naylor, S., Drabkin, H., and Heath, H. (1994). Chromosomal localization and the parathyroid hormone/parathyroid hormone–related protein receptor gene to human chromosome 3p21.1–p24.2. *J. Clin. Endocrinol. Metab.* **79,** 1046–1048.

Goff, J. P., Reinhardt, T. A., Lee, S., and Hollis, B. W. (1991). Parathyroid hormone–related peptide content of bovine milk and calf blood assessed by radioimmunoassay and bioassay. *Endocrinology* **129,** 2815–2819.

Gotoh, A., Kitazawa, S., Mizuno, Y., Takenaka, A., Arakawa, S., Matsumoto, O., Kitazawa, R., Fujimori, T., Maeda, S., and Kamidono, S. (1993). Common expression of parathyroid hormone–related protein and no correlation of calcium level in renal cell carcinomas. *Cancer* **17,** 2803–2806.

Grill, V., Hillary, J., Ho, P. M., Law, F. M., MacIsaac, R. J., MacIsaac, I. A., Moseley, J. M., and Martin, T. J. (1992). Parathyroid hormone–related protein: A possible endocrine function in lactation. *Clin. Endocrinol. Oxf.* **37,** 405–410.

Grill, V., Ho, P., Body, J. J., Johanson, N., Lee, S. C., Kureja, S. C., Moseley, J. M., and Martin, T. J. (1991). Parathyroid hormone–related protein: Elevated levels in both humoral hypercalcemia of malignancy and hypercalcemia complicating metastatic breast cancer. *J. Clin. Endocrinol. Metab.* **73,** 1309–1315.

Grone, A., Werkmeister, J. R., Steinmeyer, C. L., Capen, C. C., and Rosol, T. J. (1994). Parathyroid hormone–related protein in normal and neoplastic canine tissues: Immunohisto-chemical localization and biochemical extraction. *Vet. Pathol.* **31,** 308–315.

Grosvenor, C. E., and Mena, F. (1974). Neural and hormonal control of milk secretion and milk ejection. *In* "Lactation" (B. L. Larson and V. R. Smith, eds.), Vol. 1, pp. 227–276. Academic Press, New York.

Gutman, J. N., Burtis, W. J., Dreyer, B. E., Andrade-Gordon, P., Penzias, A. S., Polan, M. L., and Insogna, K. L. (1993). Human granulosa–luteal cells secrete parathyroid hor-mone–related portein *in vivo* and *in vitro*. *J. Clin. Endocrinol. Metab.* **76,** 1314–1318.

Hayden, T. J., Bonney, R. C., and Forsyth, I. A. (1979). Ontogeny and control of prolactin receptors in the mammary gland and liver of virgin, pregnant and lactating rats. *J. Endocrinol.* **80,** 259–269.

Hayman, J. A., Danks, J. A., Ebeling, P. R., Moseley, J. M., Kemp, B. E., and Martin, T. J. (1989). Expression of parathyroid hormone related protein in normal skin and in tumours of skin and skin appendages. *J. Pathol.* **158,** 293–296.

Heath, D. A., Senior, P. V., Varley, J. M., and Beck, F. (1990). Parathyroid-hormone-related protein in tumours associated with hypercalcemia. *Lancet* **335,** 66–69.

Hellman, P., Ridefelt, P., Juhlin, C., Åkerström, G., Rastad, J., and Gylfe, E. (1992). Parathyroid-like regulation of parathyroid hormone–related protein release and cytoplasmic calcium in cytotrophoblast cells of human placenta. *Arch. Biochem. Biophys.* **293,** 174–180.

Henderson, J. E., Shustik, C., Kremer, R., Rabbani, S. A., Hendy, G., and Goltzman, D. (1990). Circulating concentrations of parathyroid hormone–like peptide in malignancy and in hyperparathyroidism. *J. Bone Min. Res.* **5,** 105–112.

Hickey, R. C., Samaan, N. A., and Jackson, G. L. (1981). Hypercalcemia in patients with breast cancer: Osseus metastases, hyperplastic parathyroid tissue or pseudohyperparathyroidism? *Arch. Surg.* **116,** 545–552.

Hino, T., Nyby, M. D., Fittingoff, M., Tick, M. L., and Brickman, A. S. (1994). Parathyroid hormone analogues inhibit calcium mobilization in cultured vascular cells. *Hypertension* **23,** 402–408.

Holt, E. H., Vasavada, R. C., Bander, N. H., Broadus, A. E., and Philbrick, W. M. (1993). Region-specific methylation of the parathyroid hormone–related peptide determines its expression in human renal carcinoma cell lines. *J. Biol. Chem.* **268,** 20639–20645.

Honda, S., Yamaguchi, K., Moyake, Y., Hayashi, N., Adachi, N., Kinoshita, K., Ikehara, O., Kimura, S., Kinoshita, T., Shimotohno, K., Shimoyama, M., and Abe, K. (1988). Expression of parathyroid hormone–related protein mRNA in tumors obtained from patients with humoral hypercalcaemia of malignancy. *Jpn. J. Cancer Res. (GANN)* **79,** 677–681.

Hongo, T., Kupfer, J., Enomoto, H., Sharifii, B., Gainnelli Neto, D., Forrester, J. S., Singer, F. R., Goltzman, D., Hendy, G. N., Pirola, C. J., Fahin, J. A., and Clemens, T. L. (1991). Abundant expression of parathyroid hormone–related protein in primary rat aortic smooth muscle cells accompanies serum-induced proliferation. *J. Clin. Invest.* **88,** 1841–1847.

Ichinose, Y., Iguichi, H., Ohta, M., and Katakami, H. (1993). Establishment of lung cancer cell line producing parathyroid hormone–related protein. *Cancer Lett.* **74,** 119–124.

Ikeda, K., Mangin, M., Dreyer, B. E., Webb, A. C., Posillico, J. T., Stewart, A. F., Bander, N. H., Weir, E. C., Insogna, K. L., and Broadus, A. E., (1988a). Identification of transcripts encoding a parathyroid hormone-like peptide in messenger RNAs from a variety of human tumors associated with humoral hypercalcemia of malignancy. *J. Clin. Invest.* **81,** 2010–2014.

Ikeda, K., Weir, E. C., Mangin, M., Dannies, P. S., Kinder, B., Deftos, L. J., Brown, E. M., and Broadus, A. E., (1988b). Expression of messenger ribonucleic acids encoding a parathyroid hormone–like peptide in normal human and animal tissues with abnormal expression in human parathyroid adenomas. *Mol. Endocrinol.* **2,** 1230–1236.

Imagawa, W., Bandyopadhyay, W., and Nandi, S. (1990). Regulation of mammary epithelial growth in mice and rats. *Endocrinol. Rev.* **11,** 494–523.

Imamura, H., Sato, K., Shizume, K., Satoh, T., Kasano, K., Ozawa, M., Ohmura, E., Tsushima, T., and Demura, H. (1991). Urinary excretion of parathyroid hormone–related protein fragments in patients with humoral hypercalcemia of malignancy and hypercalcemic tumor–bearing nude mice. *J. Bone Min. Res.* **6,** 77–84.

Ingleton, P. M., and Danks, J. A. (1996). Parathyroid hormone-related protein in the saccus vasculosus of fishes. *In* "The comparative endocrinology of calcium regulation" (C. Dacke, J. Danks, I. Caple, and G. Flik, eds.), in press. Society for Endocrinology, Bristol, United Kingdom.

Ingleton, P. M., Hazon, N., Ho, P. M. W., Martin, F. J., and Danks, J. A. (1995). Immunodetection of parathyroid hormone–related protein in plasma and tissues of an elasmobranch (*Scyliorhinus canicula*). *Gen. Comp. Endocrinol.* **98,** 211–218.

Insogna, K. L. (1989). Humoral hypercalcemia of malignancy. The role of parathyroid hormone–related protein. *Endocrinol. Metab. Clinics N. Am.* **18,** 779–794.

Insogna, K. L., Stewart, A.L., Ikeda, K., Centrella, M., and Milstone, L. M. (1988). Characterization of a parathyroid hormone–like peptide secreted by human keratinocytes. *Ann. N.Y. Acad. Sci.* **548,** 146–159.

Ishihara, T., Nakamura, S., Kaziro, Y., Takahashi, T., Takahashi, K., and Nagata, S. (1991). Molecular cloning and expression of a cDNA encoding the secretin receptor. *EMBO J.* **10,** 1635–1641.

Ishihara, T., Shigemoto, R., Mori, K., Takahashi, K. and Nagata, S. (1992). Functional expression and tissue distribution of a novel receptor for vasoactive intestinal polypeptide. *Neuron* **8,** 811–819.

Ishikawa, M., Ouchi, Y., Akishita, M., Kozaki, K., Toba, K., Namiki, A., Yamaguchi, T., and Orimo, H. (1994). Immunocytochemical detection of parathyroid hormone–related protein in vascular endothelial cells. *Biochem. Biophys. Res. Commun.* **199,** 547–551.

Iwamura, M., di Sant'Agnese, P. A., Wu, G., Benning, C. M., Cockett, A. T., Deftos, L. J., and Abrahamsson, P. A. (1993). Immunohistochemical localization of parathyroid hormone–related protein in human prostate cancer. *Cancer Res.* **53,** 1724–1726.

Iwamura, M., Wu, G., Abrahamsson, P. A., di Sant'Agnese, P. A., Cockett, A. T., and Deftos, L. J. (1994). Parathyroid hormone–related protein is expressed by prostatic neuroendocrine cells. *Urology* **43,** 667–674.

Jahn, G. A., Edery, M., Belair, L., Kelly, P. A., and Djiane, J. (1991). Prolactin receptor gene expression in rat mammary gland and liver during pregnancy and lactation. *Endocrinology* **128,** 2976–2984.

Jelinek, L. J., Lok, S., Rosenberg, G. B., Smith, R. A., Grant, F. J., Biggs, S., Bensch, P. A., Kuijper, J. L., Sheppard, P. O., Sprecher, C. A., O'Hara, P. J., Foster, D., Walker, K. M., Chen, L. H. J., McKernan, P. A., and Kinsvogel, W. (1993). Expression, cloning and signaling properties of the rat glucagon receptor. *Science* **259,** 1614–1616.

Jüppner, H., Abou-Samra, A.-B., Freeman, M., Kong, X. F., Schipani, E., Richards, J., Kolawski, L. F., Hock, J., Potts, J. T., Kroninberg, H. M., and Segre, G. V. (1991). A G-protein-linked receptor for parathyroid hormone and parathyroid hormone–related protein. *Science* **254,** 1024–1026.

Kaiser, S. M., Laneuville, P., Bernier, S. M., Rhim, J. S., Kremer, R., and Goltzman, D. (1992). Enhanced growth of a human keratinocyte cell line induced by antisense RNA for parathyroid hormone–related peptide. *J. Biol. Chem.* **267,** 13623–13628.

Kaji, H., Sugimoto, T., Kanatani, M., Fukase, M., and Chihara, K. (1995). Carboxyl-terminal peptides from parathyroid hormone–related protein stimulate osteoclast-like cell formation. *Endocrinology* **136,** 842–848.

Kaneko, T., and Pang, P. K. T. (1987). Immunocytochemical detection of parathyroid hormone–like substance in the goldfish brain and pituitary gland. *Gen. Comp. Endocrinol.* **68,** 147–152.

Kao, P. C., Klee, G. G., Taylor, R. I., and Heath, H. (1990). Parathyroid hormone–related protein in plasma of patients with hypercalcemia and malignant lesions. *Mayo Clin. Proc.* **65,** 1399–1407.

Karaplis, A. C., Luz, A., Glowacki, J., Bronson, R. T., Tybulewicz, V. L. J., Kronenberg, H. M., and Mulligan, R. C. (1994). Lethal skeletal dysplasia from targeted disruption of the parathyroid hormone–related peptide gene. *Genes Develop.* **8,** 277–289.

Karaplis, A. C., Yasuda, T., Hendy, G. N., Goltzman, D. and Banville, D. (1990). Gene encoding parathyroid hormone–like peptide: Nucleotide sequence of the gene and comparison with the human homologue. *Mol. Endocrinol.* **4,** 441–446.

Kemp, B. E., Moseley, J. M., Rodda, C. P., Ebeling, P. R., Wettenhall, R. E. H., Stapleton, D., Diefenbach-Jagger, H., Ure, F., Michelangeli, V. P., Simmons, H. A., Raisz, L. G., and

Martin, T. J. (1987). Parathyroid hormone–related protein of malignancy: Active synthetic fragments. *Science* **238,** 1568–1570.

Kenny, A. D., and Dacke, C. G. (1974). The hypercalcaemic responses to parathyroid hormone in Japanese quail. *J. Endocrinol.* **62,** 15–23.

Kitazawa, S., Fukase, M., Kitazawa, R., Takenaka, A., Gotoh, A., Fujita, T., and Maeda, S. (1991). Immunohistologic evaluation of parathyroid hormone–related protein in human lung cancer and normal tissue with newly developed monclonal antibody. *Cancer* **67,** 984–989.

Kitazawa, R., Kitazawa, S., Fukase, M., Fujita, T., Kobayashi, A., Chihara, K., and Maeda, S. (1992a). The expression of parathyroid hormone–related protein (PTHrP) in normal parathyroid: Histochemistry and *in situ* hybridization. *Histochemistry* **98,** 211–215.

Kitazawa, S., Kitazawa, R., Fukase, M., Fujimori, T., and Maeda, S. (1992b). Immunohisto-chemical evaluation of parathyroid hormone–related protein (PTHrP) in the uterine cervix. *Int. J. Cancer* **50,** 731–735.

Kobayashi, Y., Ichikawa, T., and Kobayashi, J. (1979). Innervation of the caudal neurosecretory system of the teleost. *Gumma Symp. Endocrinol.* **16,** 81–86.

Kohno, N., Kitazawa, S., Fukase, M., Sakoda, Y., Kanbara, Y., Furaya, Y., Ohashi, O., Ishikawa, Y., and Saitoh, Y. (1994). The expression of parathyroid hormone–related protein in human breast cancer with skeletal metastases. *Surg. Today* **24,** 215–220.

Kramer, S., Reynolds., F. H., Castillo, M., Valenzuela, D. M., Thorikay, M., and Sorvillo, J. M. (1991). Immunological identification and distribution of parathyroid hormone–like protein polypeptides in normal and malignant tissues. *Endocrinology* **128,** 1927–1937.

Kriebel, R. M., Parsons, R. L., Miller, K. E. (1985). Innervation of caudal neurosecretory cells. *In* Neurosecretion and the Biology of Neuropeptides (H. Kobayashi, H. A. Bern, and A. Urano, eds.), pp. 457–463. Japan Scientific Societies Press, Tokyo.

Kukreja, S. C., Rosol, T. J., Wimbiscus, S. A., Shevrin, D. H., Grill, V., Barengolts, E. I., and Martin, T. J. (1990). Tumor resection and antibodies to parathyroid hormone–related protein cause similar changes on bone histomorphometry in hypercalcemia of cancer. *Endocrinology* **127,** 305–310.

Kukreja, S. C., Shevrin, D. H., Wimbiscus, G. A., Ebeling, P. R., Danks, J. A., Rodda, C. P., Wood, W. I., and Martin, T. J. (1988). Antibodies to parathyroid hormone–related protein lower serum calcium in athymic mouse models of malignancy–associated hypercalcemia due to human tumors. *J. Clin. Invest.* **82,** 1798–1802.

Lederis, K. (1984). *In* "Frontiers in Neuroendocrinology" (L. Martini and W. F. Ganong, eds.), Vol. 8, pp. 247–263, Raven Press, New York.

Lee, K., Deeds, J. D., Bond, A. T., Jüppner, H., Abou-Samra, A.-B., and Segre, G. V., (1993). *In situ* localization of PTH/PTHrP receptor mRNA in the bone of fetal and young rats. *Bone* **14,** 341–345.

Lee, K., Deeds, J. D., and Segre, G. V. (1995). Expression of parathyroid hormone–related peptide and its receptor messenger ribonucleic acids during fetal development of rats. *Endocrinology* **136,** 435–563.

Lee, C., Gardella, T. J., Abou-Samra, A.-B., Nussbaum, S. R., Segre, G. V., Potts, J. T., Kronenberg, H. M., and Jüppner, H. (1994). Role of the extracellular regions of the parathy-roid hormone (PTH)/PTH–related peptide receptor in hormone binding. *Endocrinology* **135,** 1488–1495.

Lehnert, S. A., and Akhurst, R. J. (1988). Embryonic expression pattern of TGF β type-1 RNA suggests both paracrine and autocrine mechanisms of action. *Development* **104,** 263–273.

Liapis, H., Crouch, E. C., Grosso, L. E., Kitazawa, S. and Wick, M. R. (1993a). Expression of parathyroidlike protein in normal, proliferative and neoplastic human breast tissues. *Am. J. Pathol.* **143,** 1169–1173.

Liapis, H., Crouch, E. C., Roby, J., and Rader, J. S. (1993b). *In situ* localization of parathyroid hormone–like protein and mRNA in intraepithelial neoplasia and invasive carcinoma of the uterine cervix. *Hum. Pathol.* **24,** 1058–1066.

Lin, H. V., Harris, T. L., Flannery, M. S., Aruffo, A., Kaji, E. H., Gorn, A., Kolawaski, L. F., Lodish, K. F., and Goldring, S. R. (1991). Expression cloning of an adenylate cyclase–coupled calcitonin receptor. *Science* **254,** 1022–1024.

McAuley, P., Asa, S. L., Chiu, B., Henderson, J., Goltzman, D., and Drucker, D. J. (1990). Parathyroid hormone–like peptide in normal and neoplastic mesothelial cells. *Cancer* **66,** 1975–1979.

Mackenzie, S. L., Gillespie, M. T., Scurry, J. P., Planner, R. S., Martin, T. J., and Danks, J. A. (1994). Parathyroid hormone–related protein and human papillomavirus in gynecological malignancies. *Int. J. Cancer* **56,** 324–330.

Mangin, M., Ikeda, K., and Broadus, A. E. (1990). Structure of the mouse gene encoding parathyroid hormone–related peptide. *Gene* **95,** 195.

Mangin, M., Ideda, K., Dreyer, B. E., and Broadus, A. E. (1989). Isolation and characterization of the human parathyroid hormone–like peptide gene. *Proc. Natl. Acad. Sci. USA* **86,** 2408–2412.

Martin, T. J. (1990). Properties of parathyroid hormone–related protein and its role in malignant hypercalcaemia. *Q. J. Med. (Series 76)* **280,** 771–786.

Martin, T. J., Moseley, J. M., and Gillespie, M. T. (1991). Parathyroid hormone–related protein: Biochemistry and molecular biology. *Crit. Rev. Biochem. Mol. Biol.* **26,** 377–395.

Martin, T. J., and Mundy, G. R., (1987) *In* "Clinical Endocrinology of Calcium Metabolism" (T. J. Martin, and L. G. Raisz, eds.), pp. 171–200. Marcel Dekker, New York.

Matias-Guiu, X., Prat, J., Young, R. H., Capen, C. C., Rosol, T. J., Delellis, R. A., and Scully, R. E. (1994). Human parathyroid hormone–related protein in ovarian small cell carcinoma: An immunohistochemical study. *Cancer* **73,** 1878–1881.

Mayo, K. (1992). Molecular cloning and expression of a pituitary specific receptor for growth hormone–releasing hormone. *Mol. Endocrinol.* **6,** 1734–1744.

Melton, M. E., D'Anza, J. J., Wimbiscus, S. A., Grill, V., Martin, T. J., and Krukreja, S. C. (1990). Parathyroid hormone–related protein and calcium homeostasis in lactating mice. *Am. J. Physiol.* **259** *(Endocr. Metab.* **22)** E792–E796.

Merendino, T. J., Insogna, K. L., Milstone, L. M., Broadus, A. E., and Stewart, A. F. (1986). Parathyroid hormone–like protein from cultured human keratinocytes. *Science* **231,** 388–390.

Moseley, J. M., Danks, J. A., Grill, V., Lister, J. A., and Horton, M. A. (1991). Immunocytochemical demonstration of PTHrP protein in neoplastic tissue of HTLV-1 positive human adult T-cell leukaemia/lymphoma: Implications for the mechanism of hypercalcaemia. *Br. J. Cancer* **64,** 745–748.

Moseley, J. M., Kubota, M., Diefenbach-Jagger, H., Wettenhall, R. E. H., Kemp, B. E., Suva, L. J., Rodda, C. P., Ebeling, P. R., Zajac, J. D., and Martin, T. J. (1987). Parathyroid hormone–related protein purified from a human lung cancer cell line. *Proc. Nat. Acad. Sci. USA* **84,** 5048–5052.

Motokura, T., Fukumoto, S., Matsumoto, T., Takahashi, S., Fujita, A., Yamashita, T., Igarashi, T., and Ogata, E. (1989). Parathyroid hormone–related protein in adult T-cell leukemia–lymphoma. *Ann. Intern. Med.* **111,** 484–488.

Motokura, T., Fukumoto, S., Takahashi, S., Watanabe, T., Matsumoto, T., Igarashi, T., and Ogata, E. (1988). Expression of parathyroid hormone–related protein in human T cell lymphotrophic virus type I–infected T cell line. *Biochem. Biophys. Res. Commun.* **154,** 1182–1188.

Mundy, G. R., and Martin, T. J. (1982). Thy hypercalcemia of malignancy: Pathogenesis and treatment. *Metabolism* **31,** 1247–1277.

Mune, T., Katakami, H., Kato, T., Yasuda, K., Matsukura, S., and Miura, K. (1993). Production and secretion of parathyroid hormone–related protein in pheochromocytoma: Participation of alpha-adrenergic mechanism. *J. Clin. Endocrinol. Metab.* **76,** 757–762.

Musso, M-J., Plante, M., Judes, C., Bartelmebs, M., and Helwig, J.-J. (1989). Renal vasodilatation and microvessel adenylate cyclase stimulation by synthetic parathyroid hormone–like protein fragments. *Europ. J. Pharmacol.* **174,** 139–151.

Nickols, G. A., Nana, A. D., Nickols, M. A., DiPette, D. J., and Asimakis, G. K. (1989). Hypotension and cardiac stimulation due to the parathyroid hormone–related protein, humoral hypercalcaemia of malignany factor. *Endocrinology* **125,** 834–841.

Nickols, G. A., Nickols, M. A., and Helwig, J.-J. (1990). Binding of parathyroid hormone and parathyroid hormone–related protein to vascular smooth muscle of rabbit renal microvessels. *Endocrinology* **126,** 721–727.

Okano, K., Pirola, C. J., Wang, H. M., Forrester, J. S., Fagin, J. A., and Clemens, T. L. (1995). Involvement of the cell cycle and mitogen-activated pathways in induction of parathyroid hormone–related protein gene expression in rat aortic smooth muscle cells. *Endocrinology* **136,** 1782–1789.

Orloff, J. J., Ganz, M. B., Ribaudo, A. E., Burtis, W. J., Reiss, M., Milstone, L. M., and Stewart, A. F. (1992). Analysis of PTHRP binding and signal transduction mechanisms in benign and malignant squamous cells. *Am. J. Physiol.* **262** *(Endocrinol Metab.* **25)** E599–E607.

Orloff, J. J., Soifer, N. E., Fodero, J. P., Dann, P., and Burtis, W. J. (1993). Accumulation of carboxy terminal fragments of parathyroid hormone–related protein in renal failure. *Kidney Int.* **43,** 1371–1376.

Orloff, J. J., Wu, T. J., Goumas, D., and Stewart, A. F. (1989). Receptors for parathyroid hormone–like peptide in vascular smooth muscle. *Clin. Res.* **37,** 457A.

Pandian, M. R., Morgan, C. H., Carlton, E., and Segre, G. V. (1992). Modified immunoradiometric assay of parathyroid hormone–related protein: Clinical application in the differential diagnosis of hypercalcaemia. *Clin. Chem.* **38,** 282–288.

Pang, P. K. T., and Yee, J. A. (1980). Evolution of the endocrine control of vertebrate hypercalcemic regulation. *In* "Hormones, Adaptation and Evolution" (S. Ishii, T. Hirano, and M. Wada, eds.), pp. 103–111. Springer-Verlag, New York.

Parsons, J. A., Gray, D., Rafferty, B., and Zanelli, J. M. (1979). Evidence for a hypercalcaemic factor in the fish pituitary immunologically related to mammalian parathyroid hormone. *In* "Endocrinology of Calcium Metabolism" (D. H. Copp and R. V. Talmage, eds.), pp. 111–114. Excerpta Medica, Amsterdam.

Paspaliaris, V., Vargas, S. J., Gillespie, M. T., Williams, E. D., Danks, J. A., Moseley, J. M., Story, M. E., Pennefather, J. N., Leaver, D. D., and Martin, T. J. (1992). Oestrogen enhancement of the myometrial response to exogenous parathyroid hormone–related protein (PTHrP) and its mRNA in the virgin rat uterus. *J. Endocrinol.* **134,** 415–425.

Pirola, C. J., Wang, H.-M., Kamyar, A., Wu, S., Enomoto, H. Sharifi, B., Forrester, J. S., Clemens, T. L., and Fagin, J. A. (1993). Angiotensin II regulates parathyroid hormone–related protein expression in cultured rat aortic smooth muscle cells through transcriptional and posttranslational mechanisms. *J. Biol. Chem.* **268,** 1987–1994.

Powell, G. J., Southby, J., Danks, J. A., Stillwell, R. G., Hayman, J. A., Henderson, M. A., Bennett, R. C., and Martin, T. J. (1991). Localization of parathyroid-hormone-related protein in breast cancer metastases: Increased incidence in bone compared with other sites. *Cancer Res.* **51,** 3059–3061.

Prosser, C. G., Farr, V. C., and Davis, S. R. (1994). Increased blood flow in the lactating goat induced by parathyroid hormone–related protein. *J. Exptl. Physiol.* **79,** 565–570.

Rakopoulos, M., Vargas, S. J., Gillespie, M. T., Ho, P. M. W., Difenbach-Jagger, H., Leaver, D. D., Grill, V., Moseley, J. M., Danks, J. A., and Martin, T. J. (1992). Production of parathyroid hormone–related protein by the rat mammary gland in pregnancy and lactation. *Am. J. Physiol.* **263,** E1077–E1085.

Ratcliffe, W. A., Bowden, S. J., Dunne, F. P., Hughes, S., Emly, J. F., Baker, J. T., Pye, J. K., and Williams, C. P. (1994). Expression and processing of parathyroid hormone–related protein in a pancreatic endocrine cell tumour associated with hypercalcaemia. *Clin. Endocrinol. Oxf.* **40**, 679–686.

Ratcliffe, W. A., Norbury, S., Stoff, R. A., Heath, D. A., and Ratcliffe, J. G. (1991a). Immunoreactivity of plasma parathyrin–related peptide: Three region-specific radioimmunoassays and a two-site immunoradiometric assay compared. *Clin. Chem.* **37**, 1781–1787.

Ratcliffe, W. A., Norbury, S., Scott, R. A., Heath, D. A., and Ratcliffe, J. G. (1991b). Development and validation of an immunoradiometric assay of parathyroid hormone–related protein in unextracted plasma. *Clin. Chem.* **37**, 678–685.

Rian, E., Jemtland, R., Olstad, O. K., Endresen, M. J., Grasser, W. A., Thiede, M. A., Henriksen, T., Bucht, E., and Gautvik, K. M. (1994). Parathyroid hormone–related protein is produced by cultured endothelial cells. *Biochem. Biophys. Res. Commun.* **198**, 740–747.

Rizzoli, R., Aubert, M. L., Sappino, A. P., and Bonjour, J. P. (1994). Cyclic AMP increases the release of parathyroid hormone–related protein from a lung cancer cell line. *Int. J. Cancer* **56**, 422–426.

Rodda, C. P., Kubota, H., Heath, J. A., Ebeling, P. R., Moseley, J. M., Care, A. D., Caple, I. W., and Martin, T. J. (1988). Evidence for a novel parathyroid hormone–related protein in fetal lamb parathyroid glands and sheep placenta: Comparisons with a similar protein implicated in humoral hypercalcaemia of malignancy. *J. Endocrinol.* **117**, 261–271.

Roskams, T., Willems, M., Campos, R. V., Drucker, D. J., Yap, S. H., and Desmet, V. J. (1993). Parathyroid hormone–related peptide expression in primary and metastatic liver tumours. *Histopathology* **23**, 519–525.

Rosol, T. J., Capen, C. C., Danks, J. A., Suva, L. J., Steinmeyer, C. L., Hayman, J. A., Ebeling, P. R., and Martin, T. J. (1990). Identification of parathyroid hormone–related protein in canine apocrine adenocarcinoma of the anal sac. *Vet. Pathol.* **27**, 89–95.

Schermer, D. T., Chan, S. D. H., Bruce, R., Nissenson, R. A. (1991). Chicken parathyroid hormone–related protein and its expression during embryologic development. *J. Bone Min. Res.* **6**, 149–155.

Schneider, H., Feyen, J. H. M., Seuwen, K., Movva, N. R. (1993). Cloning and functional expression of a human parathyroid hormone receptor. *Eur. J. Pharmacol.* **246**, 149–155.

Schipani, E., Karga, H., Karaplis, A. C., Potts, J. T., Kronenberg, H. M., Sebre, G. V., Abou-Samra, A.-B., and Jüppner, H. (1993). Identical complementary deoxyribonucleic acids encode a human renal and bone parathyroid hormone(PTH)/PTH–related peptide receptor. *Endocrinology* **132**, 2157–2165.

Seidah, N. G., Day, R., Marcinkiewicz, M., and Chretien, M. (1993). Mammalian paired basic amino acid convertases of prohormones and proteins. *Ann. N.Y. Acad. Sci.* **680**, 135–146.

Senior, P. V., Heath, D. A., and Beck, F. (1991). Expression of parathyroid hormone–related protein mRNA in the rat before birth: Demonstration by hybridisation histochemistry. *J. Mol. Endocrinol.* **6**, 281–290.

Soifer, N. E., Dee, K. E, Insogna, K. L., Burtis, W. J., Matovcik, L. M., Wu, T. L., Milstone, L. M., Broadus, A. E., Ohilbrick, W. M., and Stewart, A. F. (1992). Parathyroid hormone–related peptide: Evidence for secretion of a novel mid-region fragment by three different cell lines in culture. *J. Biol. Chem.* **267**, 18236–18243.

Southby, J., Kissin, M. W., Danks, J. A., Hayman, J. A., Moseley, J. M., Henderson, M. A., Bennett, R. C., and Martin, T. J. (1990). Immunohistochemical localization of parathyroid hormone–related protein in human breast cancer. *Cancer Res.* **50**, 7710–7716.

Spengler, D., Waeber, C., Pantaloni, C., Holsboer, F., Bockaert, J., Seeburg, P. H., and Journot, L. (1993). Differential signal transduction by five splice variants of the PACAP receptor. *Nature (London)* **365**, 170–175.

Stewart, A. F., Horst, R., Deftos, L. J., Cadman, E. C., Lang, R., and Broadus, A. E. (1990). Biochemical evaluation of patients with cancer-associated hypercalcaemia. *N. Engl. J. Med.* **303**, 1377–1383.

Strewler, G. J., Stern, P. H., Jacobs, J. W., Evelott, K., Klein, R. F., Leung, S. C., Rosenblatt, M., and Nissenson, R. A. (1987). Parathyroid hormone–like protein from human renal carcinoma cells. Structural and functional homology with parathyroid hormone. *J. Clin. Invest.* **80**, 1803–1807.

Struckhoff, G., and Turzynski, A. (1995). Demonstration of parathyroid hormone–related protein in meninges and its receptor in astrocytes: Evidence for a paracrine meningo-astrocytic loop. *Brain Res.* **676**, 1–9.

Suva, L. J., Mather, K. A., Gillespie, M. T., Webb, G. C., Ng, K. W., Winslow, G. A., Wood, W. I., Martin, T. J., and Hudson, P. J. (1989). Structure of the 5' flanking gene encoding human parathyroid hormone–related protein (PTHrP). *Gene* **77**, 95.

Suva, L. A., Winslow, G. A., Wettenhall, R. E. H., Hammonds, R. G., Moseley, J. M., Diefenbach-Jagger, H., Rodda, C. P., Kemp, B. E., Rodigruez, H., Chen, E. Y., Hudson, P. J., Martin, T. J., and Wood, W. I. (1987). A parathyroid hormone–related protein implicated in malignant hypercalcaemia: Cloning and expression. *Science* **237**, 893–896.

Takahashi, K., Inoue, D., Ando, K., Matsumoto, T., Ikeda, K., and Fujita, T. (1995). Parathyroid hormone–related peptide as a locally produced vasorelaxant: Regulation of its mRNA by hypertension in rats. *Biochem. Biophys. Res. Commun.* **208**. 447–455.

Taylor, A. C., and Kollros, J. J. (1946). Stages in the normal development of *Rana pipiens* larvae. *Anat. Rec.* **94**, 7–23.

Thiede, M. A. (1989). The mRNA encoding a parathyroid hormone–like peptide is produced in mammary tissue in response to elevations to serum prolactin. *Mol. Endocrinol.* **3**, 1443–1447.

Thiede, M. A. (1994). Parathyroid hormone–related protein: A regulated calcium-mobilizing product of the mammary gland. *J. Dairy Sci.* **77**, 1952–1963.

Thiede, M. A., Diafotis, A. G., Weir, E. C., Dreyer, B. E., Garfield, R. E., and Broadus, A. E. (1990). Intra-uterine occupancy controls expression of the parathyroid hormone-related peptide gene in pre-term rat myometrium. *Proc. Nat. Acad. Sci. USA* **87**, 6969–6973.

Thiede, M. A., Grasser, W. A., and Petersen, D. N. (1992). Regulated expression of parathyroid hormone–related protein in mammary blood supply supports a role in mammary blood flow. *Bone Min.* **17**, (*Suppl. 1*) A8.

Thiede, M. A., Harm, S. C., McKee, R. L., Grasser, W. A., Duong, L. T., and Leach, R. M. (1991). Expression of the parathyroid hormone–related protein gene in the avian oviduct: Potential role as a local modulator of vascular smooth muscle tension and shell gland motility during the egg-laying cycle. *Endocrinology* **129**, 1958–1966.

Thiede, M. A., and Rodan, G. A. (1988). Expression of a calcium mobilizing parathyroid hormone–like peptide in lactating mammary tissue. *Science* **242**, 278–280.

Thiede, M. A., and Rutledge, S. J. (1990). Nucleotide sequence of a parathyroid hormone–related peptide expressed by the 10-day chicken embryo. *Nucleic Acids Res.* **18**, 3062.

Thiede, M. A., Strewler, R. A., Nissenson, R. A., Rosenblatt, M., and Rodan, G. A. (1988). Human renal carcinoma expresses two messages encoding a parathyroid hormone like peptide: Evidence of a single copy gene. *Proc. Natl. Acad. Sci. USA.* **85**, 4605–4609.

Thorens, B. (1992). Expression cloning of the pancreatic β-cell receptor for the glucoincretin hormone glucagon-like peptide 1. *Proc. Natl. Acad. Sci. USA.* **89**, 8641–8645.

Trizna, W., and Edwards, R. M. (1991). Relaxation of renal arterioles by parathyroid hormone and parathyroid hormone–related protein. *Pharmacology* **42**, 91–96.

Van der Stolpe, A., Karperien, M., Lowick, C. W. G. M., Juppner, H., Segre, G. V., Abou Samra, A. B., De Laat, S. W., and Defize, L. H. K. (1993). Parathyroid hormone–related peptide as an endogenous inducer of parietal endoderm differentiation. *J. Cell Biol.* **120**, 235–246.

Vargas, S. J., Gillespie, M. T., Powell, G. J., Southby, J., Danks, J. A., Moseley, J. M., and Martin, T. J. (1992). Localization of parathyroid hormone–related protein mRNA expression in breast cancer and metastatic lesions by *in situ* hybridization. *J. Bone Min. Res.* **7**, 971–979.

Walker, A. T., Stewart, A. F., Korn, A. E., Shiratori, T., Mitnick, M. A., and Carpenter, T. O. (1990). Effect of a parathyroid hormone–like peptide on 25-hydroxyvitamin D-1α-hydroxylase activity in rodents. *Am. J. Physiol.* **258,** E297–E303.

Walsh, C. A., Birch, M. A., Fraser, W. D., Lawton, R., Dorgan, J., Walsh, S., Sansom, D., Beresford, J. N., and Gallagher, J. A. (1995). Expression and secretion of parathyroid hormone–related protein by human bone-derived cells *in vitro:* Effects of glucocorticoids. *J. Bone Min. Res.* **10,** 17–25.

Weaver, D. R., Deeds, J. D., Lee, K., and Segre, G. V. (1995). Localization of parathyroid hormone–related peptide (PTHrP) and PTH/PTHrP receptor mRNAs in rat brain. *Mol. Brain Res.* **28,** 296–310.

Weir, E. C., Brines, M. L., Okeda, K., Burtis, W. J., Broadus, A. E., and Robbins, R. J. (1990). Parathyroid hormone–related peptide gene is expressed in the mammalian central system. *Proc. Nat. Acad. Sci. USA.* **87,** 108–112.

Weir, E. C., Goad, D. L., Daifotis, A. G., Burtis, W. J., Dreyer, B. E., and Nowak, R. A. (1994). Relative overexpression of the parathyroid hormone–related protein gene in human leiomyomas. *J. Clin. Endocrinol. Metab.* **78,** 784–789.

White, R. P. (1989). Pharmacodynamic study of maturation and closure of human umbilical arteries. *Am. J. Obstet. Gynecol.* **160,** 229–237.

Whitfield, J. F., Chakravarty, B. R., Durkin, J. P., Isaacs, R. J., Jouishomme, H., Sikorska, M., Williams, R. E., and Rixon, R. H. (1992). Parathyroid hormone stimulated protein kinase C but not adenylate cyclase in mouse epidermal keratinocytes. *J. Cell. Physiol.* **150,** 299–303.

Whitfield, J. F., Isaacs, R. J., Chakravarthy, B. R., Durkin, J. P., Morley, P., Neugebauer, W., Williams, R. E., Willick, G., and Rixon, R. H. (1994). C-terminal fragments of parathyroid hormone-related protein, PTHrP-(107-111) and (107-139), and the *N*-terminal PTHrP-(1-40) fragment stimulate membrane-associated protein kinase C activity in rat spleen lymphocytes. *J. Cell Physiol.* **158,** 518–522.

Williams, E. D., Leaver, D. D., Danks, J. A., Moseley, J. M., and Martin, T. J. (1994). Effect of parathyroid hormone–related protein (PTHrP) on the contractility of the pregnant rat myometrium and its localization within the pregnant rat uterus. *J. Reprod. Fertil.* **102,** 209–214.

Yamamoto, M., Fisher, J. E., Thiede, M. A., Caulfield, M. P., Rosenblatt, M., and Duong, L. T. (1991). Concentrations of parathyroid hormone–related protein in rat milk change with duration of lactation and interval from previous suckling, but not with milk calcium. *Endocrinology* **130,** 741–747.

Yamamoto, M., Harm, S. C., Grasser, W. A., and Thiede, M. A. (1992). Parathyroid hormone–related protein in the rat urinary bladder: A smooth muscle relaxant produced locally in response to mechanical stretch. *Proc. Natl. Acad. Sci. USA* **89,** 5326–5330.

Yasuda, T., Banville, D., Rabbani, S. A., Hendy, G. N., and Goltzman, D. (1989a). Rat parathyroid hormone–like peptide: Comparison with the human homologue and expression in malignant and normal tissue. *Mol. Endocrinol.* **3,** 518–525.

Yasuda, T., Banville, D., Rabbani, S. A., Hendy, G. N., and Goltzman, D. (1989b). Characterization of the human parathyroid hormone–like peptide gene. *J. Biol. Chem.* **264,** 7720–7725.

Yen, T. C., Hwang, S. J., Wang, C. C., Lee, S. D., and Yeh, S. H. (1993). Hypercalcemia and parathyroid hormone–related protein in hepatocellular carcinoma. *Liver* **13,** 311–315.

Yoshikawa, T., Noguchi, Y., Matsukawa, H., Kondo, J., Matsumoto, A., Nakatani, Y., Kitamura, H., and Ito, T. (1994). Thymus carcinoid producing parathyroid hormone (PTH)–related protein: Report of a case. *Surg. Today* **24,** 544–547.

Zhou, H., Leaver, D. D., Moseley, J. M., Kemp, B., Ebeling, P. R., and Martin, T. J. (1988). Actions of parathyroid hormone–related protein on the rat kidney *in vivo. J. Endocrinol.* **122,** 229–235.

INDEX

ISBN 0-12-364570-0

9 780123 645708

90018